Digital Sound Processing for Musi

Titles in the series

Acoustics and Psychoacoustics
David M. Howard and James Angus

The Audio Workstation Handbook
Francis Rumsey

Computer Sound Synthesis for the Electronic Musician
Eduardo Reck Miranda

Digital Audio CD and Resource Pack
Markus Erne
(Digital Audio CD also available separately)

Digital Sound Processing for Music and Multimedia
Ross Kirk and Andy Hunt

MIDI Systems and Control, Second Edition
Francis Rumsey

Sound and Recording: An Introduction, Third Edition
Francis Rumsey and Tim McCormick

Sound Synthesis and Sampling
Martin Russ

Digital Sound Processing for Music and Multimedia

Ross Kirk
University of York, UK

Andy Hunt
University of York, UK

OXFORD AUCKLAND BOSTON JOHANNESBURG MELBOURNE NEW DELHI

Focal Press
An imprint of Butterworth-Heinemann
Linacre House, Jordan Hill, Oxford OX2 8DP
225 Wildwood Avenue, Woburn, MA 01801-2041
A division of Reed Educational and Professional Publishing Ltd

 A member of the Reed Elsevier plc group

First published 1999
Reprinted 2001

Transferred to digital printing 2004

British Library Cataloguing in Publication Data
A catalogue record for this book is available from the British Library

Library of Congress Cataloguing in Publication Data
A catalogue record for this book is available from the Library of Congress

ISBN 0 240 51506 4

Typeset by Avocet Typeset, Brill, Aylesbury, Bucks

Contents

Series introduction

The Focal Press Music Technology Series is intended to fill a growing need for authoritative books to support college and university courses in music technology, sound recording, multimedia and their related fields. The books will also be of value to professionals already working in these areas and who want either to update their knowledge or to familiarise themselves with topics that have not been part of their mainstream occupations.

Information technology and digital systems are now widely used in the production of sound and the composition of music for a wide range of end uses. Those working in these fields need to understand the principles of sound, musical acoustics, sound synthesis, digital audio, video and computer systems. This is a tall order, but people with this breadth of knowledge are increasingly sought after by employers. The series will explain the technology and techniques in a manner which is both readable and factually concise, avoiding the chattiness, informality and technical woolliness of many books on music technology. The authors are all experts in their fields and many come from teaching and research backgrounds.

Dr Francis Rumsey
Series Editor

Acknowledgements

This book is dedicated to our families, students and colleagues: to our wives Caroline and Charlotte for their endless patience with 'otherwise engaged' husbands at the dinner table and for their material assistance in reading the text. To Laurie and Alistair and our students for providing role models for future readers of the book. Special thanks to Rob Featherstone for extensive proof-reading of the manuscript, and to Francis Rumsey, editor for this series, for endless encouragement and suggestions. To colleagues in the Music Department at the University of York for their inspiration and insight over the past decade. Finally to the Department of Electronics at the University of York for the faith in allowing us and our colleagues to develop 'Music Technology' from an obscure predilection into a mature engineering discipline.

Introduction and background

Those of us who have an interest in digital sound processing in its various spheres – music technology, studio systems, multimedia – are witnessing the dawning of a new age. The opportunities for our own involvement in the expansion and development of sound transformation, musical performance and composition are quite unprecedented.

These opportunities are brought about by the rapid growth in the availability of competent technology in the form of powerful personal computers equipped with specialised sound-processing units. This technology is supported by software of ever-increasing sophistication, providing access for the 'man in the street' to sound processing operations which hitherto were the domain of the specialist working in expensively equipped studios.

The equally rapid development of the Internet looks set to revolutionise the distribution and dissemination of musical ideas and sound processing techniques around the world, with the result that even more of us will be able to become involved in this exciting phenomenon of growth. This may indeed give rise to a new 'electroacoustic community', where the development of studio technique and musical thought may be as much the province of the committed layperson – teenager, student or 'retired' professional – as that of the specialist and the academic.

This book is dedicated to the new community. Like many other fields associated with contemporary technologies, audio signal processing and music technology are bedevilled by a blanket of arcane technique and jargon which can appear to preclude all of those who are not in training to join the 'inner priesthood'. This need not be so. For some time now, the authors have been involved in the running of a series of courses at the University of York, UK, aimed at opening up this cross-disciplinary activity to students from a variety of backgrounds, many of a non-technical nature. This has enabled us to develop and prove an approach which we are convinced is suitable for use by any serious student who is prepared to commit effort and dedication to the task. The book is an attempt to provide a text which will demonstrate and support this approach. It can be used by students from any kind of background who are prepared to bring a modicum of logical thought to the task. The use of mathematics has been deliberately kept to a minimum, lying within the scope normally covered in schools. All other topics are dealt with from first principles.

Topics covered

Let us now consider the contents of this book. The basic technologies have been brought to a good level of maturity by specialists over the last thirty years of the twentieth century. These technologies have their roots in an exciting amalgam of telecommunication systems, computer technology, human factors (in the form of 'human–computer interaction' – HCI) and audio, all embedded within sound processing and musical contexts. This sets the framework for the book. We wish to provide enough information so that students and the dedicated layperson can understand the nature, synthesis and transformation of sound which form the basis of digital sound processing for music and multimedia. We then provide enough background in computer techniques so that you, the reader, can write computer 'algorithms' (programs) to realise new processes central to your own musical and sound processing ideas. This includes material on the structure of computer systems themselves (the 'hardware') so that you may also be able to understand how these algorithms will produce sound, as they run in the computers or on specialised sound processing units installed within the computer. It also includes use of the MIDI standard, as well as the direct synthesis of sound. Finally we include material to enable you to understand the nature of the way in which people interact with electronic instruments, so that you can ultimately *contribute* to the development of new kinds of performance and composition systems. In addition, the book will function at another level, for those who wish to know what is going on 'under the hood' in contemporary music technology systems. Taken together, these objectives are not trivial, but as we indicate above, we have shown that they are attainable by a committed student and enthusiast.

Much ground has to be covered to provide *all* of the material necessary to achieve these objectives fully, and this probably represents a lifetime of study! It is certainly beyond the scope of any one book to cover the whole field. There are plenty of texts which deal with the *detail* of the necessary techniques used in sound processing systems. However, many of them are quite inaccessible for the newcomer to the subject, and are often inappropriately structured for the purposes of an understanding of practical systems. The purpose of this book therefore is to provide a *key* to the material in these texts, so that you can approach them with the confidence that you have a firm, *non-superficial* understanding of the fundamental concepts which underpin the subject.

A considerable amount of information regarding the field of digital audio is available on the World-Wide Web. This book also provides a key to this material. A representative sample, which directly supports and extends the material in the book can be found at http://www.york.ac.uk/inst/mustech/dspmm.htm. We hope that readers will find the combined resource of the book and its associated website a valuable 'way in' to the challenging world of digital audio.

Structure of the book

In writing the book, we recognise that readers will have a variety of background, interests and points of view. Some will be particularly interested in audio engineering topics and will wish to get down to the 'signal processing' as quickly as possible. Others will come from a musical background, and may prefer a more gentle introduction, emphasising the musical context. Yet others may approach the subject from an interest in music technology, and would prefer a way into the subject from a systems-oriented perspective. We have therefore organised the text into several essentially *free standing* parts, which can be read in various orders to suit the particular needs of the reader. We suggest a number of flowlines through these parts, below, after introducing the sections of the book. In order to make the sections as free-standing as possible, in some places key information covered elsewhere in the book is summarised in outline.

We have tried in all cases to precede technical description with material which sets the musical context, so that the need for the engagement with detailed sections can be readily appreciated by the reader.

The book is split into six parts that can be consistently read from start to finish (although this is by no means the ideal route for every reader). **Part 1** introduces the reader to the ways in which sound generation and processing technology have developed and have influenced the artistic world throughout the twentieth century. Once the reader understands the context for the technological manipulation of sound, we can begin to explain the theory of sound transmission and storage. Therefore in **Part 2** we introduce the signal processing techniques that form the basis for understanding how sound can be digitally recorded, altered and played back. Equipped with these theoretical tools, we look in **Part 3** at the structure of contemporary music technology systems, considering their interconnection, basic functionality and use of signal generation and processing techniques. The reader is then acquainted in **Part 4** with the fundamentals of digital computer systems so that he/she will understand the digital 'engines' that are inside every synthesis or processing system. In **Part 5** we justify the need for learning to program these systems, and give tutorial guides on programming for MIDI, digital audio and multiple media. Finally **Part 6** considers the design of the control interface to digital instruments, and involves the reader in thinking about designing the instruments of the future.

Chapter structure

Part 1 Context

This introductory section describes the technological and artistic context within which the evolution of electronic sound and instruments have developed. Few if any human endeavours develop in isolation from related ideas, and this is as true for audio technology as for anything else. Chapter 1 briefly sets out the historical context of developments in sound generation and recording, composition and performance, which have provided the motivation for the systems described in the book.

Part 2 Sounds and signals

Chapter 2 describes the nature of the acoustic signal which is manipulated within digital audio and computer music systems. Chapter 3 then considers the way in which the signal is represented and processed in elementary signal processing algorithms. In essence, this section provides an introduction to signal processing for digital sound systems.

Part 3 Music technology systems

This section deals with the structure of contemporary music technology systems, set within the context of the way in which people have used such systems over the years. The MIDI specification is introduced and discussed in Chapter 4. This leads to a description of the essential functional components of modern music technology systems in Chapter 5. It concludes with an in-depth discussion on how signal processing techniques are used within such systems.

Part 4 Computer fundamentals

This section describes the computer systems which are used to realise the sound processing techniques described in Parts 2 and 3. It serves as an introduction to computer fundamentals for those who have not studied computers before, and who wish to understand the way in which they are used in music synthesis systems. These chapters concentrate on the issues common to all conventional processors, and do not focus on any one specific device, so that the material will remain relevant as the technology changes. Chapter 6 introduces the elementary logic gates that are fundamental to computer operations. Chapter 7 considers the structure of a typical computer system – its hardware and software. Chapter 8 then describes how the computer interacts with the outside world by considering the topic of computer interfacing.

Part 5 Programming for sound generation and processing

The processes of writing computer programs for audio, musical and multiple-media purposes are introduced in this section. It is quite feasible for the enthusiast to do this, so that he or she can contribute to the development of the subject in terms of technique and music, provided that a suitable introduction to programming is given. This is the purpose of Chapter 9 and examples are based on algorithmic composition for MIDI systems. Chapter 10 extends the programming techniques covered in Chapter 9 to show how sound can be synthesised and processed using the unit generator concept. The real-time control of signal processing and multiple-media applications is introduced. Examples of direct sound synthesis and manipulation are given using the MIDAS system as a framework.

Part 6 Interface design for the future

It was stated above that one of the objectives of this book is to assist the committed enthusiast (lay and professional) to contribute to the development of music technology and audio processing systems of the future. This section deals with the way in which people interact with such systems and considers a 'human factors' approach to their design. It concludes with a case study which describes an outline design of a new audio-visual instrument that the reader can take on and develop further.

Flow lines through the book

As stated above, the material is arranged so that the six parts of the text can be read largely independently of each other. This means that readers can choose from several flow lines through the book. Those interested primarily in sound processing and audio engineering might read the parts in the order:

1-2-3-4-5-6 or alternatively **2-3-4-5-1-6.**

Musicians might take the order:

1-3-5-6-2-4

and music technologists:

1-5-6-3-2-4.

Web-site

Access to supporting materials and further information and can be found through the web-site associated with this book. It can be found at the following address:

http://www.york.ac.uk/inst/mustech/dspmm.htm

Part 1

Context

This introductory section describes the technological and artistic context within which the evolution of electronic sounds and instruments have developed.

Few, if any, human endeavours develop in isolation from related ideas, and this is as true for audio technology as for anything else. Chapter 1 briefly sets out the historical context of developments in sound generation and recording, composition and performance, which have provided the motivation for the systems described in the book.

There is little assumed knowledge on behalf of the reader, other than an interest in sound and music, and a desire to understand the way that its technology has developed along with the artistic practices that produce it.

1 Sound generation and recording in the twentieth century

Overview

The aim of this chapter is to introduce you to the art and techniques of electronic sound creation and storage in the twentieth century and thus to encourage you to think about how these trends may develop into the next century.

This chapter is concerned with the ways in which computers and electronic systems have been used in the production and processing of sound, and examines the effect this has had on engineers, musicians and computer users. It sets the scene for later chapters by showing that during the entire history of recorded sound, the art of the recording engineer and of the musician have been inextricably intertwined – developments in the one encouraging further experimentation in the other.

The material in this book is dedicated to the furtherance of this exchange. It is therefore appropriate to review the historical context of this mutual development in this opening chapter.

Topics covered

- History of interfaces in electronic sound generation.

- Musical changes and trends in the early twentieth century.
- Early electronic instruments.
- Electronic music studios.
- How sound recording has affected music.
- Synthesisers and the live performance of electronic music.
- Digital sound and its processing and control.
- Digital computer music systems.

1.1 Setting the context

In this age of rapid technological development it is important to keep a sensible perspective by reminding ourselves that music has been a fundamental part of every human society for thousands of years. By contrast computers have only been available to the general public since the 1980s.

However, the seemingly independent disciplines of music and computing appear to be increasingly interwoven in the modern western world. The technological advances of the latter half of the twentieth century have allowed large proportions of the population to have access to music and computing facilities as an integral part of everyday life. The two subject areas are inextricably linked now that digital technology has become the prime method of storing music and an increasingly important way of producing it.

It is the purpose of this chapter to step back from the relentless pursuit of technological advancement for its own sake, and to examine the ways in which humans interact with computers when involved in musical activities. By examining some of the ways in which people perform with traditional musical instruments (methods which have developed throughout history) we can gain valuable insight into the design of interactive computer interfaces, the study of which is still, by comparison, in its infancy.

1.2 The effect of technology on sound production

The generation of sound waves requires physical vibration. As human beings we are equipped with versatile vocal systems which provide us with our most direct way of producing sound; a new-born baby wastes no time in demonstrating this!

However, throughout history, we have felt the need to use tools to create external sounds. The manufacture of such tools (or 'instruments') demands technical knowledge and technical development.

When each new technical innovation is used to make music it influences the range and style of music produced. For example, in the early 1700s Bartolomeo Cristofori invented a harpsichord with notes that could be played soft or loud (*'gravicembalo col piano e forte'*). Composers responded by writing keyboard music with expressive touch and volume.

Sometimes the reverse is true: the development of artistic thinking can set the goals for the instrument designers. At various points in musical history musicians have attempted to extend their musical language. This has often required new types of musical instrument. The twentieth century has seen an explosion of music making all over the world, and the technology of the twentieth century has been regularly called upon to produce instruments capable of new forms of expression.

Traditionally, acoustic instruments are played by one or more of the following techniques: plucking, strumming, bowing, hitting, blowing, keying or stopping (placing fingers over holes or on strings). In all of these techniques, the performer's *physical* action causes vibrations in the instrument, thus producing sound. The musician is therefore continuously in touch with the instrument as it is played. Once electricity is introduced into the range of technological tools available for making music, this situation is extended since it is possible to produce electric instruments which can be played without direct physical contact.

It is with the above points in mind that we now examine the history of the development of music and recording technology from the point where an electronic circuit was first used to make a musical instrument. Special emphasis is given to the ways in which musicians have been expected to interact with the technology as it develops.

1.3 Musical changes from 1900–1950

In this section we set the scene for the development of twentieth century music technology by considering how music itself was undergoing great changes in the early 1900s. It is important for engineers and musicians to appreciate the background to their disciplines since it is then easier to place new developments in context.

The early part of the twentieth century saw music in the western world undergoing huge changes. Symphonic works had grown to enormous lengths and involved hundreds of performers singing and playing with complex harmony in the so-called 'Romantic' period.

1.3.1 'The death of music'

The Romantic composers seemed to be running out of steam. Everything had become so large, so grand and complex that there was a feeling that everything that could be composed had already been composed. In other words there was a concern that if you pushed music any further it would just fall apart.

Austria's Gustav Mahler (1860–1911) is often acknowledged as the founder of modern music, but he saw himself as encountering 'the end of music as we know it'. His music evokes a sadness of times lost for ever. Other composers such as Jean Sibelius (1865–1957) stopped composing their traditionally romantic and nationalistic works at the turn of the century and felt compelled to write in a more abstract style.

It is interesting to note that in the field of technology there was a similar feeling of despair at the arrival of the twentieth century. Bill Gates (in his book *The Road Ahead*) mentions the US commissioner of patents who asked that his office be abolished in 1899 because 'everything that can be invented has been invented'.

1.3.2 'The re-birth of music'

Other composers saw the incoming century as a challenge to completely rethink their music. For the purposes of discussion we will classify the new music as exhibiting one or more of the following characteristics:

- impressionism (new harmonies and textures)
- rhythmic violence (complex and energetic rhythms)
- atonality (the absence of traditional harmony)
- timbral manipulation (the production of new types of sound)

Each of these characteristics is now discussed.

1.3.3 Impressionism

One of the leading composers in Paris at the end of the nineteenth century was Claude Debussy (1862–1918). Debussy's music inherited the rich harmony from Romanticism but mixed it with a new emphasis on musical 'texture'. No longer was it important to have the 'biggest and best tune', but rather to develop washes of sound and threads of melodies. This was termed 'Impressionism' – a phrase that was already in use to describe the many paintings which used washes of colour and paint texture to describe impressions of a scene (rather than a photographic representation).

It is perhaps relevant to note that the technological invention of the camera had forced painters to change their art. Artists were no longer able to match the camera's 'perfect' reproduction of a scene and thus they moved towards more abstract and representational views (which the camera could not compete with). Other composers also regarded as impressionists include Maurice Ravel (1875–1937) and Frederick Delius (1862–1934).

1.3.4 Rhythmic violence

Other composers took a different route. Rather than creating gentle textures and 'washes' of sound, these composers re-established the use of striking rhythms – particularly by using complex time signatures with unusual accents.

Igor Stravinsky (1882–1971) underwent a large compositional shift from his 1910 ballet *The Firebird* (which drew on the rich tradition of Romantic music) to his 1913 ballet *The Rite of Spring*. The premiere of this work in Paris caused a riot, as people were shocked not only by the subject matter and the appearance of the dancers, but also by the music. Chords made up of two different keys (bitonal) were thumped out in a vicious rhythm, and many passages were atonal (see Section 1.3.5). It is difficult to listen to this piece today with the same shock and horror which the Paris audience experienced (although it is worth noting that this style is often used in film music – notably John Williams' score for *Jaws*).

Other composers noted for their highly rhythmic music include Bela Bartok (1881–1945) and Serge Prokofiev (1891–1953).

1.3.5 Atonality

While many of the Romantic composers were worrying that they could not stretch musical harmony any further without totally destroying the key relationships (fundamental to music since people could remember), others were exploring a new form of music which had no central tonal base whatsoever.

Arnold Schoenberg (1874–1951) composed many works which could be described as being Romantic in style. However, he felt a growing conviction that music had to change in order to remain relevant and to have any future. He took it upon himself to explore new musical frontiers even though he knew it would severely damage his popularity.

Schoenberg first broke away from the rules of tonality by writing a series of pieces which were not in a given musical key. They used notes and melodies like other music but there was no longer a tonal 'centre'; no feeling of having to return to the home key (tonic) by the end of the piece. Schoenberg was worried by his audiences' adverse reactions to his music, but felt compelled to explore this new musical terrain.

He was concerned that the removal of tonality would leave a void in the musical language, and he set about re-inventing musical rules to take atonality into account. He proposed a new form of composition based on the assumption that all twelve semitones (in the conventional western octave) were to have equal weighting. To ensure that this basic rule was adhered to, composers were to write down a *tone row*. This was a sequential arrangement of the 12 available semitones, and the idea was to use that tone row as a 'building block' in the subsequent composition.

It is possible to make 'harmonious' tone rows, but much of the music composed by this method is dissonant. This was (and still is) the most difficult stumbling block to many listening to serial music.

1.3.6 Timbral manipulation

Many composers wanted to move beyond traditional orchestral instruments and playing styles in order to produce new timbres and textures. The idea was to create 'soundscapes' where the focus was on the sonic qualities of the sounds themselves, rather than necessarily the tune they were playing. One of the leading experimenters in this field was Edgar Varèse.

Varèse(1883–1965) was born in France but spent most of his life in America. He trained as an engineer and this was to influence his musical thinking. He urged musicians to work with engineers with the aim of producing new mechanical instruments capable of a completely new world of sound. His ideas were far ahead of the technology, and he used whatever he had to hand in order to explore sound. He used the orchestra to paint sound pictures, and some of his pieces are reminiscent of the style of composition which was to come about with computer synthesis over half a century later.

Varèse spent a lot of time and energy trying to get help from musicians, engineers and financiers to build his visionary new instruments. His inventions could only come to fruition with the development of methods of sound production which used electricity.

1.4 Early systems for electronic sound generation

Alexander Graham Bell's experiments with the first telephone systems in the 1870s showed that sound could be converted into electrical signals, transmitted to another place, then reproduced as sound in the new location. As engineers were experimenting with these new machines, they also came across features which could be used for musical purposes.

1.4.1 Accidental discoveries

In 1876 Elisha Gray was working on an alternative to Bell's telephone system. He noticed that some of his circuits oscillated under certain conditions. He converted this into sound using a home-made loudspeaker. By attaching a small keyboard to the device he invented the world's first electronic musical instrument (Figure 1.1). It was known as the 'musical telegraph' and Gray even took this device on a musical tour.

In 1900 William Duddell was trying to find a way to silence the irritating hum of the electric arc street lights of the time. He found that he could control the oscillation frequency, and he too went on tour with his 'singing arc'.

With the technological emphasis on the development of communication systems, it is no coincidence that the first large-scale electrical musical instrument was concerned with the transmission of music across the telephone network.

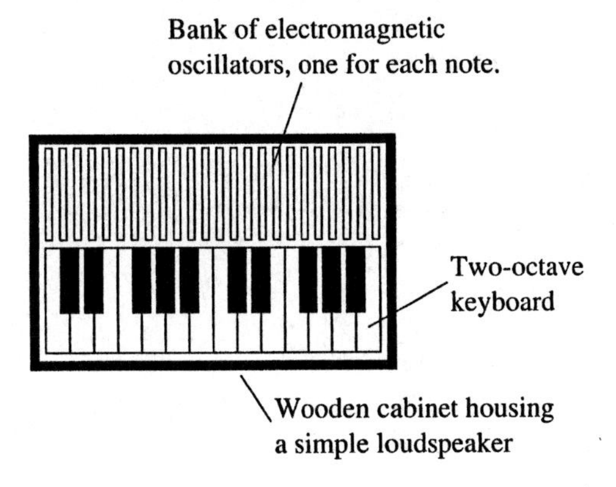

Figure 1.1 The 'musical telegraph' of 1876.

1.4.2 The telharmonium

The 'telharmonium' was a very large, electrically powered musical instrument developed in the last years of the nineteenth century, and was first demonstrated to the public in 1906 in Massachusetts, USA. This remarkable feat of engineering had several visionary design features which did not re-emerge until the latter part of the twentieth century. Its performance interface consisted of two piano-type keyboards. They were polyphonic (many notes can be played at once) and touch-sensitive. The instrument made its sound via electrical generators, giving rise to the machine's alternative name – the 'dynamophone'. The generators were tuned to produce pure tones at various frequencies.

The player could mix these tones and thus control the timbre. If this instrument had been developed further it might have yielded impressive results, but unfortunately several factors conspired to put the inventor, Thaddeus Cahill, out of business. The sheer size of the instrument (it weighed over 200 tons), its monumental power consumption and the fact that it caused major disruptive interference on the telephone network (which was intended to provide the major source of income in the form of paying listeners) meant that the telharmonium would never be used again.

1.4.3 The theremin

The development of the vacuum tube valve in 1906 had made it possible to amplify a signal for playing over a loudspeaker and to create electronic oscillators. This prompted the invention of several new electronic instruments.

The theremin was developed in Russia during the 1920s, and was the world's first non-contact instrument. It consisted of a pair of antennae (one vertical rod, one horizontal loop) which allowed continuous control of the pitch and volume of a single pure tone.

A solo performer would move his or her hands in proximity to the instrument (see Figure 1.2) and thus gestures were directly translated into pitch and volume control. Unlike the telharmonium, the theremin attracted a steady stream of interest from composers and was later used in popular music, for example by the Beach Boys (as the solo instrument in 'Good Vibrations'). In recent years there has been a theremin revival and a number of performing artists now use them.

Figure 1.2 The theremin being played by a solo performer. One hand controlled the pitch of the sound by moving close to one antenna. The other hand controlled the volume using the other antenna.

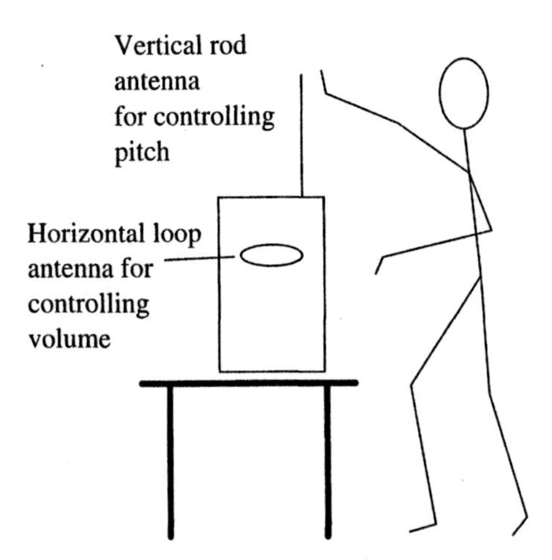

1.4.4 The domination of the keyboard

It is interesting to note that the theremin was one of the few instruments that offered a totally new player–instrument interface. However, subsequent development of electronic instruments focused increasingly on the piano-type keyboard as the standard control interface.

Organs are known to have existed as early as the third century BC. The mechanism for playing them (the keyboard) has developed through the ages into an effective way of triggering, striking or plucking individual notes.

Back in the first half of the twentieth century, electronic instruments were beginning to offer players freedom from discrete notes, and were allowing continuous control over the sound whilst it was being played. It therefore seems somewhat ironic that most of the interfaces to these instruments were restricted to the confines of a keyboard! Sounds could not be altered once the note had been struck.

There was a prolific surge of development of new keyboard instruments between the two World Wars. Some of these were the electrophon (1921), the staccatone (1923), the superpiano (1927) the dynaphon (1928), the Ondes Martenot (1928), the givelet (1929) and the trautonium (1930).

Most of the electronic keyboards were monophonic (allowing only one note at a time to be played). The exception was the givelet which permitted several notes to be played at once. It even allowed a primitive form of *sequencing* which meant that a series of notes could be stored and then automatically played back in sequence. The givelet was overshadowed by the release of the Hammond organ in 1935. This gave a much larger polyphony due to its use of spinning tone wheels (one for each note), rather like a scaled-down form of the telharmonium. The Hammond organ enjoyed astounding success as a modern replacement for the pipe organ in the growing area of popular music due to its relative portability and radically different sound quality.

The other keyboard instrument worth noting is the Ondes Martenot. It has survived through to this day (mainly in performances of Messiaen's *Turangalîla Symphony* of 1948) because of its distinctive and novel sound and its alternative method of performance. The player could put a finger through a ring (connected to a cord controlling the oscillator) and slide the hand left and right to produce continuous changes of pitch (glissandi). As with the theremin, the sound source was controllable by a device which allowed its sound to be produced in a novel way using clearly identified performance gestures.

1.5 Development of recording technology

The development of the telephone had enabled sound to be converted to electrical signals and *moved* from one place to another. The early part of the twentieth century also saw a series of innovations in the *storage* of sound. We will now explore the three main media used to store sound and to play it back.

1.5.1 Records

The '78' gramophone (so called because the disc was spun 78 times per minute) was used increasingly in the early 1900s. The explosion of interest in jazz between 1918 and 1920 was largely due to the fact that music could now be effectively 'frozen' on disc and then sold in large quantities to the general public. Record players of various types were to dominate the distribution of music to the general public until the 1980s when digital technology began to take over in the form of the compact disc (CD).

1.5.2 Tape

Magnetic recording systems (wire, tape etc.) had been available since the start of the century, but were not generally sold until the early 1930s when broadcasting companies, such as the BBC, used them to record programmes for subsequent radio transmission. By the 1950s the tape recorder was established as the primary commercial recording machine (although the general public continued to listen to music on record players – especially with the introduction of the '45' single which facilitated the explosion of rock-and-roll).

The later introduction of the compact cassette provided an alternative medium for the cheap mass distribution of music. Cassettes had the advantage of being small and less susceptible to movement – allowing them, for example, to be played in cars.

1.5.3 Optics

In the 1930s the development of cinematic sound provided a new medium for audio storage. The optical sound tracks on the edge of films could not only be used to record sound, but also allowed a remarkable form of direct synthesis.

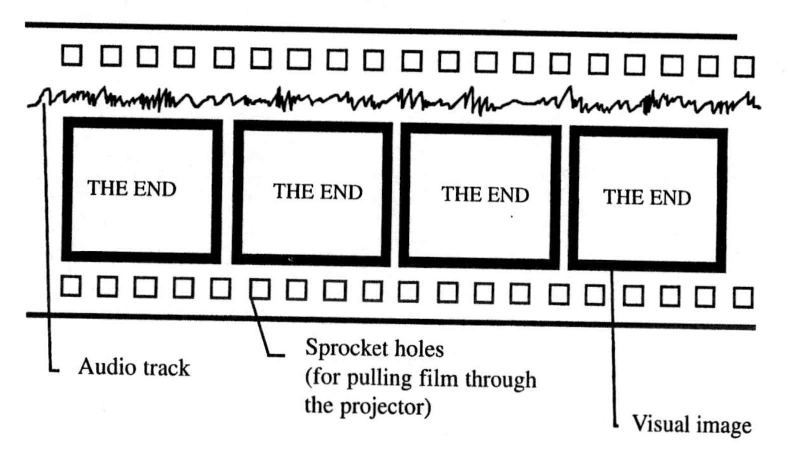

Figure 1.3 Audio signal stored as a line on a film soundtrack.

Composers such as Oskar Fischinger would use a pen to draw the soundwaves onto the film (often to accompany animation). Although this was a tedious process, and human beings are not naturally equipped with the knowledge of how to draw sound, it was (in theory) possible to compose any soundwave imaginable.

1.5.4 Comparison of the different recording media

Each of the above ways of recording sound had its own set of advantages and disadvantages. Record players were used by some composers as a new form of live performance instrument. The disc on which their sound material was recorded could be physically manipulated – spun by hand at different speeds (continuously variable), played backwards, and moved very quickly to a new part of the sound recording. In this respect, some of Pierre Schaeffer's concerts in the 1940s using several gramophones as live-performance sound sources were very similar (in terms of the technology and the human musical control involved) to the 'scratching' of records by hip-hop DJs which emerged in the 1970s. In both these musical situations the performer learns how to physically 'play' the sound source. Much time is spent in rehearsing and developing the gestural control needed to react in live performance – in the same way that conventional acoustic instrumentalists practise to improve their technique.

It should be noted that the process of recording (or etching) the disc meant that the sound must have first existed in the real world prior to being recorded. The physical structure of a disc does not lend itself to being 'edited' in any way. Therefore composers who were primarily interested in creating new sounds and editing existing ones did not find the record player a suitable tool with which to achieve their aims.

The storage of sound on magnetic tape is superficially similar to that of a record player. In both systems an analogue representation (a continuous 'trace') of the soundwaves is made on a physical medium. However, the physical forms of the different media determine the methods by which the music is reproduced. Whereas records can be used as a type of performance device (as described above) and have practically no editing potential, the situation is very different for tape. Although a reel-to-reel tape player could be 'shuttled' back and forward by hand, its live performance capabilities were regarded as inferior to the record due to the lack of random access to a particular point in the music and the lower physical robustness of the tape.

Records were an inappropriate medium for editing, but tape could be chopped up, rearranged, copied to another tape, looped and sections removed, thus providing a completely new way of organising and transforming musical sounds.

Thus tape recorders freed composers from the range of sounds available on acoustic instruments and simultaneously removed the live performance element that had been the central part of music making for thousands of years.

1.6 Electronic music studios

With the end of the World War II came a new focus on arts and technology around the world. People continued to experiment with new art forms including new forms in music. Various individuals were interested in creating fresh sounds, and buildings were provided so that this experimentation could take place.

1.6.1 Paris

The first experimental music studio took shape in Paris in 1948 where Pierre Schaeffer was working at French National Radio. His early experiments with magnetic tape recording led him to produce sound collages made up of tape transformations of sounds (such as the hissing of steam engines) recorded in the real world. Audience responses to these unusual sound collages seemed to suggest that there was too much 'association' with the original source (i.e. 'It sounds like a collection of steam engines!').

When Schaeffer was joined soon afterwards by composer Pierre Henry, the studio began producing tape-based compositions which were much more abstract, but were still taken from recordings of sounds found in the real world. Their first full piece was *Symphonie pour un homme seul* (1949/50) which used modified voice sounds. They termed this type of composition *'musique concrète'* and it was very popular with some composers, including Messiaen and Varèse (Sections 1.3.6 and 1.4.4).

1.6.2 Köln

Meanwhile, at West German Radio, Köln, a collaboration of engineers and composers was established and by 1952 the Köln Electronic Music Studio was producing music by electronic means alone. Composer Herbert Eimert developed a new methodology of music making known as *'Elektronische Musik'*. In this, only newly synthesised sounds were used, not sounds from the real world as used by the Paris studio. To do this they assembled a collection of oscillators and noise generators. They were very large, very expensive and extremely basic by today's standards but the composers used them to produce sounds which had never been heard before.

1.6.3 Rivalry

From today's viewpoint it is quite strange to hear of the intense rivalry which existed between these two studios. Presumably it was because of the desire to be first at producing totally original sounds. People working at each studio tended to praise the work of that studio and denounce the work of the other.

The two studios were actually exploring complementary areas of sound generation. The French studios were pioneering the art of *sampling* – taking sounds from the real world and processing them to make new sonic collages. At the same time the German studio was developing *synthesis* techniques – making sound from electronic components.

It took one particular young composer (who had worked at both studios) to point out the fact that the work of each studio could potentially enhance that of the other. That composer was Karlheinz Stockhausen.

1.6.4 Stockhausen

Stockhausen (b.1928) joined the Köln studio in 1953. He put the studio though its technological paces by composing a series of studies. These are considered to be the earliest pieces of purely electronic music. Stockhausen built each piece out of sinewaves, adding them together to form new timbres. He invented his own form of graphical notation because conventional musical notation (which represents note pitches and rhythms) was no longer valid for pieces where the timbre was the central element.

He soon realised that these pieces all sounded acoustically 'dead', and he wanted to use sounds from the real world as well as reverberation techniques for processing the sounds he had generated. He was then able to blend for the first time the techniques of the two opposing music studios. His next piece, *Gesang der Jünglinge* (1956), used both concrete sounds (a boy's voice) and electronic sounds. It was also performed on an array of loudspeakers as Stockhausen believed that sound spatialisation was very important to electronic music.

By 1960 during the composition of *Kontakte* Stockhausen developed some radical theories about the nature of sound and music. In this piece he explored one of those theories – that the concepts of 'pitch', 'rhythm' and 'timbre' are all essentially equivalent. At one key point in *Kontakte* a buzz-like timbre descends in pitch so deeply that the listener perceives it as a rhythm. Then the individual clicks which make up the rhythm slow down even further so that the listener can perceive the timbre and pitch 'hidden' within one click.

1.7 Live electronic music

While all the above developments in the synthesis and storage of new sounds were evolving, there was a gradual revolution taking place in the popular music scene. In order to play to larger audiences, several acoustically quiet instruments (such as the guitar) were amplified and played over loudspeaker systems.

1.7.1 Electric guitar

The earliest electrically amplified guitar was prototyped in 1927. As guitarists explored their newly amplified guitars they discovered that they could play the instruments in new ways. They could produce long sustained notes and audible solo lines rather than simply having to strum chords loudly in order to be heard. In addition, it became obvious that the timbral quality of the guitar could be radically altered. In a cycle of development in which the guitar, amplification systems and outboard effects (such as tape-based reverberation systems) underwent continuous alterations, the electric guitar evolved into a versatile instrument of fundamental importance to the emerging rock and pop music (from the 1950s onwards).

Other conventional instruments were also electrically amplified – for example the radiopiano (1931) and the violectric (1936).

1.7.2 'Sonic microscope'

Karlheinz Stockhausen wrote a series of exploratory pieces called *Mikrophonie* in 1964–65. He used a microphone to pick up and amplify sounds that cannot normally be heard – a 'sonic microscope'. *Mikrophonie I* requires six performers – two to play a large tam-tam in a variety of ways, two to move the microphones (to act as the sonic microscope) and two to operate the live electronic amplifiers and filters.

The physical gestures of each performer contribute to the final sound, but each of the three groups operates in a different way. The first group actually makes the sound by physical contact with the tam-tam (a generation process), the second group 'homes in' on various sounds (a selection process) and the final group manipulates the sound via filters and amplifiers (a transformation process).

1.8 Synthesisers

Nowadays the term 'synthesiser' tends to imply a keyboard-based instrument that can generate sound. The early keyboard-based devices and the theremin could be considered to be primitive synthesisers in that sound was produced from rotating wheels or oscillating electric currents. However, the first instrument to form the basis of the modern synthesiser was the 'monochord'.

The monochord was one of the sound sources available in the Köln studio in the early 1950s. It was a live performance instrument that was played by a keyboard which could react to key pressure to control the dynamic response of the sound. Volume could also be controlled by a foot pedal. The monochord's single oscillator could produce sawtooth, square and triangle waveforms to give a range of timbres. It is interesting to note that this instrument, which offered live performance of electronic music but limited control of sound, was generally ignored by the studio's composers in favour of constructing sounds on tape using a sinewave generator.

1.8.1 The RCA synthesiser

The first large-scale synthesiser offering composers complex control of sound was built in 1956 by the Radio Corporation of America (RCA). This massive machine (which occupied an entire room) was programmed by punched paper tape and produced its sound output by directly engraving a gramophone disc (Figure 1.4).

The designers, Harry Olson and Herbert Belar, were mainly interested in imitating conventional instruments and reproducing synthesised versions of the classical repertoire. The two things they could not overcome were the sheer complexity of

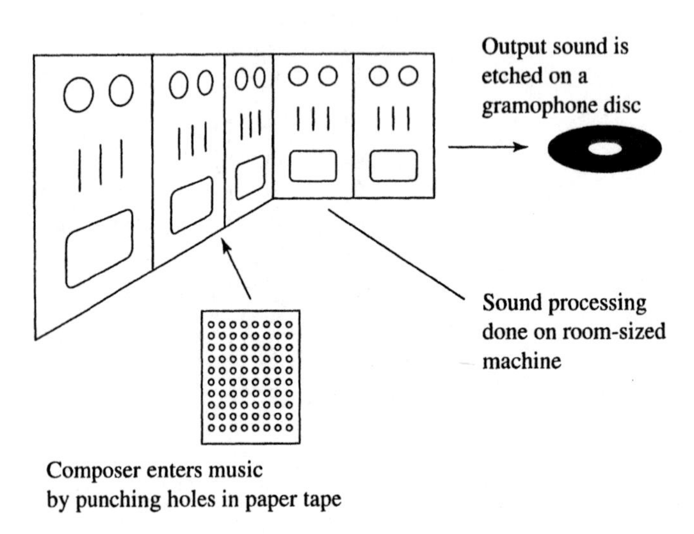

Figure 1.4 The RCA synthesiser (1956).

timbre in acoustic instruments and the fact that an acoustic performer constantly varies the pitch, volume and timbre of an acoustic instrument in a way that is impossible to imitate by programming pitches in non-real-time on punched tape.

Composers such as Milton Babbitt relished the use of this machine as it gave him total control over the compositional timbres as well as the rhythmic structure of the music. Babbitt explored highly 'robotic' rhythms, for example in *Ensembles for Synthesiser* (1963), as he saw this as being a musical niche which could not be filled in any other way.

Thus a new form of composition, sequencing or 'programming' was born. The model of composer as 'computer programmer' was thus firmly established. Many of today's composers follow this model by producing music which has no need of further interpretation by a human performer.

1.8.2 Synthesisers and sound generation for rock and pop

As the academic music world was exploring serialism and electronic music during the 1950s, the outside world was seeing a musical revolution on an enormous scale. Just as the world had experienced an explosion of jazz music around 1917 due to the popularisation of the 78 record, so the introduction of the 45 'single' record coincided with a new explosion of popular 'rock' music.

Rock music thrived on live performance at a loud volume, so it was vital to have good amplifiers and instruments which could be easily amplified. The electric guitar became the defining symbol of the rock band. What had started out as a simple attempt to amplify the classical guitar developed into a complex art-form in its own right, with a range of performance techniques and sound processing requirements.

However, the search was on for new forms of sound generation (synthesis) devices that could be used on stage. This meant that synthesisers had to be portable, and playable in real-time; so obviously room-sized paper-tape reading devices like the RCA synth were unsuitable.

1.8.3 The transistor and voltage-controlled synthesis

The use of synthesisers as live performance instruments (particularly in the area of popular music) was helped dramatically by the introduction of transistor circuits in the late 1950s. Various engineers worked with the new circuits in order to reduce the size of electronic musical instruments, but the most successful and influential design came from Robert Moog.

Moog's first transistor-based synthesiser was released in 1966 and was based on techniques whereby various aspects of the sound generation process could be controlled by a voltage. The advantage of this method of working was that this 'control voltage' could come from anywhere – a keyboard, a joystick or the output of another part of the synthesiser such as an oscillator (Figure 1.5).

This is considered to be an important development in instrument design. No longer were sound generation functions controlled from permanently assigned dials or sliders. Instead, one could now control any aspect of an instrument's sound (e.g. its pitch or timbre) from a variety of devices, simply by 'patching' them together.

One of the early uses of the Moog synth was in film soundtracks, and in 1970 Walter Carlos composed a synthesised backing to the film *A Clockwork Orange*. He used a monophonic Moog synth with a multi-track tape recorder to create complex, orchestral-type, synthesised textures.

The Moog synthesiser and its competitors were generally played via a monophonic keyboard. As the prices reduced, they became so popular that various bands began to use the synthesiser increasingly in preference to the electric guitar, or alongside it. Keith Emerson became famous for playing (and abusing) Hammond organs and various synthesisers in the early 1970s. Some bands began to use the increasingly sophisticated polyphonic synthesisers as the main sonic focus of the band. Groups such as Tangerine Dream used guitars as supplements to the synthesised sound.

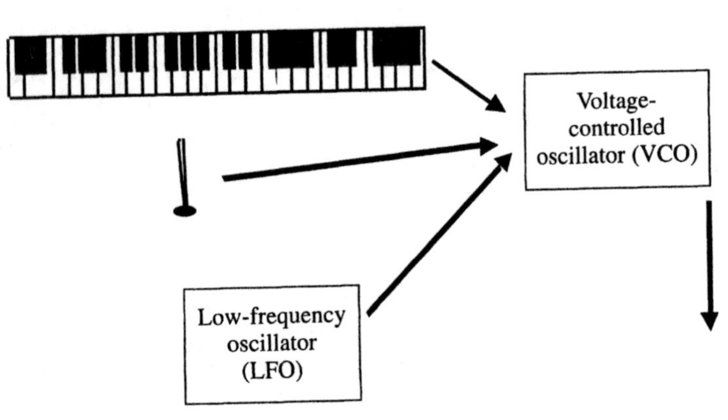

Figure 1.5 The principle of voltage-controlled synthesis. All control devices (keyboards, joysticks, oscillators) generate voltages. These are indicated on the diagram as arrows. They can be fed into other synthesis components which in turn produce voltages. Only at the last stage are they turned into sound.

1.8.4 Visual control

One musical synthesis device from this period, using a non-keyboard interface, was developed at the BBC Radiophonic Workshop in the early 1960s by composer Daphne Oram. She developed a system of 'drawn sound' synthesis known as 'Oramics'. The composer specified various sonic parameters by drawing them onto a transparent plastic sheet which was moved across a strip of photocells. These cells reacted to the pen-strokes on the film and subsequently controlled a monophonic voltage-controlled synthesiser.

It should be noted that despite the superficial resemblance of Oramics to the 'optical film track' compositions of the 1930s, the paradigms in which the composer worked were very different. On film the *soundwaves* themselves were drawn, whereas with Oramics the composer was effectively drawing a series of *control* codes. In this respect an Oramics user worked in a manner similar to that of a composer writing a traditional score.

1.9 Digital sound

The development of the digital computer and its subsequent application to music synthesis, recording, storage and playback is one of the most important and influential developments in the technology of music in the twentieth century.

In 1957 Max Mathews produced the world's first computer program (MUSIC I) which could synthesise sound. Mathews was working at the Bell Telephone Labs on the digitisation of sound. His goal was to store sound as a stream of numbers, send it as a stream of pulses down a telephone cable, then reconstruct it at the far end. It was a natural step to try transmitting music, and then to experiment with the generation of sound from algorithms running on the computer.

MUSIC I was very limited in its scope as it could only produce a single triangular waveform, but it prompted the rapid development of a series of increasingly more complex programs. When musicians from nearby universities heard Mathews' attempts to play simple tunes on his new program they soon saw its potential. They urged him to work on more powerful versions of the program which could play several sounds at once on a variety of waveforms. Soon Mathews had written MUSIC IV which was used by composers in several different American universities. They subsequently developed their own versions of the MUSIC series of programs, using various programming languages on different hardware platforms.

In 1986 at the Massachusetts Institute of Technology (MIT) the latest version of the MUSIC program was translated into the 'C' programming language by Barry Vercoe. The advantages of C were that it was widely available on many different hardware platforms and that programs written in C tended to run faster than those written in other languages. This new program was known as Csound and is today the most widespread and popular direct synthesis program – mainly because it was made freely available to anyone who wanted to use it.

The use of Csound demanded a very different way of composing as numbers had to be typed in to specify *everything* about the final sound. No longer could composers rely on the innate musical sense of a set of human instrumental players to make the notes 'musical'. Composers had gained the ability to *totally control* the sound of their compositions, but had to work extremely hard with a lot of numbers to produce anything of interest.

1.9.1 Parametrical control

The majority of the variations on MUSIC IV (including Csound) followed its fundamental design very closely and differed only in syntactic detail and in the range of commands available. Mathews established the concept of a computer-controlled 'SCORE' and 'ORCHESTRA' by which the composer could specify musical activity. The ORCHESTRA was a text-based file that specified how sounds should be synthesised. This was done in terms of a number of simple algorithms, known as unit generators. The composer would describe how these units should be connected together and where their inputs should come from. The SCORE, also a text file typed in by the composer, was a time-based list of numbers which formed the inputs to the network of unit generators. Figure 1.6 shows diagrammatically how this works.

Thus the composer was given the role of 'algorithmic instrument designer' and of 'parameter specifier'. The program took both the SCORE and the ORCHESTRA and did the 'performance' itself by writing each generated sound sample to a file. This process usually took many minutes to produce each second of sound.

So influential were Mathews' programs that even today, when it is technologically possible to perform with complex synthesis algorithms in real-time, many programs still expect the composer to specify streams of numbers.

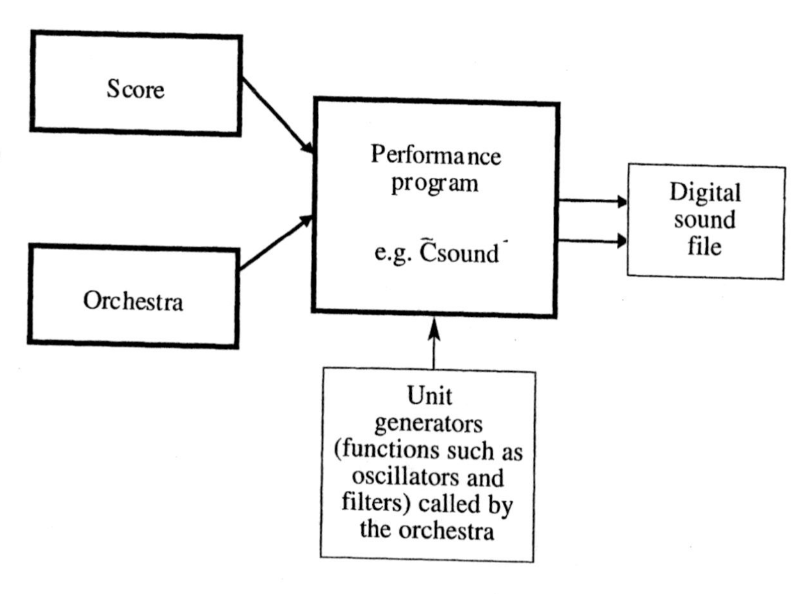

Figure 1.6 The concept of the non-real-time computer instrument. The composer defines the *orchestra* and *score*, and these are processed by the computer to produce a digital sound file.

1.10 The performance interface

While many composers around the world were busy following the MUSIC IV philosophy and were effectively turning into computer programmers, Max Mathews became concerned with the lack of a performance interface. Musicians using the MUSIC series of programs had no interactive feedback and so could not actually 'perform'. This was left entirely to the computer.

The main reason for MUSIC-type programs being unsuitable for live performance is their need to have pre-prepared files of textual and numerical information. This information cannot be effectively generated or altered by the composer/performer *as the sound is produced* – even if the computer is fast enough to instantly process the information and make the sound.

Various designers and composers therefore deemed it necessary to explore alternative methods of giving the computer synthesis unit the information it needed in *'real-time'* – to make it an instrument capable of performance.

1.10.1 Real-time manipulation of a computer score

In 1970 Mathews worked on a system which used a computer to help the composer to specify musical instructions by displaying a visual representation of the instructions on a monitor. The sound was generated in real-time by an analogue synthesiser. It was called 'GROOVE' – Generated Real-time Output Operations on Voltage-controlled Equipment. It was known as a 'hybrid' system because it used both analogue and digital technology.

The score was specified by typing a series of codes into the computer. The computer then controlled the analogue synthesis unit by translating the score into a series of instructions to drive a set of relays and DACs (digital-to-analogue converters). In addition to the score being played back, the performer could modify various parameters in real-time by using a joystick or a small piano-type keyboard.

At this point in time the direction of Mathews' work became strongly focused on the idea of conducting a computer-stored score, a philosophy which has continued in most of his subsequent work to date.

1.10.2 MUSYS

MUSYS was an alternative hybrid digital-analogue system developed in 1970 by Peter Zinovieff (later to form the company EMS). Its two computers were used to control an extensive set of analogue synthesisers, filters and modulation units. Unfortunately for the composer every single instruction had to be preprogrammed either by typing in each register value or by making use of an elaborate array of spinning wheels and binary light patterns to program the registers. No performance interaction was possible.

1.11 Digital computers with performance interfaces

The advent of the affordable microprocessor in the 1970s launched a growth spurt in the production of portable computers and associated musical devices.

1.11.1 Synclavier

In 1976 a new musical instrument called the synclavier was designed around a special 16-bit microprocessor. The performer played a piano-type keyboard and had a bank of buttons which controlled the synthesis. The buttons operated sinewave generators to allow interactive additive synthesis and frequency modulation (see Chapter 5 for a description of these synthesis methods).

The keyboard was polyphonic, but in addition there was a strip-sensor for continuous variations in pitch (similar to the Ondes Martenot). There were also two foot pedals which enabled continuous changes of volume and any other parameter during performance. Soon after its release, additional hardware became available which extended the editing facilities. A visual display unit (VDU) allowed the user to see the parameter settings more clearly, and the user's synthesis configurations could be stored on a floppy disk within the machine.

1.11.2 Fairlight

Towards the end of the 1970s another system emerged which also addressed the needs of an ever increasing market for reasonably flexible but instant synthesis. The 'Fairlight Computer Music Instrument' (CMI) set the standard for all subsequent commercial computer music development. Its novelty was the digital storage and playback of sound (sampling) combined with an interactive computer display.

Like the synclavier, the Fairlight also used a keyboard and foot-pedals for the performance interface, but the editing functions were dramatically improved. Users could 'draw' their own waveforms on the screen by using a light pen (similar to the optical film soundtrack composers in the 1930s), or specify sound parameters via a QWERTY keyboard using a series of control codes.

Peter Gabriel and Kate Bush are amongst those in the pop music world who saw the potential of this instrument. They have built up such a range of sounds and processing techniques on the Fairlight that they still use it today. The reason that more people did not use it was the prohibitive price. When it was released in 1979 one could buy a quite reasonable house for the same price as this machine!

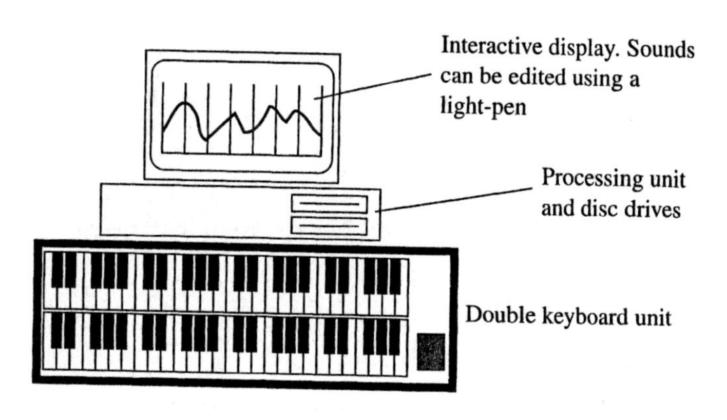

Figure 1.7 The Fairlight computer music instrument (CMI).

1.11.3 Structured Sound Synthesis Project (SSSP)

Further advances in the performer interface were made by Bill Buxton and his team in their Structured Sound Synthesis Project (SSSP) at the University of Toronto, Canada. The hardware consisted of a set of digital oscillators which could be configured for different types of synthesis (including additive synthesis, and 'VOSIM' – a vocal-sound algorithm). The novel performer interface required players to hold onto a set of plastic bands and move their hands around in space, with the option of watching the display on an interactive graphics unit.

The information on the screen could also be accessed using a tracker-ball. The layout of the screen display would take on the characteristics of the current editing option; for example dragging notes onto a stave when in 'score' mode, or drawing amplitude envelopes on a grid, or moving graphical objects (such as sliders) when in 'synthesis editing' mode.

Such forms of graphical interaction were to become much more widespread with the advent of microcomputers.

1.12 The digital revolution

The Fairlight and synclavier systems were widely used in studio and live performance work, particularly in the popular music industry where fast turnover of musical material was of utmost importance. Many of the original systems are still in use, and their design philosophies have had an enormous influence on many subsequent systems.

By the early 1980s a revolution had occurred in the world of music technology. With microprocessors widely available and increasingly powerful, there was an explosion in the number of computer-based musical instruments and processing systems.

An example of the developments taking place at this time was a range of hardware known as the 'IRCAM-4' series. Composers and engineers in Paris at the Institut de Recherche et Coordination Acoustique/Musique (IRCAM) worked together with the aim of producing a powerful and flexible synthesis engine. The '4C' contained a bank of configurable oscillators and envelope shapers of which various parameters could be adjusted in real-time using a set of sliders. However, composers could only achieve real-time complex synthesis by setting everything up in advance in low-level assembler code.

The '4X' machine was also developed at IRCAM. It used an extra microchip known as a Digital Signal Processor (DSP), running alongside the microprocessor, to handle, in real time, the vast amounts of addition and multiplication operations required by the synthesis engine.

1.12.1 Real-time digital synthesis on VLSI chips

While the academic research institutes were developing flexible (but costly and unfriendly) devices, the commercial music manufacturers were producing an unprecedented number of keyboards to supply the ever increasing demand for synthesisers. With more pop music bands using synthesisers, the requirement for cheaper 'home' and 'school' keyboards drove down market prices and brought in a whole new range of large companies (such as Yamaha, Roland, Oberheim and Casio) which produced competing products.

The production of these cheaper instruments was only made possible by the recent advances in VLSI (very large scale integration) technology which allowed companies to customise their own fast circuit designs onto a single silicon chip.

The Casio 'VL-tone' of 1981, though regarded by many as a toy (and resembling a large calculator), was the first synthesis and sequencer unit which could be purchased for under £30. It launched a torrent of new keyboard synthesiser devices from an ever increasing number of manufacturers. These products were electronically incompatible with each other as every device's internal digital operation was custom designed.

1.12.2 The effect of digital technology on pop music

From 1980–82 there was a flourishing use of synthesisers in pop groups, very much at the expense of the electric guitar. By this time there was a variety of sequencers available which enabled pre-programmed drum-tracks, basslines (and sometimes entire accompaniments) to be played in extremely accurate rhythms. Listeners began to *expect* to hear this rhythmic precision and for a while conventional bands (with drums and guitars) seemed to be out of fashion.

Whilst the 'punk' and 'heavy metal' scene continued to revolve around guitar and drums, the new electronic sound of synthesisers launched a whole new popular music emphasis on synthesisers. The 'New Romantic' movement was epitomised by entirely synthesiser-based bands (and yet another style of haircut). It seemed to support the general view that western society was becoming increasingly computer-oriented. Indeed Orwell's *1984* prediction of 'Big Brother' and the take-over by machines was echoed in the increasingly synthetic sounds and robotic rhythms of the early 1980s synthesiser/sequencer music. Artists such as Kraftwerk, Gary Numan, John Foxx and the Human League popularised this style of music and sound production. Some even dressed as robots or got robots to take their place on stage as part of their statement about music and machines.

1.12.3 MIDI – the Musical Instrument Digital Interface

Players of voltage-controlled analogue synthesisers had been able to make rudimentary connections between different devices by taking the control voltage of one keyboard and making it drive the synthesis unit of another. This allowed a form of 'keyboard coupling' similar to the function with which church organists are familiar, allowing one keyboard manual to slave to another. However, the connections and voltage levels were not standardised and, more importantly, this form of analogue voltage linking could not apply to the new digital synthesisers. Between 1981 and 1983 an international group of electronic musical instrument manufacturers met together to discuss ways of implementing a standard protocol for the interconnection of digital music devices. The resulting 'v1.0 MIDI Protocol' was adopted by practically all major electronic musical instrument manufacturers and has had a profound influence on the world's electronic music making ever since.

The MIDI specification is a description of the manner in which electronic instruments are to be physically connected (i.e. the electrical description of the interface) and perhaps more importantly a list of the messages which can be transmitted between devices. MIDI allows devices to communicate at the real-time 'control' level only; it does not transmit audio data. It was designed originally for keyboard synthesiser devices. Thus communication consists primarily of key-press data, supplemented by optional 'continuous' controls that transmit information such as the position of the pitch-bend wheel.

MIDI has its critics and its champions, but no-one doubts the enormous effect it has had on the way in which electronic musical instruments have developed in the 1980s and 1990s. The critics tend to look at the limitations of the MIDI specification from both the 'electronic connection' and 'message format' points of view.

Many of the developments outlined in the remainder of this chapter were possible primarily because MIDI had been accepted as an international standard of control-rate musical data. Many people were given access to music making for the first time because of the widespread acceptance of MIDI.

The MIDI standard is of such importance in the history of electronic music that we devote Chapter 4 to examining it in more detail.

1.13 Performance instruments in the MIDI age

Inevitably, after the invention of MIDI, the majority of new instruments were keyboard based because that was the medium which originally prompted the MIDI specification and subsequently influenced its message format.

1.13.1 The Yamaha DX-7

Yamaha released one of the earliest MIDI-compatible keyboards in 1983. It also happened to be one of the most advanced digital synthesisers available at the time. The DX-7 was capable of producing a wide range of very 'bright', timbrally rich sounds and it became very popular.

It introduced the general public to FM synthesis and it became known for 'not sounding like other synthesisers'. As well as its polyphonic velocity-sensitive keyboard with an 'aftertouch' pressure bar and pitch and modulation wheels, the DX7 allowed various parameters to be controlled by a breath-controller. Although not often used, this offered another option for shaping note timbre or volume during live performance whilst the player's hands were in use on the keyboard.

This keyboard had a huge effect on the synthesiser market. Not only did more bands buy them, but home users could finally hope to afford their own electronic keyboard. Other manufacturers were quick to spot this and, coupled with the popularity of the MIDI specification, an enormous range of keyboards flooded onto the market. Digital synthesis was now firmly established as the way forward.

1.13.2 Sequencing, sampling and synthesis

By the mid-1980s several very important developments had taken place. Digital sound samplers (such as the Ensoniq 'Mirage' and the Akai 'S' series) were available at a fraction of the cost of the 'Fairlight'. Sound sources were now 'multi-timbral', allowing several different voicings to be played on one physical device, each voice assigned to its own MIDI channel.

The techniques for making novel sounds (*synthesis*), recording and editing existing sounds (*sampling*) and accurately playing them back without human input (*sequencing*) were now available to a large number of people – not just researchers in universities. Consequently whole new musical directions opened up in popular music as people explored these techniques.

As equipment got cheaper more and more people were making music in their own home studios. Commercial studios suddenly had to take notice and to offer new services since many people were composing and recording entirely in their bedrooms on their computer systems and sampling synthesisers.

The phenomenon of 'dance' and 'house' music gained popularity during this time – almost to the surprise of conventional bands and record companies. Like much popular music the *beat* was very prominent in this new music, but there was an increasing tendency to experiment with the sound *timbres* in a manner reminiscent of some academically based electroacoustic composers.

1.13.3 MIDI modules and home studios

It was possible to produce small sound modules – synthesisers and samplers without keyboards – that could be triggered from any MIDI source. Therefore it finally became possible to break away from the domination of the piano-type keyboard. One could use completely novel performance interfaces as long as the messages being sent to the sound module obeyed the MIDI standard.

Most of the major manufacturers produced specialised percussion-based units which were known as 'drum machines'. Some were simply dedicated sound modules with a range of pads for programming a drum sequence, others (such as the 'Linn Drum' series) offered a well-crafted set of percussion pads which would be familiar to a drummer.

1.13.4 Guitar interfaces

Guitarists were very anxious to enter the world of synthesised sound, and so various manufacturers produced guitar-to-MIDI converters. Some guitar interfaces were simply modified 'pick-ups', fitted to an electric guitar, which sent MIDI messages in response to the pitch of each string. Whilst they allowed guitarists to use the traditional playing interface, they became the subject of much criticism regarding their slow response and the lack of performance subtlety which was possible via MIDI. Many guitarists have found that an acoustic instrument, put through an assortment of effects units, is their favoured way of accessing new sounds whilst retaining full performance intimacy.

1.13.5 Wind instrument interfaces

Wind instruments have also been the subject of innovation in the age of MIDI. These have varied from the simple Yamaha 'Breath Controller' via easy-to-use instruments with limited flexibility such as the Casio 'MIDI Horn', to quite complex instruments such as the Yamaha WX7. This latter instrument could be customised to the requirements of an individual user. For example, the mapping of the fingering to the notes produced can be altered, the sensitivity to air movement can be adjusted to suit the player's breath, and the lip pressure sensor can be changed to suit the performer's preferences.

1.14 The microcomputer

During the mid-1980s the concept of the 'desktop microcomputer' became a reality. It was now possible for powerful computing facilities to be available in every home, office and school. There was consequently a great need for an improvement in user interfaces. The Apple Macintosh was the first widely available microcomputer to be supplied with a standard graphical interface, driven by the user with the 'pointing, clicking and dragging' actions of a 'mouse'. Atari followed suit with the 'ST' range which were not only cheaper, but came with a built-in MIDI interface. Atari clearly had high expectations of the market for music software, and they were not disappointed.

In recent years the PC market has expanded enormously and there is a bewildering range of computer configurations and sound cards available to today's computer user.

1.14.1 Types of music software

Over the last ten years there has been an explosion in the range of software available for music on microcomputers. It would be impractical to list the full range of programs available, but the main focus has been on the following functional categories:

- sequencing (recording, editing and playback of MIDI information)
- notation (producing conventional printed musical scores)
- sampling/hard-disk recording (digitally-based recording, editing and playback of sound)
- performance instruments (devices which produce sound in real time under the direct control of a human player)
- instrument editors (graphical control or storage of synthesis parameters, enabling sound modules to be edited)
- composition (software for aiding the composer in the process of composition or 'algorithmic composition' where the computer produces new musical material from mathematical formulae)
- education/instruction (music tutors, reference information)
- studio automation (routing of signals and control of complex MIDI set-ups)

1.14.2 Manipulation of sound using computers

In the late 1980s digital signal processing (DSP) chips became increasingly powerful and cheap. As manufacturers began to incorporate these devices as part of standard computer hardware, it was possible to perform fast digital sound operations on home computers.

The release of the 'NeXT' computer in 1988 offered a real leap forward in terms of what composers could achieve on their own computer systems. It contained two powerful microprocessors which could handle compact-disc quality sound in real time, and allow the composer to interact with the program using complex graphics (e.g. displays of musical notation or images which represented the sound being worked on).

By 1991 the 'IRCAM Musical Workstation' had been designed around the NeXT computer by adding on a series of extra processing cards. All the major computer manufacturers developed high-quality graphical interfaces for their machines, and produced extra hardware (where necessary) to process sound in real time. Much of the software developed for MIDI applications was now adapted to handle the digital recording of sound alongside the MIDI data.

1.14.3 Integrated systems

The increasing power and affordability of PC-based microprocessing power during the 1990s have led to the widespread development and acceptance of *integrated* systems where MIDI sequencing coexists in one package alongside audio recording and processing.

Composers such as Mike Oldfield now regularly record many tracks of acoustic instruments alongside their computer-based digitally synthesised sounds. The equipment offers a range of sound processing techniques (from reverberation, echo, and chorusing through to 'quantising' an acoustic signal to a particular time signature). Many albums and movie soundtracks are *entirely* edited digitally.

1.15 Systems for capturing performance gesture

Human beings take part in musical performance by controlling musical instruments with physical gestures. We have already looked at some of the interfaces which have been made available because of the introduction of MIDI (Section 1.13). This section outlines a number of ways in which computers have been given a gesture-based performance interface.

1.15.1 Drawing

The 'UPIC' system, developed during the early 1980s and associated with composer Iannis Xenakis, was a drawing interface (not unlike the 'Oramics' system of two decades earlier). A light pen was used to draw waveshapes, frequency profiles and envelopes on a special screen. These would then be formed into a score where sound events could be freely drawn and listened to immediately. Because of its direct approach to score and sound specification it was very popular as a teaching aid.

1.15.2 Conducting

One school of thought, pioneered by Max Mathews, views the orchestral 'conductor' as the model on which interfaces should be designed. The 'GROOVE' system (see Section 1.10.1) was modified to produce the 'Conductor' program which could interpret the movements of a hand-held wand and control various parameters of a 'score' held in the computer.

Mathews developed a variety of systems which gradually became more physical, culminating in the popular 'Radio Drum'. This drum-like sensor can be struck by the performer in various places and with a range of force in different strokes. It is also capable of detecting the position of the beater in a 3D space above the drum using an array of small radio antennae contained in the drum's surface. The Conductor program interprets these signals as perturbations to the score, and the music is played via MIDI.

1.15.3 Hand gestures

Many designers have investigated the possibility of using the subtlety of movement a human hand can produce to control a variety of musical parameters.

- *Computer mouse:* MidiGrid is a software environment (developed at the University of York UK) which turns the computer into a musical performance instrument based upon the computer 'mouse' controller (Figure 1.8). A 'grid' of boxes appears on the computer screen and the user moves a cursor (by using the computer mouse) which instantly plays the notes in each box. Users quickly discover that different gestures produce different musical results. The 'grids' can be stored on disk and can consist of hundreds of boxes full of user-definable layouts of MIDI notes, chords and sequences.
- *Sensors and beams:* Various non-contact devices have been invented in recent years. Generally these are 'beams' (of light or ultrasound) which are used to detect the player's movement in space. The device senses the point at which the beam is broken, and this is converted into a MIDI message, so that sound can be made using a MIDI sound module (see Figure 1.9). The EMS *Soundbeam* uses a series of ultrasonic sensors to convert the movements of dancers (for example) into MIDI

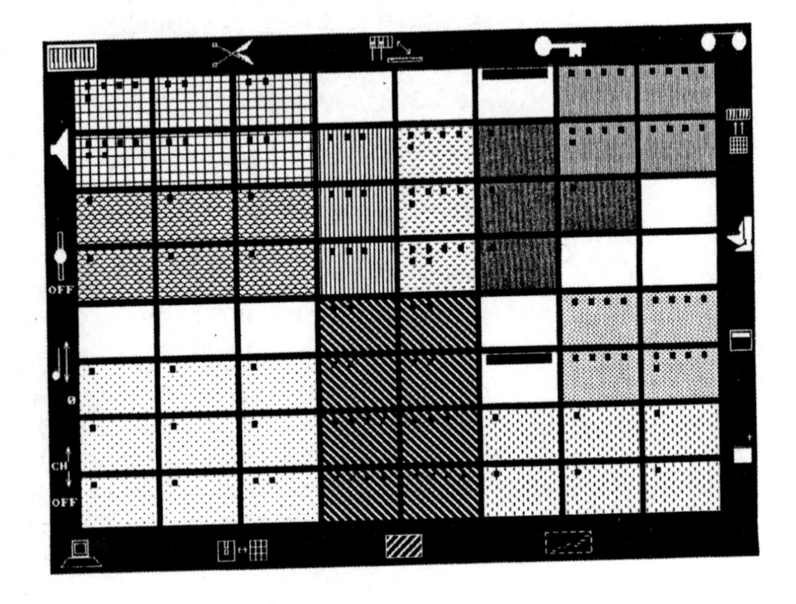

Figure 1.8 An example screen from the *MidiGrid* computer instrument. Boxes contain notes, chords and sequences and are played using a computer mouse.

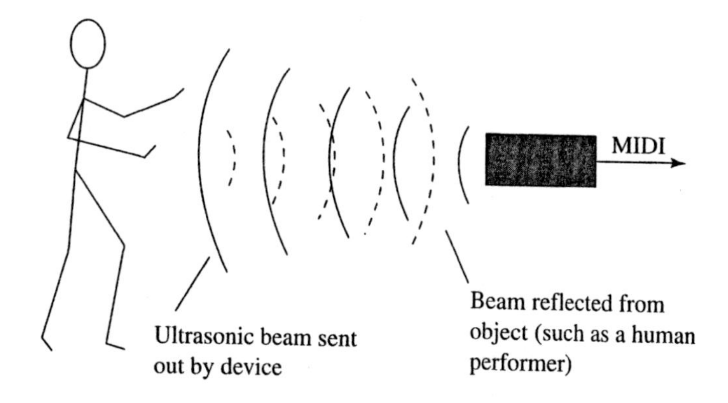

Figure 1.9 A device that sends out an ultrasonic beam. It also measures the time taken for the sound to be reflected. This is then converted into MIDI messages and can be used to play sound.

messages. The University of York's *MidiCreator* system will turn any electrical switch or resistance control into MIDI. Thus dials, levers, switches, even bits of wire immersed in water can be used as a musical instrument. The *MidiGesture* is an ultrasonic beam which plugs into the *Creator*. All these devices are used widely in helping people with disabilities to gain access to music. Switches, sensors and beams can be configured to suit particular physical problems, and can be played without the combination of fine control and considerable physical effort required by many acoustic musical instruments. People are even investigating how to use variations in skin resistance and brain-waves to drive MIDI and computer systems.

- *Gloves and wires:* The STEIM team in Amsterdam, led by Michael Waisvisz, produced a two-gloved device called 'The Hands'. The performer wears the gloves which sense finger position and hand separation and convert these to a stream of control signals. A further development consists a lattice of wires which are under tension. 'The Web' allows players to subtly pull the wires about and disturb the tension distribution and thus influence the musical signals. Both of these developments show much promise for 'explorative' control over a complex set of parameters.

1.16 Interactive music environments

Many systems exist nowadays which allow the user to construct 'networks' of musical processing objects. Often these objects are represented by graphical icons on a computer screen and are connected together using a computer mouse. This type of software package allows the user to devise interactive musical environments, where certain gestures produce specific musical results. Two specific examples of this are 'Max' and 'MIDAS'.

1.16.1 Max

'Max' is a program, written in the 1980s by Miller Puckette, that allows composers to define interactive musical environments. It offers a fast way of prototyping performance environments and sound-processing techniques. The composer uses the computer's graphical screen to connect together a set of boxes that each represent a simple process. For example it is possible to connect a 'get a note from a MIDI keyboard' object to a 'transpose a note up one octave' object. The output of this example network would then collect notes from a MIDI keyboard and transpose the pitch up an octave. More complex networks can be built up to achieve a variety of interactive musical situations.

1.16.2 MIDAS

The University of York UK's 'MIDAS' system is an audio-visual toolkit which is designed to run on any future computing system. Composers can use it to design interactive multimedia compositions and installations in the confidence that these works will be performable even if the current computing technology goes out of date. MIDAS can also operate on many computers at the same time (in the form of a computing *network*). This networking ability means that extra processing power (which is vital for real-time performance) can be added into the system without changing the composer's interface. Composers can work with MIDAS in a variety of ways (from graphically connecting together boxes that represent audio-visual functions, to programming the system in computer code). We refer to it in more detail in Chapter 10.

1.16.3 The Internet

The exponential growth in recent years of the world-wide interconnection of computers – known as the Internet – is transforming the way that people communicate. Two visible signs of this are the growth of personal *e-mail* addresses and the necessity for a company to have its own 'home-page' on the World Wide Web (WWW).

Some people think that it will not be long before the entire commercial music industry is overturned or transformed by the Internet. The present sluggish (if not tedious) response of the Internet when downloading sound files may make us suspicious about the feasibility of some of these ideas. However, many people believe that in a few years' time we will no longer buy tapes or CDs, but will download the tracks of our choice from on-line catalogues or even download them in *real-time* and pay for them for each time we listen.

1.17 Summary

Electronics and computer technology have had an ever increasing and transforming effect on the storage, distribution, and production of music in the twentieth century. Each new invention brings its own new method of operation which must be learnt by its users. Tape, and more recently computers, have created an entirely new mode of operation in the musical world; the complete preparation in advance of a piece of music which is 'performed' at the touch of a button. However the essence of music for thousands of years has been its interactive human quality and given the recent technological advantages (which have made real-time audio computing possible) it is no wonder that the search is now on for new methods of musical *performance* and *control*.

It is easy to become distracted by the glamour of the latest technology, though what does *not* change is the mysterious human compulsion to communicate feeling and thoughts via the sonic language we call music.

The twentieth century has seen an enormous expansion and proliferation of musical ideas and the technology to enable them. It is an exciting prospect to consider what the next century might bring, and the role which *today's* musicians might have in shaping this future. This book is dedicated to those starting out on this path.

Suggested further reading

The web-site for this book (http://www.york.ac.uk/inst/mustech/dspmm.htm) contains links to related sites on the Internet covering these topics. Other books for recommended reading are listed below :

Dodge, C. and Jerse, T. (1985) *Computer Music: Synthesis, Composition and Performance* (especially Chapters 1, 8 and 9), Macmillan, New York, ISBN 0-02-873100-X.

Manning, Peter (1993) *Electronic and Computer Music*, Clarendon Press, Oxford, ISBN 0-19-816329-0.

Roads, Curtis (1995) *The Computer Music Tutorial* (especially Part V), MIT Press, Cambridge, Massachusetts, ISBN 0-262-68082-3.

Rumsey, F. and McCormick, T. (1997) *Sound and recording: an introduction*, Butterworth–Heinemann, Oxford, ISBN 0-240-51487-4.

Part 2

Sounds and signals

The purpose of this section is to understand the nature of the sound signal as it is encountered in digital audio and music technology systems – its structure, genesis and transformation. Part 2 also deals with the important topic of sampling, and the representation of a signal as a sequence of binary numbers – the basis of digital audio. In essence, this section provides an introduction to signal processing for digital sound systems. No prior knowledge of these topics is assumed on the part of the reader.

Chapter 2 describes the nature of the acoustic signal which is manipulated within digital audio and computer music systems. Chapter 3 then considers the way in which the signal is sampled and processed in elementary signal processing algorithms.

2 Sound and signals in music technology and digital audio

Overview

Digital audio and music technology are concerned with the recording, transmission and manipulation of the acoustic signal within electrical systems. Chapter 1 outlined the experiments of the pioneers of sound processing, and we can see that an understanding of the sound signal is an important first step in following in their work. In order to be able to develop and exploit these processes, it is necessary to understand the nature of the acoustic signal as it is represented in the electrical world, and to recognise the effects which transmission through electrical systems will have on these signals. This is the purpose of this chapter. Chapter 3 will then describe how the signal is converted into a form which can be processed and stored in computers and other digital audio systems.

Topics covered

- The electrical analogue of acoustic signals.
- The nature and content of signals.
- The propagation of signals through electrical systems.

2.1 The electrical analogue of acoustic signals

In modern society, we are surrounded by so much 'consumer' electrical audio equipment that we might be forgiven for thinking that the existence of an acoustic signal in an electrical form is one of the fundamental properties of the universe! In fact, the validity of the process of *encoding* an acoustic signal into an electrical *analogue* would have been far from obvious to most of our forefathers of even two or three generations back.

As shown in Figure 2.1, an acoustic signal, such as the sound produced by a drum, travels as pressure waves through the air. When the drum is struck by a player, the vibrations produced in the skin of the drum head alternately compress and rarefy the air immediately surrounding the drum. These pressure waves travel through space, rather like ripples on the surface of a pond. When these waves fall on our ear, they cause vibrations to be set up in a membrane located at the boundary of the outer and middle ear and these vibrations are perceived as sound by the auditory system. However, as also shown in Figure 2.1, it is possible to transmit these acoustic signals via an electrical system.

In this case, the sound pressure waves fall on a moveable diaphragm in a microphone and this causes the diaphragm to move in direct sympathy with the sound waves. There are many different kinds of microphone, and it is not necessary to consider them all here. However, the *magnetic microphone* shown in

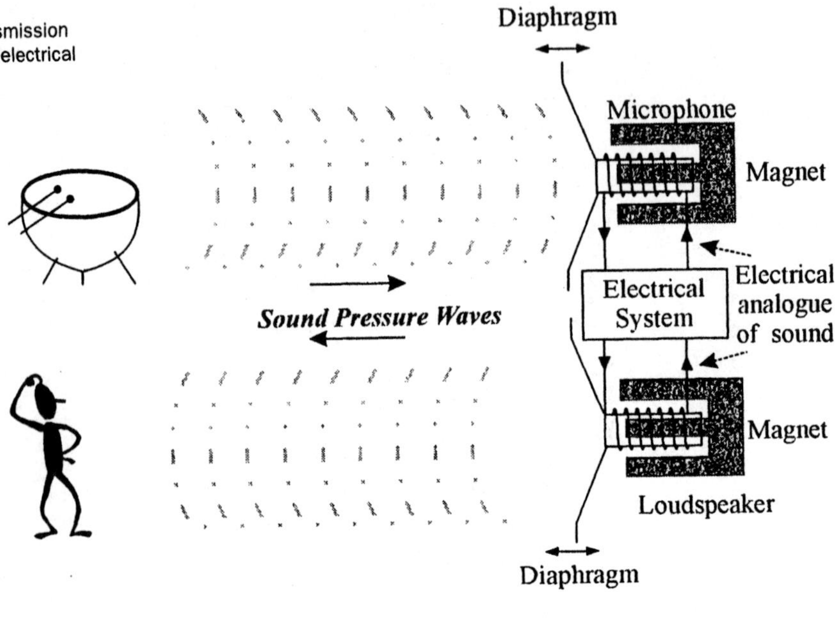

Figure 2.1 The transmission of sound through an electrical system.

Figure 2.1 will assist our understanding of the nature of signals. A small coil of wire is wound on a bobbin attached to the microphone's diaphragm, and this coil is placed in close proximity to a magnet, so that it sits in a strong magnetic field. The magnetic microphone works on the same principle as the cycle dynamo: if a wire is moved in the presence of a magnetic field, then a current is induced in the wire. When our bobbin moves with the diaphragm as the sound waves fall on the microphone, a current is produced in the wire on the bobbin in exact step with the sound waves. Within the mechanical capabilities of the diaphragm/coil assembly, this current is an exact *electrical analogue* of the incident sound waves. We may manipulate and store this electrical analogue, and any changes which we make upon it will have an equivalent effect on its analogue in the sound world.

Figure 2.1 also shows the process of converting the electrical analogue back into sound waves using a *loudspeaker.* A loudspeaker is like a magnetic microphone worked in reverse. The cycle dynamo principle above is a reversible effect: if a current is *forced* to flow in wire which is placed in the presence of a magnetic field, then the wire will tend to move. Therefore if the wire is wound on a bobbin attached to a diaphragm, then the diaphragm will move in step with the electrical analogue (the current), which is itself a close analogue of the original sound. The movement of the loudspeaker's diaphragm compresses and rarefies the air surrounding it, setting up sound waves which we perceive as sound signals. These signals are an accurate analogue of the original signal, within the electrical and mechanical capabilities of the sound reproduction system.

Having established the analogue nature of the electrical signal, we are now in a position to discuss all of the operations and transformations which we can impart on this electrical analogue, understanding that there will be an equivalent effect on the sound wave emanating from the loudspeaker. These operations and transformations form the substance of the rest of this book!

2.2 The nature and content of signals

As a first step in understanding the nature of the electrical analogue of sound signals, we might think that some system such as that shown in Figure 2.2 could help us to visualise the construction of sound. An oscilloscope is an electrical instrument which displays variations in voltage (vertical axis) against time (horizontal axis). It therefore traces out fluctuations in voltage over time on its screen. Voltage is a measure of the electrical 'pressure'

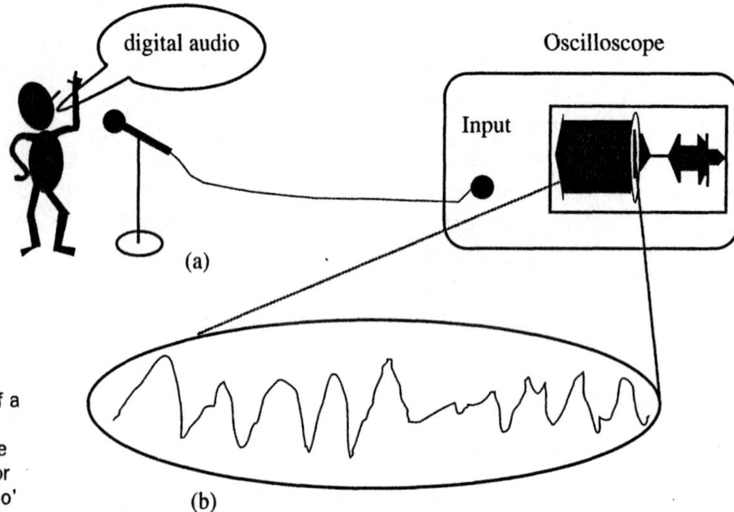

Figure 2.2 (a) An oscilloscope used to visualise the output of a microphone; (b) a closer view of the oscilloscope screen for the words 'digital audio' spoken into the microphone.

which forces current along the wires of a circuit, and so it is another analogue of sound in our microphone example. The figure shows the waveform which might be observed on an oscilloscope for the words 'digital audio' spoken into the microphone.

Whilst the information on the oscilloscope screen is a completely accurate description of the *information content* of the sound 'digital audio', (if we reproduced the information exactly in an electrical system it would say 'digital audio') we would be probably hard-pressed to *classify* it in any meaningful form, except to describe it as a 'wiggly line'. This indicates to us that we are likely to need to consider simpler signals than the sound of a spoken phrase in our journey of understanding and also that we should develop simpler models, or ways of analysing the content of a signal. These are the main topics of this section.

Figure 2.3 shows a much simpler signal which can also form a *building block* to construct more complex signals. It is known as a *sinusoid*. If you listen to the sound which a sinusoid makes when it is played through a loudspeaker, it has a very pure, almost mellow tone. Perhaps the nearest instrumental sound to this would be that produced by the ocarina, a small flute-like instrument.

Since the sinusoid has an important role in the construction (and analysis) of more complex signals, we need to introduce a few terms which describe its essential characteristics.

Figure 2.3
A sinusoidal signal

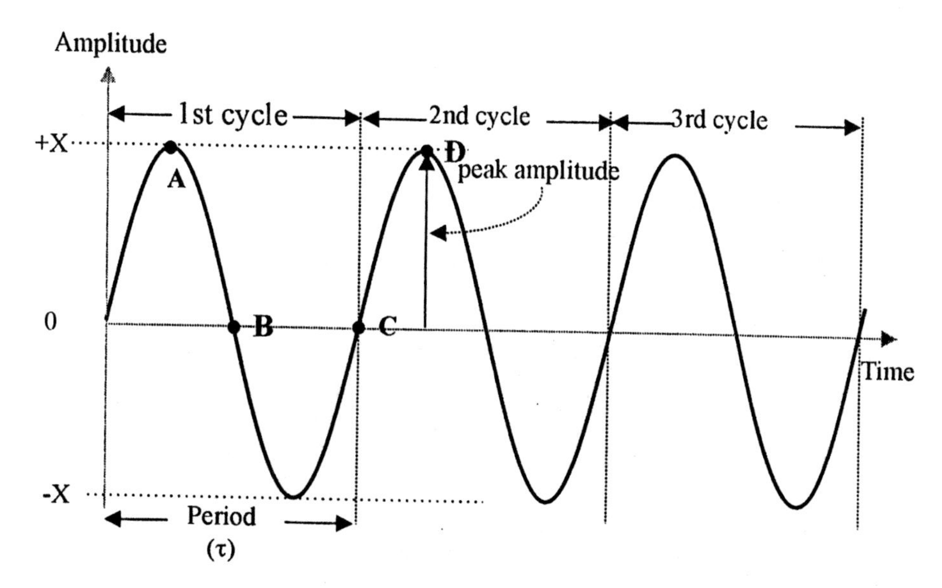

The first thing to notice is that the sinusoid is a wave, continuous in time (time is the horizontal axis in Figure 2.3) which consists of endless repetitions of a basic waveform shape known as a *cycle*. The first few cycles are identified in the figure. The number of cycles which occur per second is known as the *frequency* of the wave. Frequency is sometimes measured in numbers of cycles per second (cps), although this a unit of measurement is now known as the Hertz (Hz). Frequency corresponds to pitch in musical terms. When an oboe tunes the orchestra by playing the note 'concert A', it is playing a pitch of frequency 440 cycles per second (440 Hz).

The vertical axis in Figure 2.3 is known as the *amplitude*. Broadly speaking, the amplitude corresponds to the strength of the signal, so the waveform is a plot of the way in which 'signal strength' varies with time. In our sound world containing a drum, amplitude could describe the variation in strength of the pressure waves set up in the air by the vibrations of the skin of the drum. The louder the note, the greater the peak amplitude of the resulting pressure wave. The higher the pitch, the higher the frequency, or numbers of cycles per second in the number of sound pressure waves. Notice that in the case of the sinusoid, the amplitude swings either side of the horizontal axis. If we take the horizontal axis as the zero line of amplitude, the sinusoid has positive and negative peaks of equal peak amplitude, X, as shown in Figure 2.3.

Figure 2.3 also identifies something called the period (t) of the waveform. This is simply the length of time which one cycle of the waveform occupies. This is linked to the frequency of the waveform. If there are (for instance) 10 cycles per second, then each cycle lasts 1/10 s. Generalising this into an equation:

$$\tau = 1/f$$

and from this,

$$f = 1/\tau$$

where f is the frequency of the waveform.

It is also useful to have a measure of whereabouts *within* a cycle a point such as A, B or C in Figure 2.3 is located. This is known as the *phase* of such a point. Since the waveform exactly replicates itself on every cycle, a measurement of rotation (degrees or radians) is useful for describing position within a waveform. This analogy is useful since a point travelling around the circumference of a circle finds itself in the same position after travelling through a rotation of 360° (or 2π radians) around the circle. By analogy, when a point travels along a waveform, it is said to have reached a phase of 360° when it reaches the same point on the *next* cycle to that which it started. Thus point A is located at phase 90° (π/2 rads) relative to the beginning of the cycle, point B 180° (π rads), point C 360° (2π rads), and D 450° (5π/2 rads) relative to the beginning of the first cycle.

Phase can also be used to describe the position in time of one waveform relative to another, as shown in Figure 2.4. Waveforms 1 and 2 have the same frequency and amplitude, but Waveform 2 is delayed in time by a phase shift of –90° (the '–' sign means a

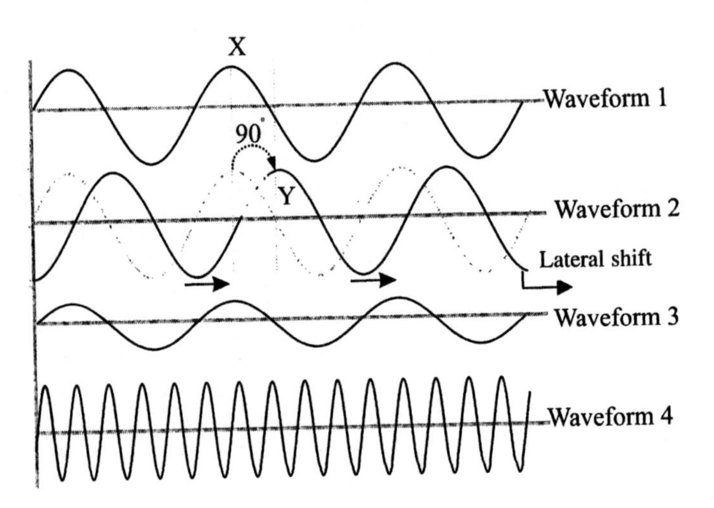

Figure 2.4 Illustrating phase, amplitude and frequency of a waveform.

phase shift delay). Waveform 2 falls behind Waveform 1. This is because point Y lies 90° in phase behind the corresponding point (X) in Waveform 1. A phase shift of a waveform therefore corresponds to a lateral shift of the waveform along the time axis, as Waveform 2 in Figure 2.4 shows. Waveform 3 has the same frequency and phase as Waveform 1, but has half the *amplitude* (it would sound quieter). Waveform 4 has the same amplitude as Waveform 1, and the same initial phase, but has a higher *frequency* (it would sound with a higher pitch).

Now that we understand some of the basic properties of our 'building block' (sinusoidal) waveform, we are ready to understand how it can be used to characterise more complex waveforms. Figure 2.5 shows what happens when we *add* (or to give it its technical term, *superpose*) three sinusoids whose frequencies just happen to be related by simple integer multiples. That is to say, Waveform 2 has a frequency exactly three times that of Waveform 1, and Waveform 3 has a frequency five times that of Waveform 1. The result is shown in Waveform 4. It is known as a *composite waveform* because it is made up of simpler (sinusoidal) components. In the technical parlance, Waveform 1 is known as the 'fundamental', and the others are known as 'harmonics'. A harmonic whose frequency is three times that of the fundamental is known as the 'third harmonic', and one five times the frequency of the fundamental is known as the 'fifth harmonic' and so on.

Waveform 5 shows what happens when the phase (only) of Waveform 2 is changed to that shown in the dotted line. There is a radical change to the shape of the composite.

We could, again, describe the shape of Waveform 5 as a 'wiggly line' (as we did with the 'digital audio' waveform above), but now we have a much greater understanding of its nature and

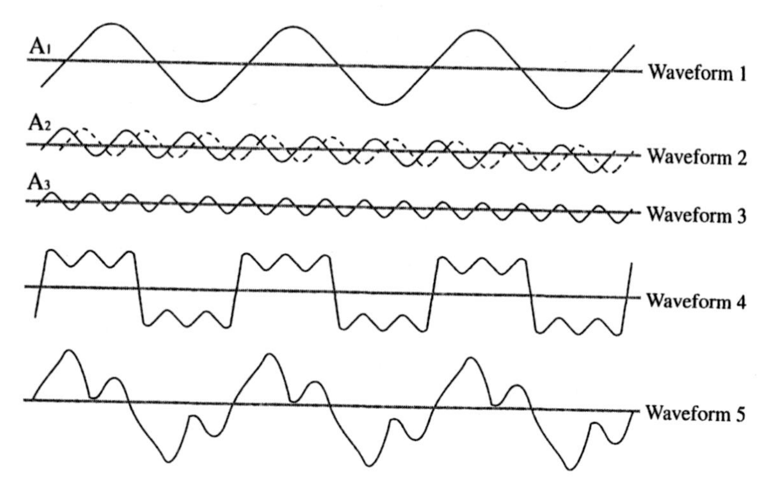

Figure 2.5 The superposition of a fundamental, third and fifth harmonics.

construction, in terms of its 'frequency content' (the superposition of various sinusoids of appropriate frequency, amplitude and phase).

If we were to listen to Waveform 4 or 5 (shown in Figure 2.5) through a loudspeaker, they would have the same pitch as the fundamental, after all they have the same number of cycles per second. However, the higher frequency harmonics add a different tone colour, or 'timbre' to the sound. The *harmonic content* of a sound produced by an instrument is an important factor in our recognition of the instrument. For instance if an oboe and a violin played the same sustained note, the two sounds would have the same fundamental (because they have the same pitch), but the differing harmonic content of the two sounds enables us to identify one sound as violin, and the other as oboe.

It is interesting to note the audible effect of changing the phase of Waveform 2 in Figure 2.5. If we were to listen very acutely, we might detect a slight change in pitch whilst the phase was being shifted to a new position (lateral translation), but thereafter we would hear under normal circumstances no detectable difference (e.g. between Waveform 4 and Waveform 5). This is because to a first approximation, the ear is insensitive to phase. The phase might affect the operation of electronic equipment in some way (because the waveform shape changes for different phase relationships for instance, as Figure 2.5 shows), and this might create certain electronic artefacts which our ears might detect, but these are secondary effects.

The slight change in pitch we detect whilst the phase is changing is because frequency is related to phase – frequency is the rate of change of phase. After all, phase is a way of measuring the number of cycles in a waveform (by having a point such as D in Figure 2.3 trace along the waveform), and frequency is the rate of production of cycles of the waveform.

A composite waveform such as those shown in Figure 2.5 is sufficient to completely identify the signal. Just as an electronic instrument replaying the wiggly line in Figure 2.2 over a loudspeaker would enable it to say 'digital audio', the same system playing the wiggly line corresponding to Waveform 4 or 5 in Figure 2.5 might produce a sound approximating to a simple organ stop.

It would sound a note whose pitch is set by the fundamental, with a timbre appropriate to third and fifth harmonics. However, trying to *classify* the sound of an electronic instrument by defining the waveform shape (e.g. by drawing Waveform 5 in the figure) would be exceedingly tedious, if not impossible to carry

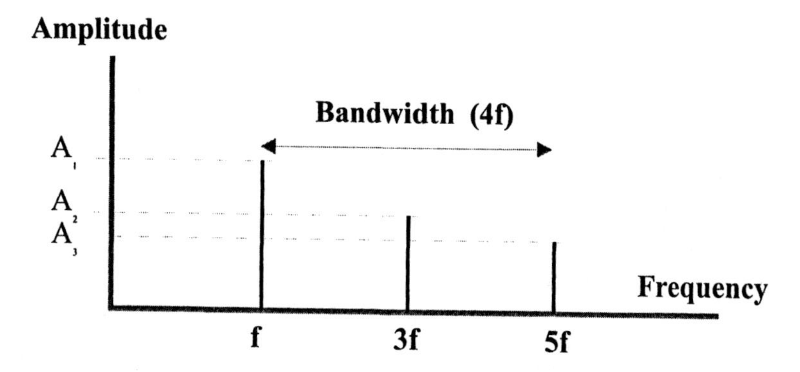

Figure 2.6 The frequency domain representation of the signal in Figure 2.5.

out. We need a more convenient way of representing the same information as that shown in Figure 2.5. The diagram shown in Figure 2.6 does this.

In Figure 2.6, a line represents a sinusoid which forms one of the frequency components of a signal. Its frequency is indicated by the position on the horizontal axis and its amplitude by the length of the line. The signal described by Figure 2.6 therefore has three frequency components: fundamental, third and fifth harmonics with amplitudes of A_1, A_2, A_3. This is *exactly the same information* as that in Figure 2.5, but drawn more conveniently. Phase can also be represented in Figure 2.6, by showing a rotation of a frequency component around the horizontal (frequency) axis, but this would convert the diagram into a three-dimensional image, and again make it difficult to draw. Phase is therefore not usually shown in such diagrams, unless it is particularly relevant.

We now have *two* ways of representing the structure of a signal: Figure 2.5, known as the *time domain* representation, since it is a graph of amplitude versus *time*, and Figure 2.6, known as the *frequency domain* representation because it is a graph of amplitude versus *frequency*.

Any *signal* or *process applied to a signal* therefore can be represented in either of these two ways. They are like two different sides of the same coin. Sometimes a signal or process will be easier to describe and understand in the time domain, sometimes in the frequency domain. Needless to say, we shall always choose the simpler in the chapters which follow! However, it is important to realise that any process applied in one domain will always have an equivalent in the other, it is just that it may be more difficult to follow.

The line structure of a signal shown in a frequency domain diagram is often known as the *frequency spectrum* of the signal.

The width of the frequency spectrum, measured in Hz, is known as the *bandwidth* of the signal, as indicated in Figure 2.6. It is one of the important sonic characteristics of an audio signal. Just as there are instruments which display the time domain behaviour of a signal (the oscilloscope), there are others, known as spectrum analysers which display the frequency domain, in a similar form to that shown in Figure 2.6.

It is actually a rather gross simplification to imply, as we have so far, that the frequency spectra of natural acoustic sounds are entirely harmonic in nature. A device such as the reed of a woodwind instrument vibrates at many different frequencies depending upon its precise physical dimensions, and these frequencies are not necessarily harmonically related. Even though the resonating bore of the instrument (the 'tube') may favour the harmonic components, to give a well pitched sound, there will still be these inharmonic components present in the sound. They are important, as they give a certain timbral edge or 'bite' to the tone produced. If they are absent in a sound which is created for instance in an electronic instrument, the ear will quickly interpret the sound as 'dead' or 'artificial'. Unfortunately it is rather easy to produce such dead sounds as by-products of the sound synthesis process in electronic instruments, as we now show.

Returning to Figure 2.5, we notice that successive cycles of the composite waveform (4 or 5) have identical shape. This is because each of the frequency components starts from the same point in its cycle for the beginning of each cycle of the composite. This in turn is because there is an exact integer number of cycles of each component within each cycle of the composite (i.e. they are exactly harmonically related). The components are said to be *phase locked*.

Baron Fourier recognised this in his famous statement that a stationary wave (i.e. one whose successive cycles have the same shape) can be resolved into *harmonically related* components. So-called 'Fourier analysis' is a mathematical technique which resolves a composite into its frequency components – that is, which effects a transformation of a time domain signal into its frequency domain equivalent. Various computer programs used in digital audio and music technology carry out this transformation process, examples being the Fast Fourier Transform (FFT) and the Phase Vocoder.

Contrast this phase locked behaviour with the situation shown in Figure 2.7, which is again a superposition of frequency components (two this time), where the frequencies are *not* harmonically related – the higher frequency is about 2.8 that of the lower.

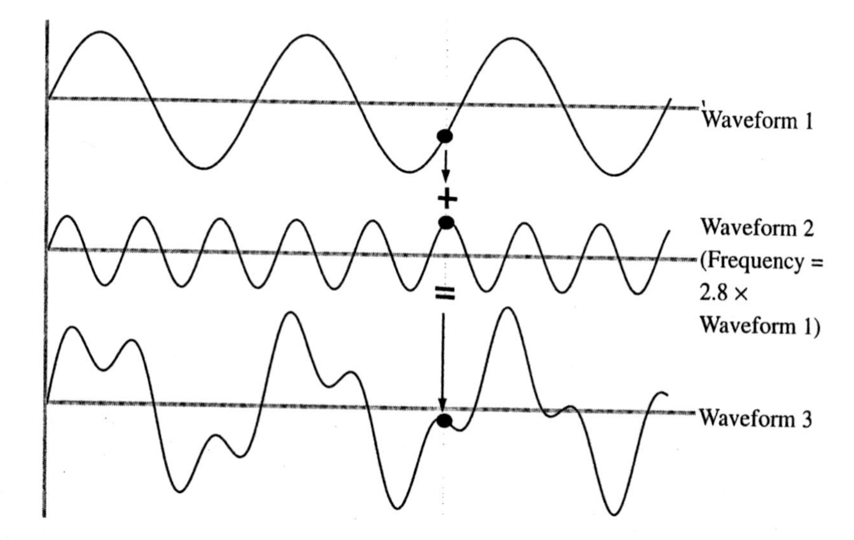

Figure 2.7 A composite wave formed from inharmonic components.

Notice that the resulting composite still has the same pitch as the fundamental, but that each successive cycle is slightly different in shape. This is because the components are not phase locked. In musical parlance, both kinds of frequency components (harmonic or inharmonic) are called 'partials'.

The consequence of this behaviour is that any electronic instrument which reproduces the same waveform shape on each cycle, as a result of a particular synthesis process, will produce a sound devoid of any inharmonic content. Many synthesis processes behave in this way and have a detrimental effect on the acoustic quality of the sound as noted above. This is one reason why electronic samplers will not replace natural acoustic instruments in the foreseeable future (fortunately!).

Any sound such as speech or music can be resolved down into a superposition of harmonic and inharmonic partials. If the speech or music changes through time (as of course it does, continuously), then the frequency content will correspondingly change on a continuous basis.

We now have an adequate understanding for our purposes of the nature of the signal which occurs in digital audio and music technology. We are ready to consider what happens to that signal when it is transmitted through an audio system. This is the subject of the next sections.

2.3 The effect of linearity and gain on the transmission of signals

The different media through which signals are transmitted all have their own effect on the frequency, phase and amplitude content of the signals. Examples of such media include electronic equipment such as amplifiers, CD and tape players, the tapes themselves, loudspeakers and microphones, and in the natural acoustic world, the body of an instrument such as a violin (even the varnish it is coated in!), the air and our ears.

All of these components in the audio chain affect our perception of the signal through their *modification* of the characteristics of the signals propagated through them. This section looks at effects imposed by these media on the amplitude characteristics of the signal. Section 2.4 will look at effects on frequency and phase.

Figure 2.8 shows a typical 'gain' or 'transfer characteristic' of a device such as an amplifier. It is a graph of how input signal amplitudes (positive and negative) are transformed into corresponding output amplitudes. 'Gain' is a parameter which describes how the *amplitude* of a signal is transformed when it is transmitted through a system. When the gain is greater than 1, the output signal amplitude is greater than the input, and the system therefore provides *amplification*. When the gain is less than 1, the system provides *attenuation*. The system *multiplies* the input amplitude by the gain, to produce the output amplitude.

Figure 2.8 Typical gain characteristics of a non-linear system.

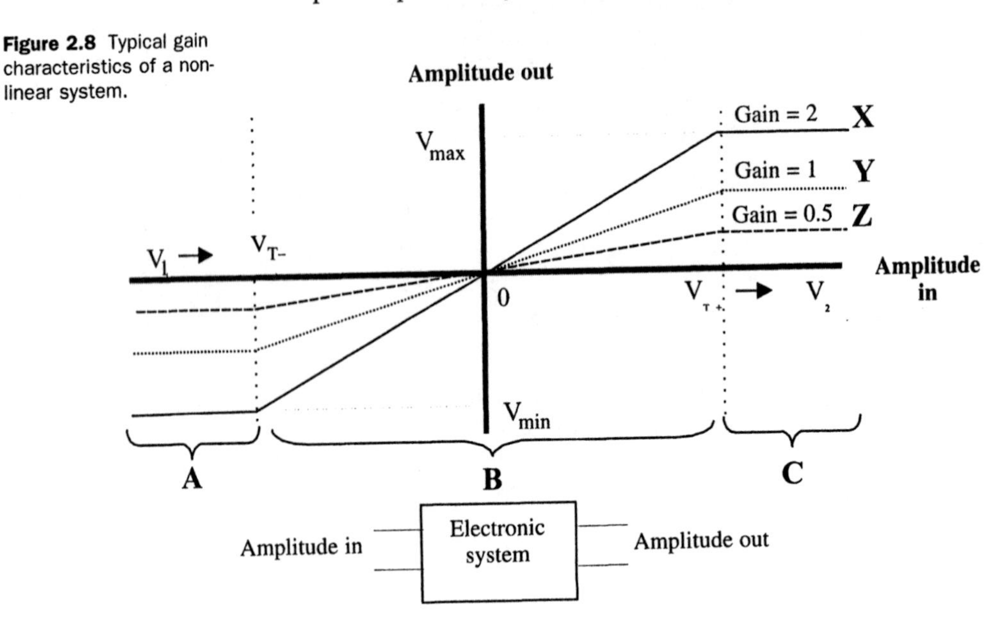

The transfer characteristics shown are rather typical of electronic systems. Within the region marked 'B' in Figure 2.8, signal amplitudes are scaled (multiplied) by a simple number. In our example this is 2 for characteristic X, 1 for Y and 0.5 for Z. This is the region of 'linear gain'. However in the regions A and C, the amplitude of the signals at the output of the system have a fixed value, no matter what the amplitude of the input signal. These are called regions of 'gain saturation'. They are regions where the amplifier (or whatever it is) cannot provide any more gain for signals greater than V_{T+} or less than V_{T-}. The amplifier has just 'run out of steam'. This is often caused by the limitations of power supplies, etc. in the amplifier.

If you imagine that you had a 'volume control knob' which allowed you to select an input amplitude ranging through values from a large negative value, V_1 to a large positive value, V_2 (see Figure 2.8) then assuming that the system has characteristic X, the output amplitudes returned by the system as you turned the knob would be as follows. There would be a fixed negative value, V_{min} for an input up to a value V_{T-}, then a value where the input amplitude is multiplied by 2 (a gain of 2) for all input values falling between V_{T-} and V_{T+}. For all input amplitude values above V_{T+} the system would output a fixed value V_{max}.

Let us suppose that the threshold values V_{T-} and V_{T+} are set at values –3 and +3 respectively, and let us apply sinusoidal signals to the input of the system, rather than our 'manual' volume control knob values. Figure 2.9 shows the effect of this transfer characteristic on input sinusoidal signals of various amplitudes. The first three Waveforms (a) to (c) show the output of the system if an input sinusoid of amplitude 1 is applied to a system with gain characteristics X, Y and Z respectively. In these cases, the input signal amplitude lies within the region of linear gain, so the system simply multiplies the input signal everywhere by the fixed gain value (2, 1, 0.5 respectively). Importantly, since all input values are treated equally, if the input is a sinusoid, *then the output will also be a sinusoid*. The shape of the signal is not changed within the region of linear gain. Its amplitude is simply multiplied by the gain value.

The case of Waveform (d) however is different. Let us suppose that a sinusoidal input signal of amplitude 4 is now applied to gain characteristic Y. Here the input strays at the peaks of the input signal beyond the threshold values ±3. In between the output amplitudes ±3, the output signal is a faithful copy of the input signal (the linear gain is 1 here). However at the peaks of the signal, the output is clipped – it is limited to the saturation

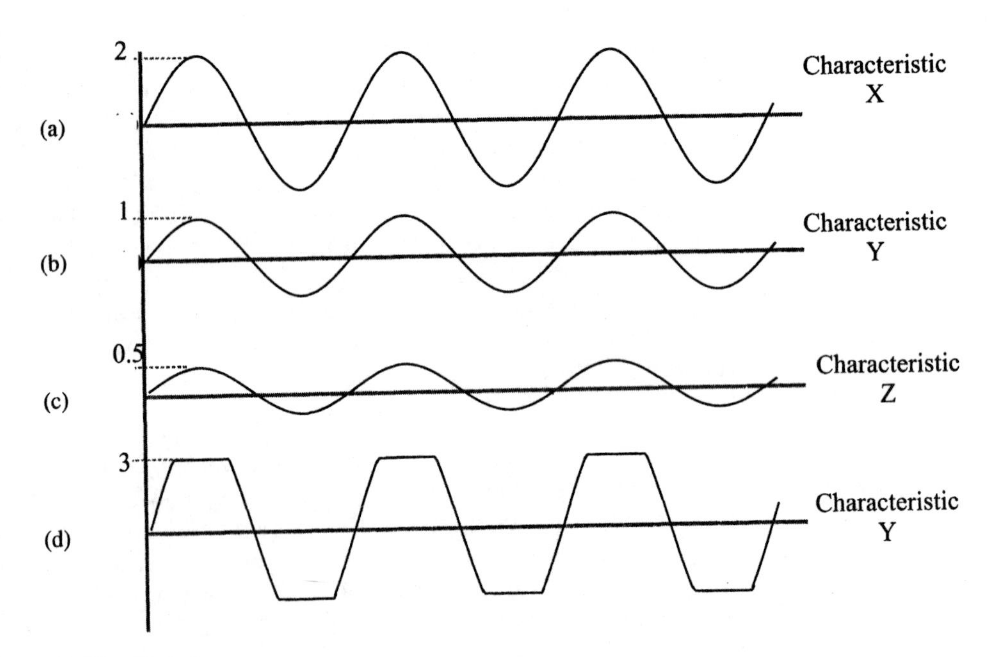

Figure 2.9 The effect of non-linear gain on signal amplitudes.

value as the amplifier runs out of steam. We would hear this as a distortion. This is the rather fuzzy sound which we hear at the loud parts of the signal when a bass guitar is played, too loud, through a poor amplifier.

Now, as far as we are concerned, there are only two types of signal: sinusoids and composites. Waveform (d) is clearly a composite, departing from the pure sinusoidal shape when the input signal hits the non-linearities in the transfer characteristic. Waveform (d) therefore consists of a fundamental whose frequency equals that of the input signal (the composite has the same pitch as the input, since it is directly produced by it), plus a number of harmonics. These harmonics are produced directly by the non-linear transfer characteristic. They are known as *harmonic distortion*.

Audio engineers go to great lengths to avoid harmonic distortion by making their audio equipment *linear.* One technique used which some readers may have heard of is *feedback*. Here, some of the gain of the system (e.g. an amplifier) is sacrificed to make the transfer characteristic linear. The degree of success of this procedure is quantified in a measurement of *total harmonic distortion (THD)*. To measure this, a very pure sinusoid is input to the system and the component present in the output at this frequency (the fundamental) is removed by a process known as filtering (see Section 2.4). This leaves only the harmonic components. The amount of energy in these components is then measured and compared with that of the fundamental. The result is expressed as a percentage figure – we are obviously aiming for a low figure for the THD, perhaps of the order 0.01 per cent, or lower.

It is typical of a musician's use of audio systems that what would be otherwise regarded as an imperfection (harmonic distortion) is often exploited and even accentuated! For some musical applications, a non-linear gain characteristic is deliberately chosen because the resulting harmonic distortion provides a timbral enrichment to the sound produced. This is known as *waveshaping synthesis.* It has a long and honoured history amongst electric guitarists, who learned to work creatively with the (distorted) sound produced when their amplifiers became faulty, and therefore suffered non-linearities. In waveshaping synthesis, a problem exists in controlling the *precise shape* of the gain characteristic to produce the desired harmonic structure. This problem can be solved by using a series of non-linearities produced by mathematical functions known as Chebychev polynomials. The harmonic generation properties of these polynomials are controllable and predictable.

2.4 The effect of frequency response on the transmission of signals

The discussion in Section 2.3 ignored the *frequency* of the signals being propagated through audio systems; it concentrated purely on *amplitude* effects. The implication was that all frequency components are treated equally by the gain characteristic – that the gain was not selective in any way with respect to frequency components present in an input signal. This is an unrealistic simplification.

In practice, virtually all systems have a frequency-selective behaviour in addition to non-linearities – they have more amplification for certain frequencies than they do for others. They also impart phase changes on signal, where the amount of phase shift varies with the frequency of the signal or harmonic. The amount of variation imparted by the system on a signal's amplitude and phase, plotted against frequency is known as its *frequency response.*

An electrical component which amplifies or attenuates certain frequencies in relation to others is known as a *filter.* Although the description above suggests that the frequency selective (or filtering) action of systems may be unavoidable, there is no implica-

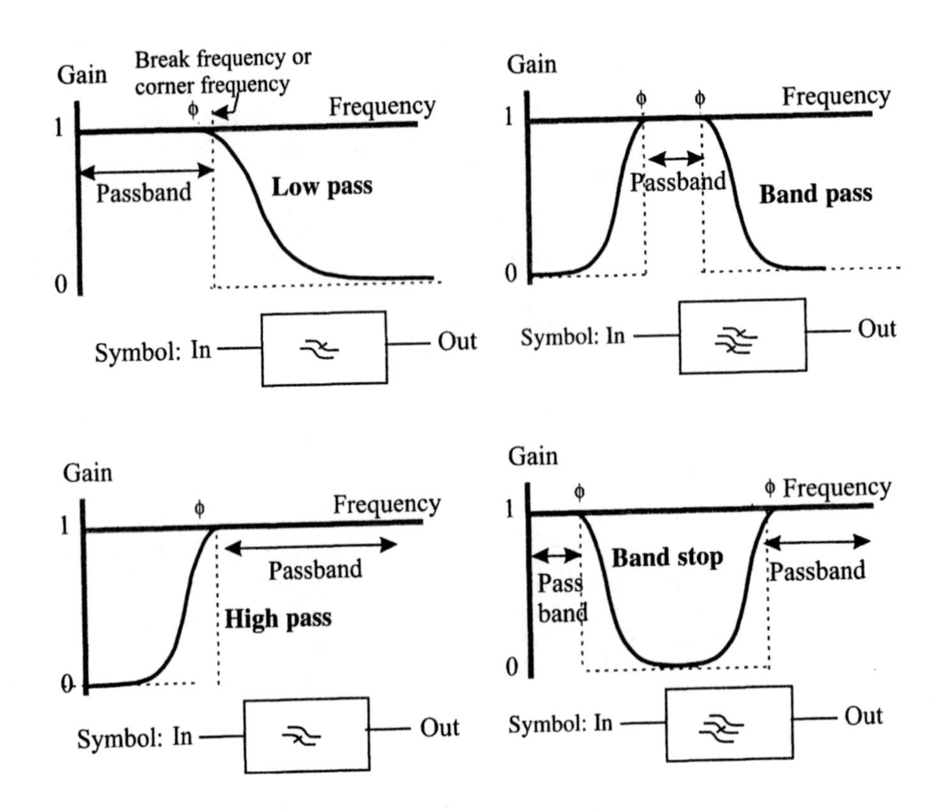

Figure 2.10 The amplitude characteristics of the four major classes of filters.

tion that such action is undesirable. The application of filters are essential in digital audio and music technology systems. We have already seen one application in the measurement of harmonic distortion in the previous section, and we shall see many others later in the book.

There are, in fact, four major classes of filter, as shown in Figure 2.10. The low pass filter transmits (passes) and possibly amplifies the lower frequencies, and removes the higher frequencies by attenuation (the gain falls to 0 at high frequencies). The opposite occurs with the high pass filter. High frequencies are retained, and possibly amplified, whilst the lower frequencies are removed. The bandpass filter transmits signals within its passband, and removes those frequency components above and below this passband. The bandstop filter has the converse frequency characteristic. It removes frequencies within a band and transmits those lying outside the band. Figure 2.10 shows the symbols usually given to these filters in block diagram schematics of audio systems such as mixing desks and synthesisers.

The solid lines in Figure 2.10 show the amplitude responses of typical filters. The gain approximates to 1 in the passband (the signal amplitude is multiplied by 1, and so is unchanged), whilst it is multiplied by less than 1 (i.e. multiplied by a fraction, progressively reducing the amplitude) outside the passband.

Figure 2.10 also shows the amplitude responses of *ideal* filters of the four classes. These are shown as dotted lines. Within the passband of these characteristics, there is no change of amplitude (gain = 1), and then infinite attenuation abruptly applied outside the passband (signal multiplied by 0).

Anyone who is familiar with audio systems is also familiar with the use of filters, even though s/he may not be explicitly aware of it. The use of bass and treble tone controls on amplifiers is a direct example of the use of low pass and high pass filters (respectively) in audio systems. These days, such tone controls are often replaced by graphic equalisers. However these are in essence just a cascaded bank of filters: low pass, band pass, and high pass, as shown in Figure 2.11 where the output of one filter directly drives the input of the next in the chain. The corner frequencies (ϕ) of the filters are all set to different values, and can be adjusted by control knobs on the front panel of the audio system. This is indicated by the arrows in the figure.

The overall gain of the graphic equaliser, V_{out}/V_{in} is given by:

$$\frac{V_{out}}{V_{in}} = \frac{V_{out}}{V_4} \times \frac{V_4}{V_3} \times \frac{V_3}{V_2} \times \frac{V_2}{V_1} \times \frac{V_1}{V_{in}}$$

where the signal amplitude at a certain point is given the symbol *V*. Interpreting the equation, the overall gain or frequency response of the graphic equaliser is given by the *product* of the frequency responses of the individual elements in the cascade of filters. Now each of the individual frequency responses is typically expressed as a graph of gain versus frequency, such as those shown in Figure 2.10. Therefore in order to obtain the overall frequency response, we would have to carry out frequency point by frequency point *multiplication* of the frequency responses of all of the elements in the cascade. Whilst this is feasible, at least in

Figure 2.11 A cascaded five element graphic equaliser.

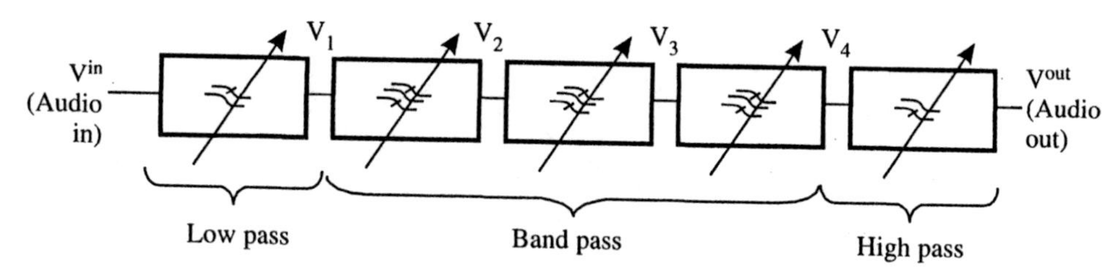

concept, it would be easier in practice if this point by point multiplication could be converted into a point by point *addition*. A *logarithmic* expression of gain would enable us to do this. The use of logarithms converts the process of multiplication into a process of addition:

$$\log(A \times B) = \log(A) + \log(B)$$

For this reason, amongst others, gain is often expressed in *decibels* (dB), which is a logarithmic expression of gain.

$$\text{Gain (in dBs)} = 20 \log_{10}(V_{out}/V_{in})$$

If our gain curves, such as those in Figure 2.10, now have their gain axes expressed in units of dB, then the overall gain of a cascade can be obtained by simple *addition* of the individual gain characteristics.

This is illustrated by the example shown in Figure 2.12, where the individual frequency responses, amplitude (dB) and phase, of each element of a two-element cascade is shown plotted against frequency, where the frequency is itself shown on a loga-

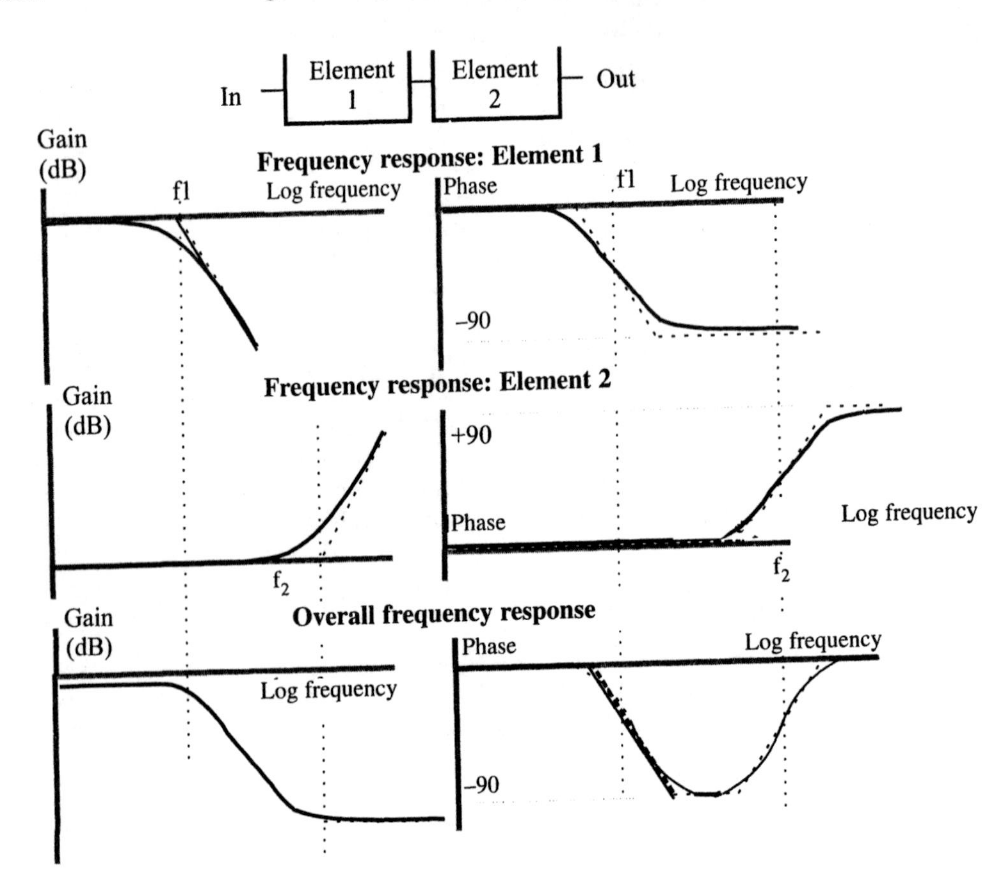

Figure 2.12 Addition of gain curves expressed in dB, to form the overall frequency response of a two element cascade.

rithmic scale. The reason for the use of this logarithmic frequency scale is described below.

The overall frequency response in Figure 2.12 (amplitude and phase) is simply formed by the addition of the two individual frequency responses. The reason for the addition of the amplitude responses (in dB) is explained above. In terms of phase, if the first element for instance gives a phase shift of –90°, and the second +90°, then the overall phase will be 0° (addition of phases).

It is worth noting a few points about dB plots of gain. If $V_{out} = V_{in}$, then $20 \log (V_{out}/V_{in}) = 0$, since $\log 1 = 0$. That is, if a dB plot shows an amplitude response of 0 dB, then there is no change in the amplitude of the signal propagating through the system. If $V_{out} < V_{in}$, (the signal is attenuated), then the corresponding dB plot will show negative dBs, since log of a fraction is a negative number. Similarly a positive dB plot of gain indicates amplification of the signal.

Logarithmic scales of frequency, as shown in Figure 2.12 are often used because in most natural systems (including music!) the *octave* is the natural measurement of frequency. An octave is a doubling of frequency. *Wherever* an octave interval is located on the log frequency axis, it has the same, uniform length. Log frequency paper therefore shows frequency as a *linear* measurement of frequency in the natural unit (the octave), even though this happens to show the 'cycles per second' measurement of frequency (the Hz) on a logarithmic, non-linear scale.

2.5 Summary

In this chapter we have discussed the essential nature of signals, and the ways of representing these essential characteristics. We have met the concepts of superposition, sinusoidal and composite waveforms, time and frequency domains. We also have analysed the important effects which propagation through electrical systems imparts on the signal characteristics: linearity, gain and frequency response. We are now ready to see how these signals and characteristics are captured and represented in digital audio systems.

Suggested further reading

Chamberlin, Hal (1980) *Musical Applications of Microprocessors*, Hayden, ISBN 0-8104-5753-9.

Dodge, Charles and Jerse, Thomas A. (1985) *Computer Music: Synthesis, Composition and Performance*, Schirmer Books, ISBN 0-02-873100-X.

Howard, David M. and Angus, James (1996) *Acoustics and Psychoacoustics*, Butterworth–Heinemann, Oxford, ISBN 0-240-51428-9.

Roads, Curtis (1995) *The Computer Music Tutorial*, MIT Press, Cambridge, Massachusetts, ISBN 0-262-68082-3.

3 Digital audio

Overview

The sound signal in multimedia and music technology systems is almost universally stored and manipulated in 'digital format'. The same is becoming true of the general audio world, with the ubiquitous use of CDs, DATs and digitally based studio systems. This chapter explains what is meant by 'digital format', how it works, and why it is used, including its advantages and disadvantages. The chapter also describes the 'sampling theorem', which is a necessary aspect of the format. The material forms an essential prerequisite to later chapters which describe the techniques used to synthesise and manipulate sound.

Topics covered

- Preparation: frequency translation.
- The sampling process.
- Pulse code modulation (PCM) systems.
- Characteristics of digital audio systems.
- Synthetic audio systems.

3.1 Preparation: frequency translation

In order to understand the sampling process, which is an essential aspect of digital audio systems, we need to understand what happens when signals are *multiplied* together. Signal multiplication is used in its own right for the transformation of timbre, (for instance in various forms of ring modulation). However, its role in this section is really a preparation for our understanding of the consequences of sampling a signal; something which will be dealt with in the next section. Essentially, we need to derive some concepts which will be useful to us later, so please bear with us as we derive this result, and mentally carry it forward to Section 3.2.

We can approach the necessary concept by considering the simple *superposition* of the two signals shown in Figure 3.1. This process of superposition is exactly the same process which we considered in Chapter 2: the simple point-by-point *addition* of Waveform 1 in Figure 3.1 to Waveform 2, to produce the resulting waveform shown as Waveform 3.

The shape of Waveform 3 may be a little surprising on first sight, but you could confirm the result for yourself simply by adding two such signals for yourself using a piece of graph paper. The reason why the *amplitude* of the result (Waveform 3) varies *with time* is because the relative phases of the two constituent waveforms (1 and 2) are gradually but continuously moving apart.

Figure 3.1 The superposition of two (co)sinusoids of slightly differing frequencies. A cosinusoid is a sinusoid with a 90° phase shift.

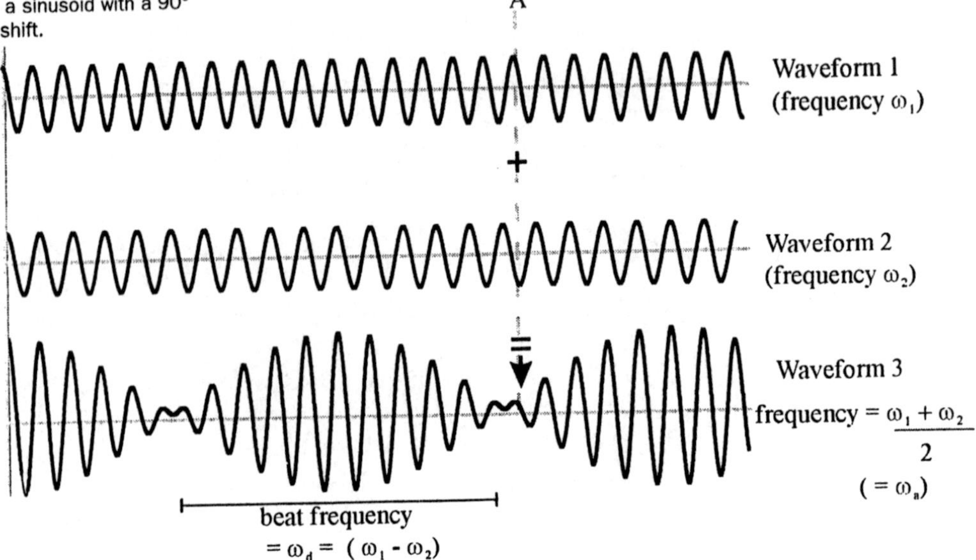

This is because they are of slightly different frequencies. Consequently, at a point such as A, we have a positive amplitude value in Waveform 1 added to a negative amplitude in Waveform 2, to produce a net zero amplitude at the corresponding point in Waveform 3. This kind of process is happening all along the waveforms, to produce a fluctuating 'envelope' (the shape of the amplitude) as shown in Waveform 3.

Let us make some closer observations of the result. If we were to measure the frequency of Waveform 3, say using an oscilloscope, we would find that it was the *average* of the constituent waveforms, 1 and 2. That is to say that the addition of two sinusoids of frequencies ω_1 and ω_2 produces a result whose frequency is at neither of these two frequencies, but is instead of frequency $(\omega_1 + \omega_2)/2$. Let us call this average frequency, ω_a. The rate at which the amplitude envelope fluctuates (known as the 'beat frequency') is equal to the *difference* in frequency of Waveforms 1 and 2. Let us call this the difference frequency, or beat frequency, ω_d.

This 'beating' effect is familiar to musicians who are tuning their instruments. Two musicians (attempt to) play the same note and listen to the rapidly fluctuating (or beating) amplitude of the result. The more rapid the beating effect, the greater the difference in the two individual frequencies, i.e. the more they are playing out of tune. The musicians adjust their playing pitches, until the beats disappear; that is until the difference (ω_d) in their respective pitches is zero, at which point they are playing in tune.

The process of superposition described above is a perfectly correct and adequate description of what happens when we add these two signals together. However, it will suit our purposes in the future to devise a very simple *mathematical model* which describes Waveform 3. By saying that we have a model, we are not saying that the processes described in the model are actually happening in physical reality; rather that the results produced by the model are *directly equivalent to* (or indistinguishable from) the *results* observed in physical reality. The reason for devising such a model is that it will turn out to be more convenient in describing what is going on in other processes (in our case sampling) than the directly observed result (superposition in this case).

Figure 3.2 shows that *multiplying* Waveform 1, a (co)sinusoid of frequency w_b by Waveform 2, another (co)sinusoid of frequency ω_a .This produces a result, Waveform 3 which is *indistinguishable* from that of Waveform 3 in Figure 3.1. This is absolutely the case, if we make the frequency of Waveform 2 in Figure 3.2 (ω_a) equal to the average of the two tones in Figure 3.1, as before, and

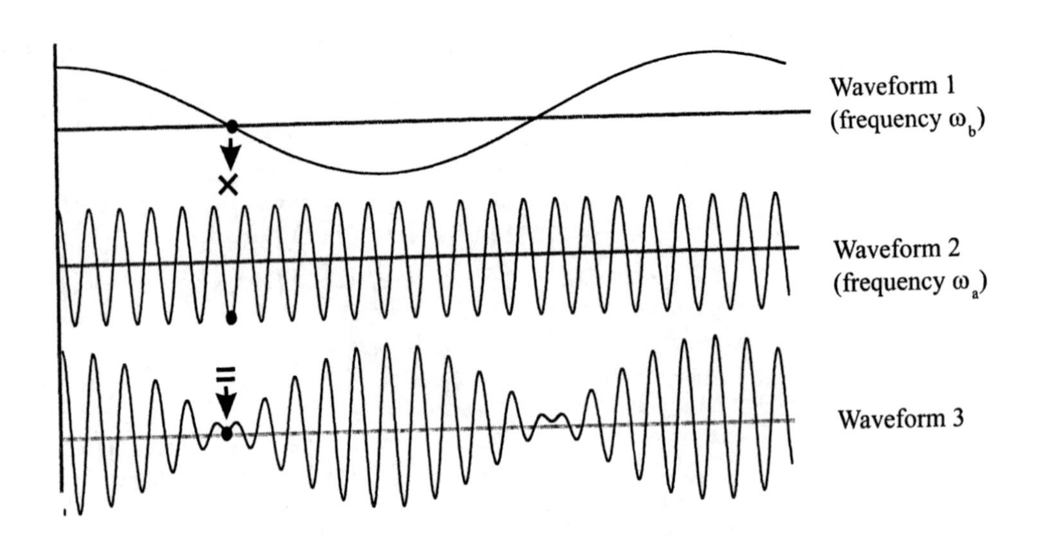

Figure 3.2 Time domain representation of the product of two (co)sinusoids.

ω_b equal to half the *difference* between the frequencies of the two constituent Waveforms 1 and 2 in Figure 3.1 (see also Figure 3.3). This gives the result in the frequency domain shown in Figure 3.3.

To summarise the frequency domain result shown in Figure 3.3, multiplying two cosinusoids of frequencies ω_b and ω_a together produces a result which is equivalent to *adding* two cosinusoids of frequencies ω_1 and ω_2 in Figure 3.2. We obtain a copy of the

Figure 3.3 Frequency domain representation of the product of two (co)sinusoids.

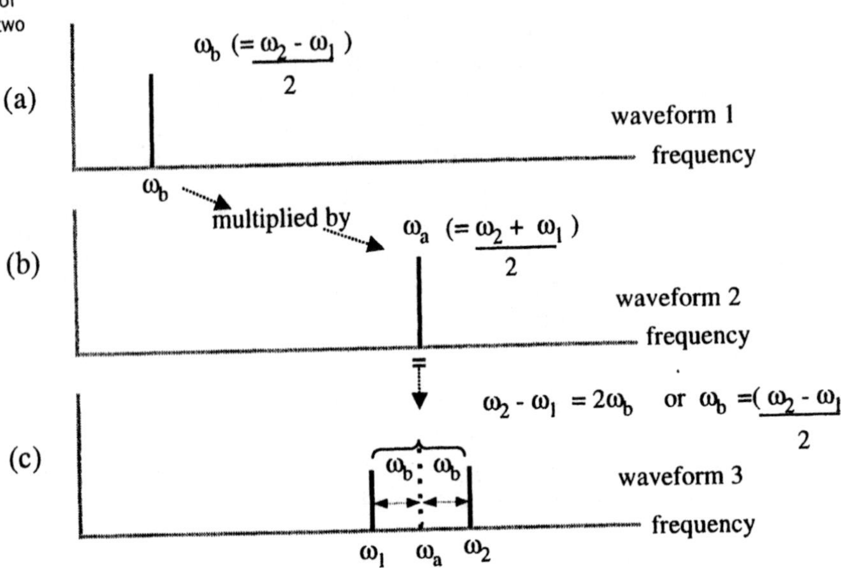

frequency spectrum of Waveform 1 placed either side of the location of the frequency component of Waveform 2, separated from it by frequency of the component in Waveform 1. This process of *multiplication* is the mathematical model described above which is equivalent to the process of superposition shown in Figure 3.1.

For those who remember a little high-school mathematics, this is directly equivalent to a standard trigonometrical result:

$$\cos \omega_a t \cos \omega_b t = 1/2\cos(\omega_a + \omega_b)t + 1/2\cos(\omega_a - \omega_b)t$$

You can rely on your memory of mathematics, or on the evidence of Figures 3.1 and 3.2: our model of multiplication of signals of appropriate frequencies is equivalent to our process of superposition, producing the spectra shown in Figure 3.3.

If Waveform 1 in Figure 3.2 was *itself* a composite, consisting of a number of frequency components, as shown in Figure 3.4(a), *multiplied* by a single frequency component Waveform 2, then the result would be a spectrum shown in Figure 3.4(c). This result could be proved by substituting $[\cos \omega_x t + \cos \omega_y t + \cos\omega_z t]$ (a multi-component composite) for $\cos w_b t$ in the equation above:

$$[\cos \omega_x t + \cos \omega_y t + \cos\omega_z t] \cos \omega_a t =$$

$$\mathbf{1/2\cos(\omega_a + \omega_x)t + 1/2\cos(\omega_a - \omega_x)t}$$

(a pair of components at positions (1) in Figure 3.4)

$$\mathbf{+ 1/2\cos(\omega_a + \omega_y)t + 1/2\cos(\omega_a - \omega_y)t}$$

(a pair of components at positions (2) in Figure 3.4)

$$\mathbf{+ 1/2\cos(\omega_a + \omega_z)t + 1/2\cos(\omega_a - \omega_z)t}$$

(a pair of components at positions (3) in Figure 3.4).

Again, we have a copy of the spectrum of Waveform 1 arranged either side of the component of Waveform 2 (ω_a), with each component being separated from ω_a by its frequency in Waveform 1. This means that the highest frequency components in Waveform 1 (ω_z in this case) are separated (by the greatest frequency) from ω_a in Waveform 3, and this gives the mirror image aspect to the spectrum of Waveform 3. This spectrum is known as 'double sideband suppressed carrier' (DSBSC) in the telecommunications literature. In Figure 3.4 there are clearly two sidebands, or groups of frequency components: an upper sideband (USB) and a lower sideband (LSB). The suppressed carrier refers to the *absence* of the component ω_a (known as the 'carrier') in the figure.

DSBSC is an example of *frequency translation* since we have a copy of the input frequency spectrum (spectrum (a) in Figure 3.4)

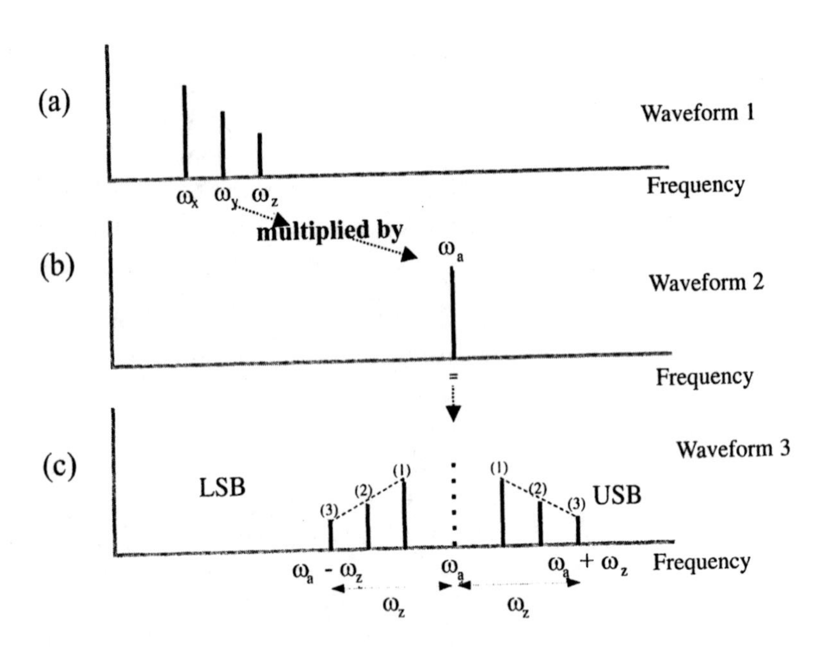

Figure 3.4 Frequency domain representation of the product of two signals: one a composite, the other a (co)sinusoid.

arranged either side of the higher frequency component ω_a – the carrier. (The lower sideband is a mirror image). If we make the carrier, ω_a, a higher frequency, the sidebands would move up the frequency axis, following it.

We can now go on to use these results in our study of sampling, in Section 3.2.

3.2 The sampling process

Sampling is the process of taking 'snapshots' of a signal and transmitting (or storing) the snapshots, rather than dealing with the continuous signal itself. Surprisingly, if we take snapshots at a sufficient rate, *we do not lose information about the signal*. This means that we can use the interval between snapshots for a variety of useful purposes, without compromising the integrity of the signal itself. This, as we shall show, makes multi-channel digital audio systems and also many modern telecommunications systems possible. Sampling is clearly an important technique for our purposes.

This section will address the process of sampling, together with the point about information content and sampling rate. The next sections will deal with the application of sampling in digital audio systems.

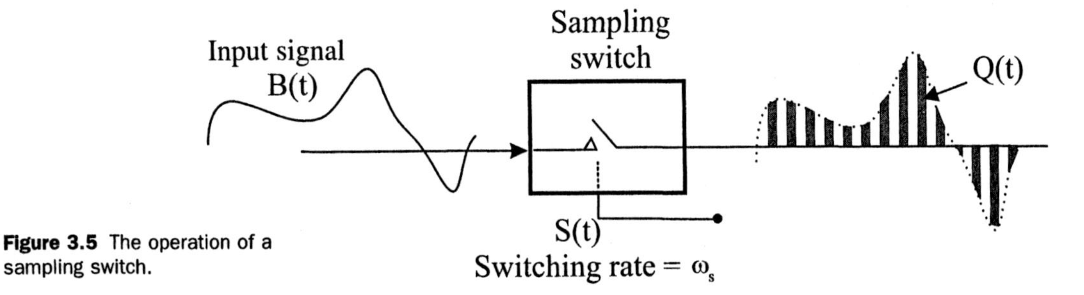

Figure 3.5 The operation of a sampling switch.

Figure 3.5 shows the archetypal sampling system, which will help us in our understanding of the sampling process. An input signal B(t), (for example an audio signal) is applied to the input of a very fast electronic switch. A *ring modulator,* often found in studios is an example of such a switch. The switch can open (break contact) and close (make contact) hundreds of thousands of times per second – or even faster – under the control of a 'switching signal', S(t); otherwise it operates just like any other simple switch. In particular, when the switch is closed, the input signal appears at the output, and when it is open, nothing appears at the output. The time domain output of the switch is therefore a 'chopped up', or *sampled* version of the input signal, shown as the shaded signal and labelled as Q(t) in the figure.

In order to learn anything more useful about the properties of the sampling process, we need to explore the frequency domain representation of the sampled signal, Q(t), the output of the switch. Another of our simple mathematical models will help here. As illustrated in Figure 3.6, Q(t) can be described mathematically as the *product* of the input signal, B(t), and a square wave switching signal Sq(t) which runs between amplitude values 0 and 1. When the switching signal is 1, we get B(t) out, and when the switching signal is 0, we get nothing out. This is our simple mathematical model – it behaves in *exactly* the same way as the output of the sampling switch, Q(t). The output of the mathematical model is shown in Figure 3.6(c), and it is identical to the output waveform of the switch in Figure 3.5.

We now need to consider the frequency domain content of the switching signal. It is a composite wave, and therefore consists of the superposition of a fundamental of frequency ω_s, together with its harmonics $2\omega_s$, $3\omega_s$ etc. of appropriate amplitude and phase. The superposition also needs to include a constant number, shown as K in the figure. This is because without it, the switching signal Sq(t), constructed from this superposition, would run between values of ±1, since the other components in

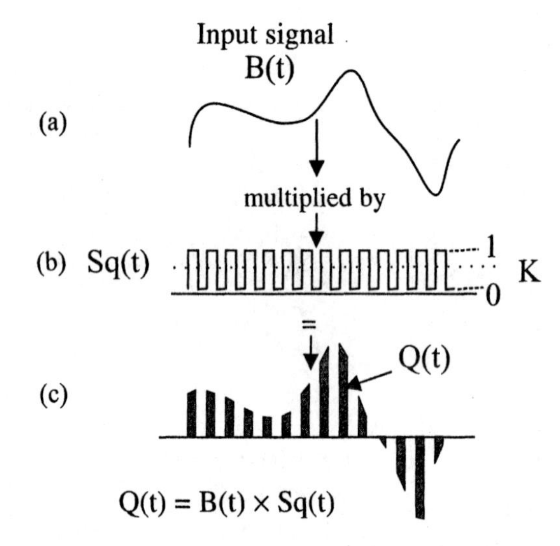

Figure 3.6 A mathematical model of the sampling process.

the superposition are sinusoids of maximum amplitude 1, and these run between equal positive and negative values. The constant value, K, 'lifts' (or biases) the waveform Sq(t), so that its minimum value is 0, and not some negative value. The amplitudes of the other components in the superposition would be adjusted so that the positive value of Sq(t) becomes 1. The entire Sq(t) signal therefore consists of positive values (or 0). This factor, K, is critical in our model, as we shall see shortly.

We can therefore write the switching signal, Sq(t) as:

$$Sq(t) = K + A_1 \cos \omega_s t + A_2 \cos 2\omega_s t + A_3 \cos 3\omega_s t + A_4 \cos 4\omega_s t + \ldots..$$

The amplitudes and phases of the harmonics, $A_1 \ldots A_n$ are not particularly important here, and do not affect the generality of the argument. Taking our model of the output of the sampling switch:

$$\begin{aligned} Q(t) &= B(t) \times Sq(t) \\ &= B(t) \times K && \text{(Term 1)} \\ &+ B(t) \times A_1 \cos \omega_s t && \text{(Term 2)} \\ &+ B(t) \times A_2 \cos 2\omega_s t && \text{(Term 3)} \\ &+ B(t) \times A_3 \cos 3\omega_s t && \text{(Term 4)} \\ &+ B(t) \times A_4 \cos 4\omega_s t && \text{(Term 5)} \\ &+ \ldots \end{aligned}$$

We can now use our result from Section 3.1. Each of Terms 2–5 represents a pair of DSBSC sidebands at the switching frequency ω_s and each of its harmonics. This is because each of these terms

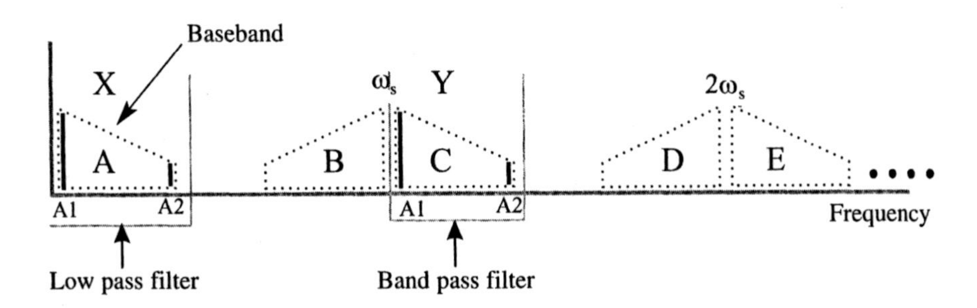

Figure 3.7 The output frequency spectrum of a sampling switch.

represents a multi-component input signal B(t) multiplied by a high frequency (co)sinusoid. Term 1 is somewhat different. It is simply the input signal B(t) multiplied (scaled) by a constant K. This has the (only) effect of changing the amplitude of the input signal, it does not affect its frequency content.

These results are shown as a frequency domain diagram in Figure 3.7. We see the input spectrum to the switch (known as the *baseband* in the jargon), plus several DSBSC spectra.

Some important consequences arise from Figure 3.7. Imagine that the sampling switch was followed by a bandpass filter whose frequency characteristic is located at Y. The output of the filter would be a frequency translated version of the baseband, C. Frequency translation such as this is an important way of manipulating timbral relationships. Suppose that the baseband B(t) had only two frequency components, A_1 and A_2, of frequencies 60 and 180 Hz, that is, a fundamental and third harmonic. If the switch is operated at 400 Hz, then the corresponding frequency components in the translated baseband, C_1 and C_2 would become 460 Hz and 580 Hz. *These are no longer harmonically related.* The act of frequency translation can be used to manipulate harmonic relationships in sound using a device such as a ring modulator.

The output of the lowpass filter at X in Figure 3.7 has even more momentous consequences. It is the baseband. The situation can be interpreted in the following way: if *samples* of a baseband are passed into a low pass filter, the output of the filter is a reconstructed version of the baseband. In particular, *the frequency spectrum characteristics of the baseband are retained.* Its overall amplitude may be changed (reduced) through the sampling process, but this is of little consequence, since this can be restored through a simple process of amplification. The retention of the frequency spectrum is the important factor.

This result is counter-intuitive to the point of being astounding!

One would have thought that if a signal was chopped up, leaving only snap-shots or samples of the signal, then information about the content of the signal (including its frequency spectrum) would have been lost. Figure 3.7 indicates that this is not the case. The whole of the baseband spectrum is produced at the output of the lowpass filter.

The consequences of this are far-reaching. Specifically, the 'dead-time' between samples can be used for other useful purposes without significantly affecting the spectrum of the baseband. One such important use is that the time can be employed to convert the sample into a binary number (a finite length of time is required to do this), and this gives rise to the whole world of digital audio. The way this is done, and the reasons for doing it are described in the next section. Another use of the 'dead-time' would be to interleave samples from one (or indeed many) additional signal channels, so that many individual signals can be sent down one signal path, such as a single pair of wires. This is possible as long as the samples of the individual signals can be separated out at the receiving end, and applied to their own individual reconstruction lowpass filters. This is known as 'time division multiplex', or TDM and it is widely used in modern telecommunications systems, particularly when the signal exists in digital (binary) form, as described in Section 3.3. The use of individual satellite links to transmit thousands of telephone conversations and TV programs is an example of the use of sampled digital TDM. The transmission of digital stereo signals through a CD player is another.

Although we have established that a signal can be reconstructed from samples of it, using a lowpass filter, we need to give some consideration to the effect of the *sampling rate* (the frequency at which the sample switch is operated) on this process. Returning to Figure 3.7, we see that the frequency of the switching signal (ω_s) determines the position on the frequency axis of each DSBSC sideband pair. The position of the baseband (X) is not affected by the switching frequency.

In particular, if we *reduce* the sampling frequency ω_s, then the pairs of sidebands move down the frequency axis until the situation shown in Figure 3.8(a) might occur. The lower sideband of the first DSBSC sideband pair has encroached into the passband of the reconstruction filter X. It is not possible for the filter to remove this vestige of the lower sideband, and we would hear the result as an objectionable distortion known as *aliasing distortion.* It is particularly objectionable because all of the frequency components in a normal baseband are turned upside down (the

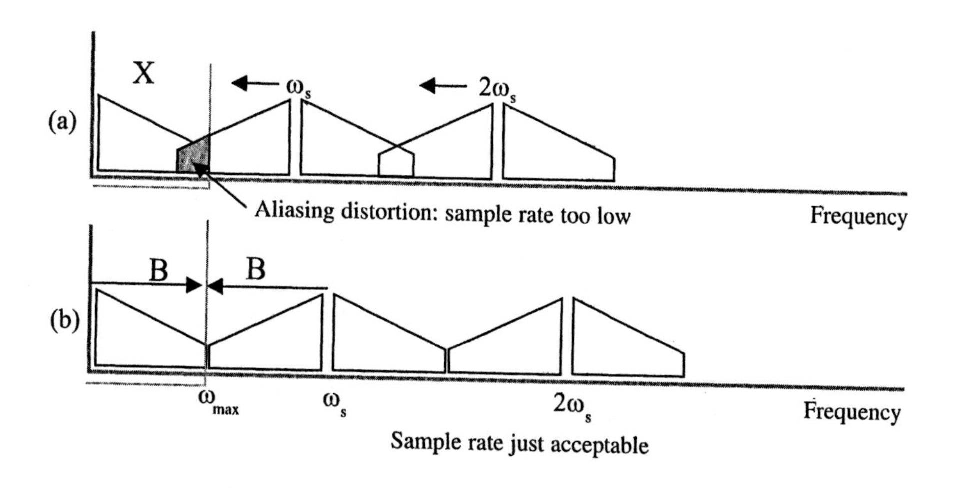

Figure 3.8 (a) and (b) Aliasing effects in sampled systems.

mirror image aspect of the lower sideband), and this contradicts our normal perception of sound.

Figure 3.8(b) illustrates the use of the lowest sampling frequency, where aliasing is just on the point of occurring. Here the lower sideband is located immediately above the baseband, and is just on the point of encroaching into the baseband region. If the bandwidth of the baseband is B, then in this marginal situation the sampling rate, $\omega_s = 2B$ (recall that the lower sideband has the same bandwidth as the baseband). To put this another way, the bandwidth of the baseband (B) is set by the frequency of the highest frequency component in the baseband (ω_{max}). Aliasing will therefore occur if the sampling rate is less than twice the frequency of the highest frequency component in the baseband. This gives rise to the following statement of the *sampling theorem:*

> *A signal can be recovered from samples of that signal using a low-pass filter, providing that the sampling rate is at least twice the highest frequency component in the signal.*

ω_{max} in Figure 3.8(b) is known as the *Nyquist frequency.* In practice, a sample rate somewhat higher than twice the Nyquist frequency would be used, to accommodate the non-ideal characteristics of realistic filters. Only a non-realistic ideal filter would be able to separate the immediately adjacent baseband and lower sideband in Figure 3.8(b). For example, high quality audio is reckoned to have a total bandwidth extending up to 20 kHz (the upper limits of the human hearing system). Typical sample rates in digital audio systems are therefore 44.1 and 48 kHz.

We are now ready to consider the important topic of digital audio.

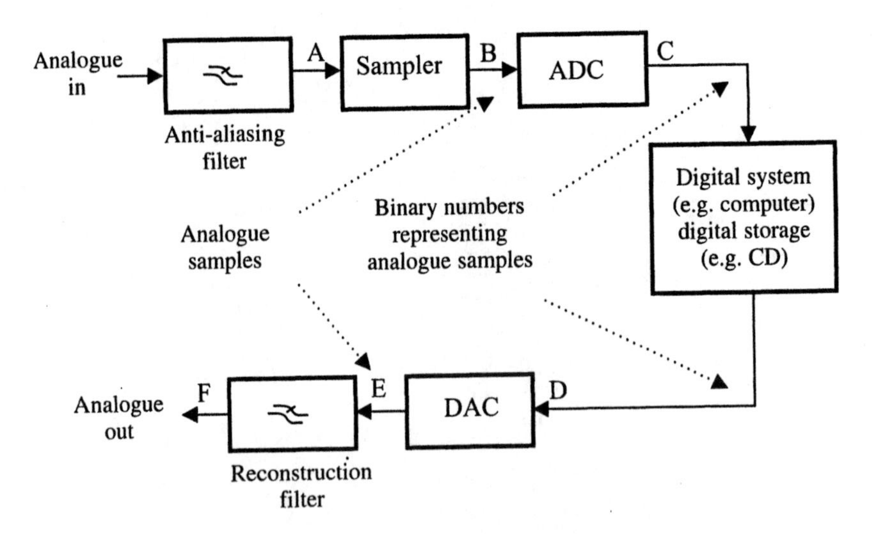

Figure 3.9 The outline of a PCM system.

3.3 Pulse code modulation (PCM) systems

This section is split into two parts: first the outline operation of a digital audio system is described. This is then followed in the second part by an explanation of *why* one would want to use digital audio systems. Section 3.4 carries on this theme by describing some of the costs associated with these systems.

3.3.1 Digital audio system operation

Figure 3.9 illustrates the block diagram of a pulse code modulation (PCM) system, which forms an archetypal example of a digital audio system. The analogue input signal from a microphone or analogue tape deck is first of all filtered through a low pass filter. This produces an analogue signal, A, with a well-defined maximum bandwidth. This is important for the next stage, where signal A is *sampled*. This sampling process produces a signal B, with a spectrum similar to that shown in Figure 3.7. This is why signal A needs to be bandwidth-limited, otherwise aliasing effects such as those shown in Figure 3.8 would occur.

Each analogue sample (signal B) is converted into a *binary number* whose magnitude is proportional to the amplitude of the sample. Thus, for example, if the amplitude of a sample were 4 units, this would be converted into a binary number such as 00000100. This operation is carried out by the analogue to digital converter (ADC). It takes analogue samples as its inputs and produces pulses at its output (signal C) which encode the binary numbers corresponding to the sample amplitudes. A low amplitude pulse might encode a binary 0, a higher amplitude pulse a binary 1. This is the origin of the 'pulse code' aspect of the name of this scheme. An example is illustrated in Figure 3.10. The operation of such a converter is described in Chapter 8, although the detail is not required here. The binary numbers can then be stored in a digital system such as a computer, DAT or CD.

When we wish to play the signal back, the binary numbers are retrieved from the digital system and converted back into analogue samples using a digital to analogue converter (DAC). Again, the operation of a DAC is explained in Chapter 8. As far as we are concerned here, it carries out the reverse process to the ADC: it converts the binary numbers into analogue samples. All that remains is to use a lowpass filter to convert the samples back into the original analogue input. The sampling theorem, described in Section 3.2, ensures that no information is lost through the sampling process.

Figure 3.10 illustrates how the ADC/DAC system might convert the amplitudes of analogue samples into binary numbers and vice-versa. The ADC compares the analogue samples such as A, B, C and D against a mesh of *quantisation levels* (the horizontal

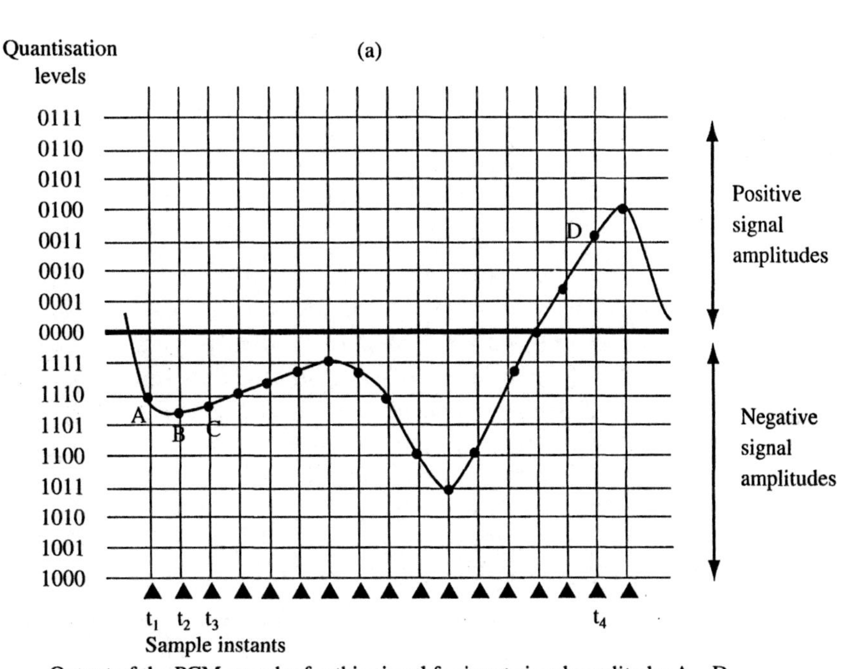

Figure 3.10 The PCM encoding process.

lines in the figure). It then outputs a pulse code representing the nearest quantisation level *below* the amplitude of the sample value. (Some systems might output the nearest quantisation level *above* – it makes little difference.) These codes are shown on the left-hand sides of the quantisation levels in the figure.

In Figure 3.10, the code groups are shown as signed 2s complement binary numbers, and this is often the case, although other forms of coding are sometimes used. 2s complement encoding is explained in Chapter 7, although for our purposes here, it is sufficient to regard this as a scheme for representing positive and negative amplitudes with binary numbers (i.e. 1s and 0s).

Therefore, taking the example shown in Figure 3.10, the signal amplitudes at A–D would be converted into bursts of pulses shown in part (b) of the figure. It can be seen clearly that the amplitudes of the pulses encode the 1s and 0s of each PCM binary code group.

Each of the code groups in Figure 3.10 is shown as consisting of four binary digits. These four digits are capable of encoding 2^4 = 16 unique quantisation levels. (There are 16 quantisation levels in part (a) of the figure.) In general, if there are n binary digits in each code group, then there are 2^n quantisation levels in the system. The number of quantisation levels doubles for each additional binary digit in the code group. Digital audio is often encoded with n = 14 or 16. You may have seen 16-bit PCM audio systems described as '16-bit sound'. This indicates that there are 2^{16} = 65 536 possible amplitude levels.

When the code groups are reconstituted back into analogue samples at the DAC, it outputs an amplitude 'step' proportional to the magnitude of the binary PCM code group, as shown in Figure 3.11. This form of output is not quite the same form as the output of the sampling switch shown in Figure 3.5, where Q(t) follows exactly the input signal when the switch is closed. In Figure 3.11 the output remains at the same, fixed value until the next sample interval. This 'stepped' form of output is known as a 'zero-order hold'. It makes only a slight modification to the spectrum of the sampled signal (Figure 3.7), and this can be ignored for our purposes.

Figure 3.11 also shows the amplitude samples of the original signal, for the purposes of comparison. It can be seen that the output samples (the stepped amplitudes) are at slightly different amplitudes compared with the originals. This is of course an inevitable consequence of attempting to represent a continuous signal by a set of discrete quantisation levels in the PCM encoding mesh.

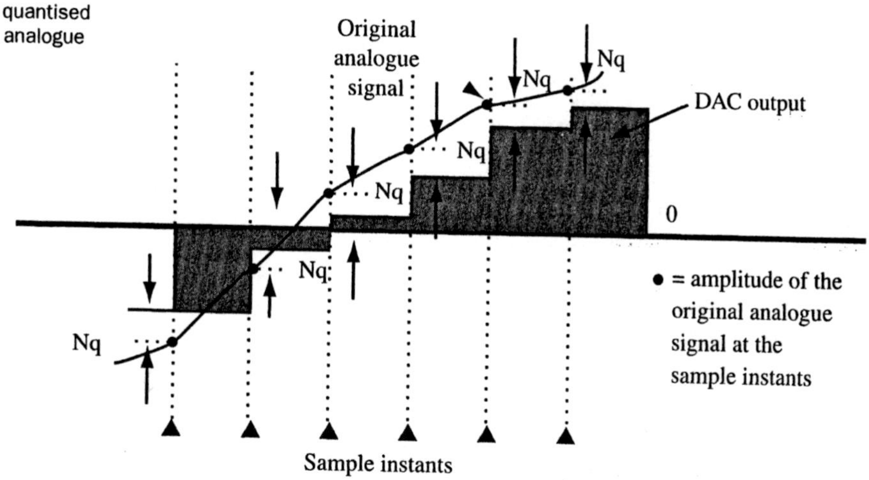

Figure 3.11 PCM quantised approximations to analogue signal samples.

The difference between the PCM approximated sample amplitudes and those of the real sample points (shown as Nq in the figure) sounds like a form of noise when it is listened to in the output of the PCM system. It is known as *quantisation noise* or *quantisation error*. The size of the error Nq will vary from sample to sample. We shall return to the effects of this noise later.

3.3.2 The benefits of digital audio

The sampling of a signal and converting to and from PCM code groups sounds like a complex business, when compared with simply recording the signal directly, for instance in an analogue tape recorder. Yet contemporary high quality audio systems (for example CDs) are almost exclusively digital in nature, based on some form of PCM. Why is this?

One reason, important in music technology and multimedia, is that the signal is transmitted (and stored where required) as binary numbers. This is the form of information processed by computers, as explained in Chapters 6 to 8, and this means that the (digital) audio can be easily manipulated by computer systems. Computer programs or 'algorithms' exist to filter signals (digital filters) and to add artificial reverberation for instance. Digital representation of sound also allows computers to *generate* artificial audio signals by *synthesising* a stream of binary numbers which will be converted into sound with a DAC and reconstitution filter. This is the basis of most contemporary music synthesisers.

Another important reason is that it is possible to store and transmit digital audio (PCM) with much reduced susceptibility to *noise*. Electrical noise is present everywhere in electronic

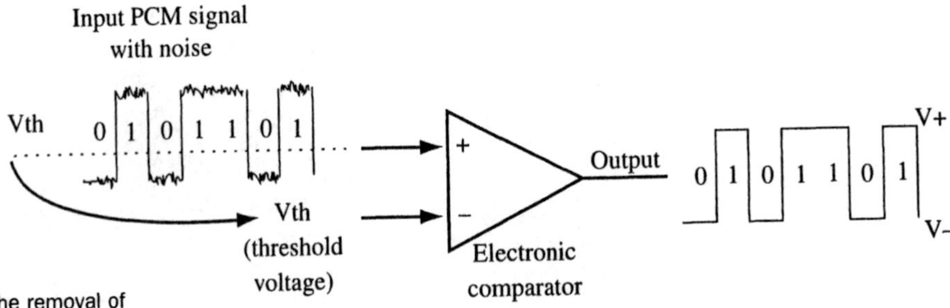

Figure 3.12 The removal of electrical noise on a PCM signal using an electronic comparator.

systems, and it cannot be avoided. In conventional analogue systems such as tape cassette decks and FM tuners, it sounds like 'mush' or 'hiss', and is difficult to reduce. The digital nature of the signals makes this much more straightforward, as Figure 3.12 shows.

The figure shows a PCM code group 0101101, with electrical noise riding on it. This is then applied to an *electronic comparator.* This is a relatively simple electronic device which works in the following way: if the signal on the '+' input is greater in (amplitude) level than that on the '–' input, the comparator outputs a fixed voltage, V+. It doesn't matter how much bigger the '+' input is, the output voltage remains fixed at V+. If on the other hand, the '+' input signal is smaller in level than the '–' input, then the comparator outputs a fixed voltage, V–. Again, it doesn't matter how much less the '+' input is compared with the '–' input. Returning to our noisy PCM signal, if this is applied to the '+' input, and a fixed amplitude level Vth is applied to the '–' input, then the output shown in the figure will be produced. Vth is typically set to a *threshold* level about half-way up the input PCM signal. Notice that the noise is removed from the PCM signal by the operation of the comparator!

Use of a simple device such as a comparator can remove transmission and storage noise from PCM systems. This is because we are only interested in the *absence* or *presence* of a PCM pulse, not in its precise amplitude. This is quite different to conventional analogue signals, where the information content of a signal is carried in the precise variation in the amplitude.

This is an important result. Its consequence is that PCM systems can be made to be much less noise-prone than conventional analogue audio systems. Transmission and storage noise (the 'furry' tops and bottoms to the PCM code groups in Figure 3.12) can be removed by the comparator. For example, in PCM telephone systems, a call from around the world can sound as though it were coming from next door, so clear is the sound. A similar clarity can be found on CD recorded audio. In PCM systems, transmission/storage noise is traded for quantisation noise (an unavoidable consequence of the operation of PCM), but this quantisation noise can be reduced to almost arbitrarily low levels, as Section 3.4 will show.

3.4 Characteristics of digital audio systems

Given that we can remove nearly all of the electrical noise surrounding a PCM signal, to be replaced by quantisation noise, the question arises as to the factors which influence our ability to reduce this quantisation noise. In principle, the task is straightforward to comprehend. Referring to Figure 3.10, the quantisation noise arises from the *limited resolution* with which a particular quantisation level can approximate an analogue sample such as A.

In particular, if we want to reduce the error, we need to reduce the *spacing* between quantisation levels, so that a sample such as A is measured against a finer mesh of levels. The maximum error between the true value of A and the nearest quantisation level will then be reduced.

The way in which the number of levels increases with PCM code digits is helpful here. As noted earlier, the number of levels *doubles* with each extra PCM binary digit (thus halving the interval between quantisation levels), so in this sense, there is a nice scaling of noise with each additional PCM bit. What stops us however from increasing the number of PCM digits *indefinitely* to reduce the quantisation noise to an arbitrarily low value?

A number of factors reduce our ability to do this, chief amongst them are limitations in storage capacity, system bandwidth and dynamic range. These are now discussed.

3.4.1 Storage capacity of PCM systems

PCM systems generate a lot of data. In high quality audio systems we need to sample at high rates, getting on towards 50 000 samples per second. Let us say that each sample has 16 bits, so that there are $2^{16} \approx 65\,000$ quantisation levels (not uncommon in contemporary systems). A common unit of storage in computer systems is the 'byte', which consists of 8 bits, so our PCM system will generate approximately 100 000 bytes (100 kbytes) *per second*. If we have a stereo TDM system, we will double this number of bytes per second. Therefore, 5 seconds of stereo sound will occupy approximately 1 Mbyte of computer disk space, if the sound were to be stored there. This approximates to something less than one and a half hours for a 1Gbyte disk given over entirely to the storage of sound – (a similar density of information) which would be stored on one CD.

Adding extra digits to each PCM code group will increase this storage requirement, especially since capacity tends to come in quanta of 1 byte. For instance, adding an extra bit in the example above (17 bit PCM), could increase the storage capacity by 50 per cent in some implementations (since it would require three bytes). Storage capacity could therefore limit our quest for reduced quantisation noise.

Various encoding schemes are used as an alternative to the simple linear PCM scheme described above, in order to pack information into the binary digits more efficiently so that storage requirements can be reduced. Instead of starting the coding of each sample from scratch each time a sample was taken, it might be more efficient for instance to use the binary digits to encode the number of quantisation levels *up or down* from the *present* level, to get to the next sample amplitude. Then, information about this present sample would contribute to the encoding of the next sample, and this would take less binary digits. Various alternative schemes such as 'differential PCM (DPCM)' or 'delta-sigma modulation' use these kinds of ideas. Unfortunately some of these encoding schemes are less tolerant to code group binary digit errors, so the design of coding schemes often becomes a compromise between information capacity and performance in electrically noisy environments.

3.4.2 Bandwidth requirements of PCM signals

Even if we manage to store all of our PCM information in some suitable medium, the shifting of this information from one point to another in a system requires a certain bandwidth, and this can become a limiting factor. The restricted bandwidth available in radio and TV channels has prevented the use of digital audio and video on standard broadcast channels for many years for instance. Another example would be the limited bandwidth in the reading head of a DAT or CD player. The question therefore arises: what is the minimum bandwidth which may be used for the reliable transmission of PCM code digits, and how does this influence an attempt to reduce quantisation noise by increasing the number of digits in each PCM code group?

Figure 3.13 PCM code groups within sample periods.

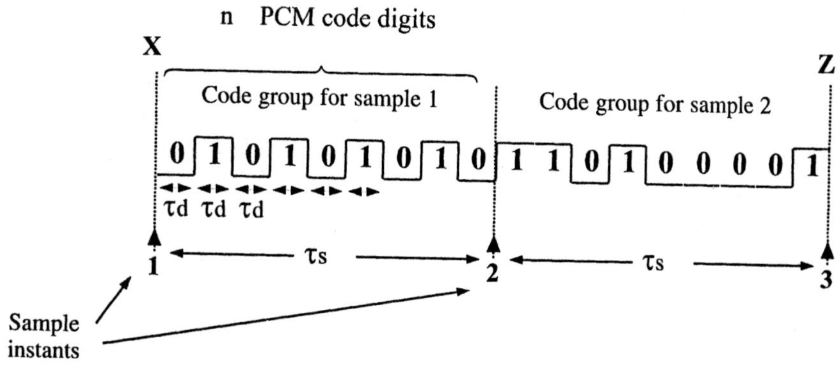

Consider the situation shown in Figure 3.13. Some sampling instants are shown, marked X, Y, Z. The time interval between these instants, τ_s, is set by the sampling frequency f_s: $\tau_s = 1/\ f_s$. Putting a few figures in: if there are 100 samples per second, then the interval of each sample is 1/100th of a second.

Notice that the sample period, τ_s, has nothing to do with the number of code digits per PCM code group. It is set entirely by the sample rate.

If there are n binary digits per code group, then each digit will occupy a time interval of $\tau_d = \tau_s\ /n$. The question is: what bandwidth is required to transmit a pulse stream such as that shown for Sample 1, where each digit lasts for τ_d seconds? There is no single 'correct' answer to this question. It depends on the system in which the code digits are used and the way they are detected

Figure 3.14 (a) Time domain and (b) frequency domain representation of digit stream for Sample 1.

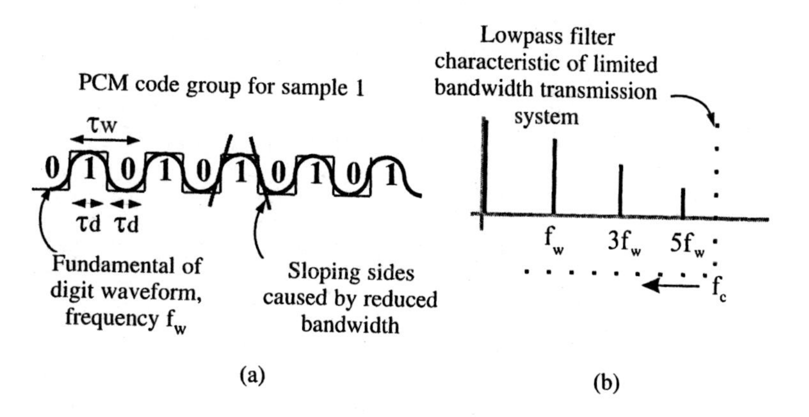

to some extent. However, let us develop a simple model for this bandwidth, to give us an idea of the factors at play here.

Consider the waveform of the stream of binary digits representing Sample 1, which exists between time instants X and Y. It is a composite wave, consisting of a fundamental related to the bit rate, plus harmonics of that fundamental. This is illustrated in Figure 3.14. In the figure, the fundamental is superimposed on the digit stream to show how it relates to the digit interval, τ_d.

The 0 Hz frequency component is the average value of the waveform, which arises because the signal is *unipolar,* taking only positive values. It is not particularly important here. Let us now consider the effects of transmitting this code group through ever-decreasing bandwidths. Assuming that the bandwidth limitation takes a low pass characteristic (which is the normal case, for example for a DAT reading mechanism), as the cut-off frequency f_c is successively reduced, the higher harmonics of the waveform are removed. If this cut-off frequency, which defines the available bandwidth, falls below the frequency of the code group fundamental, f_w, then any chance of discerning the code group waveform will be lost. We will have filtered out the fundamental, leaving only the 0 Hz component, which has no discernible shape (it is just a constant level). To put it another way, if our filtering action leaves just the fundamental, ($f_c = f_w$) we still have a chance of discerning the individual 1s and 0s of the waveform, as Figure 3.14 shows. If the bandwidth falls below f_w then we lose this information, and we are left with nothing useful. f_w therefore represents the minimum usable bandwidth.

How does f_w relate to the sample rate and the number of digits per PCM code group? Looking at Figure 3.14, the period of the fundamental,

$$\tau_w \ (= 1/\ f_w) = 2\tau_d.$$

Hence the minimum usable bandwidth, f_w is given by:

$$f_w = 1/2\tau_d = \ n/2\tau_s \text{ (using } \tau_d = \tau_s/n).$$

And since f_s (the sample rate) = $1/\tau_s$

$$f_w = n\ f_s\ /2$$

In practice, a greater bandwidth would be used since the use of restricted bandwidth causes the waveform to have sloping sides (as Figure 3.14 shows). Any electrical noise present in the system will ride up these sloping sides, and interfere with the detection threshold in the receiving device, causing a mistrigger of the detection process, and subsequent errors. For this reason, the bandwidth requirement of sample 2s code group in Figure 3.13 is similar to that of Sample 1, even though Group 2 could be considered to have a lower fundamental frequency content, and would therefore perhaps be more tolerant to reduced bandwidth.

3.4.3 Dynamic range of PCM systems

Dynamic range is simply the ratio of the maximum to minimum signal amplitude which can be accommodated by a system. It usually expressed in a logarithmic form:

$$\text{Dynamic range (DR)} = 20 \log_{10} (V_{max} / V_{min}) \text{ dBs}$$

It is a measure of the signal *amplitude range* which can be safely transmitted through the system. V_{max} is the largest signal amplitude which can be handled by the system before it starts to overload or distort. V_{min} is usually set by the amount of noise present in the system. If the signal falls below the general noise level, then it will be lost, or overcome by the noise.

In the specific case of PCM, the maximum signal amplitude is given by the number of quantisation 'slices', times the amplitude interval between them. Let us call this 'a'.

$$V_{max} = (2^m - 1)a,$$

where m is the number of binary digits in each PCM code group. (There are 2^m quantisation levels, so there are $(2^m - 1)$ intervals between them.)

V_{min} is also set by the amplitude interval between quantisation levels. A signal has to be large enough to break through at least one quantisation level, otherwise it will not be 'noticed' by the PCM system. The code groups will not change unless the signal spans at least one quantisation interval. Hence $V_{min} = a$. Therefore, dynamic range (DR)

$$= 20 \log_{10} (V_{max} / V_{min}) = 20 \log_{10} (2^m - 1) \text{ dBs}$$

Since 2^m is usually a large number (greater than 65,000 for m = 16), the '–1' is usually ignored, to leave:

$$DR = 20 \log_{10} 2^m = 20m \log_{10} 2$$

Since $20 \log_{10} 2$ has an approximate value of 6, then the dynamic range for a PCM signal is approximately 6m dBs.

The dynamic range is therefore directly proportional to the number of bits in a PCM code group. If we wish to increase the dynamic range, so that the system can accommodate a greater range of signals, from loud to soft, we need to increase the number of PCM code digits. Unfortunately, as we saw in the previous sections, this will cost us more in terms of storage capacity and system bandwidth, so that there is a direct trade-off between these system characteristics.

3.5 Synthetic audio systems

In our discussions so far about digital audio systems, we have assumed that an analogue signal existed at the outset and we have asked ourselves what the effect of sampling and digitisation of this pre-existing signal would be. We now turn to an important class of application where a digital system *synthesises* a digital audio signal from scratch, where the original sound may not have existed earlier in the acoustic world.

Such systems are known as *synthetic audio* systems, especially in multimedia applications. Synthetic audio (perhaps by another name) is also used extensively in digital computer music systems. In this section we look at the components from which these synthetic audio systems are constructed. These components go by various names such as 'unit generators', 'opcodes' and 'operators', depending upon the environment in which they are used – whether this is a MIDI controlled synthesiser, a computer music language or a multimedia application.

Some of these unit generators carry out functions which we are already familiar with: filtering and mathematical operations such as addition and multiplication for example. Later we shall see other operators such as reverberation and flanging. In this section we shall discover how an *oscillator* can be made into such a unit generator. This will form the basis of networks of unit generators which carry out various forms of timbral synthesis and transformation, such as frequency modulation and ring modulation, described earlier.

The basis of such a digital oscillator is in most circumstances a set of PCM samples of one cycle of a waveform stored in a digital memory. Such memory devices are described in Chapters 6 and 7. For the moment, it is sufficient to say that they are capable of storing the 1s and 0s of PCM code groups in addressable memory locations, each location holding one PCM code group.

If we specify the address of the location (typically as a binary 'address' number), then the memory responds by providing the code group stored in that location. If this code group is applied to a DAC, then we have a voltage sample whose amplitude is proportional to the magnitude of the binary number. If we then

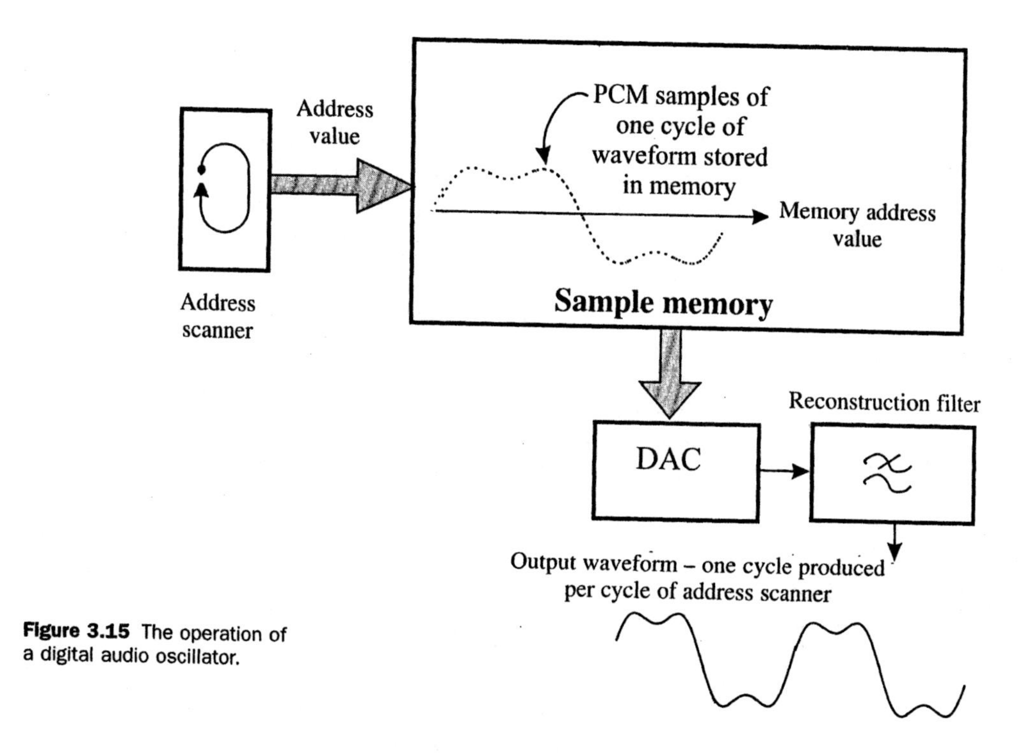

Figure 3.15 The operation of a digital audio oscillator.

provide a mechanism to *repetitively* scan through the addresses in the memory, then the samples will be played out through the DAC, to form consecutive cycles of a continuous sampled waveform. This is illustrated in Figure 3.15.

Notice that each cycle of the waveform must be identical, since it is constructed from a common set of samples held in the memory. It is therefore not possible for this waveform to contain inharmonic components, following the discussion given in Section 2.2, and this will limit its timbral interest. Nevertheless, this is a very commonly used synthesis process.

Supposing that we wanted to use this 'wavetable mechanism' as the basis of a music synthesiser where we wished to be able change the *pitch* of the output waveform. What methods would we use to control this pitch? A number of options are available to us, but the process illustrated in Figure 3.16 is the one most commonly used. The idea is to have the synthesiser *emulate* the operation of a natural sampler, as the diagram shows.

Imagine we had an instrument called a 'harmophone' which produced a timbre consisting of a fundamental plus third harmonic. In part (a) of Figure 3.16, we have a harmophone player playing into a microphone attached to a sampler. The samples are sent to a listener in another room, where the sound is recon-

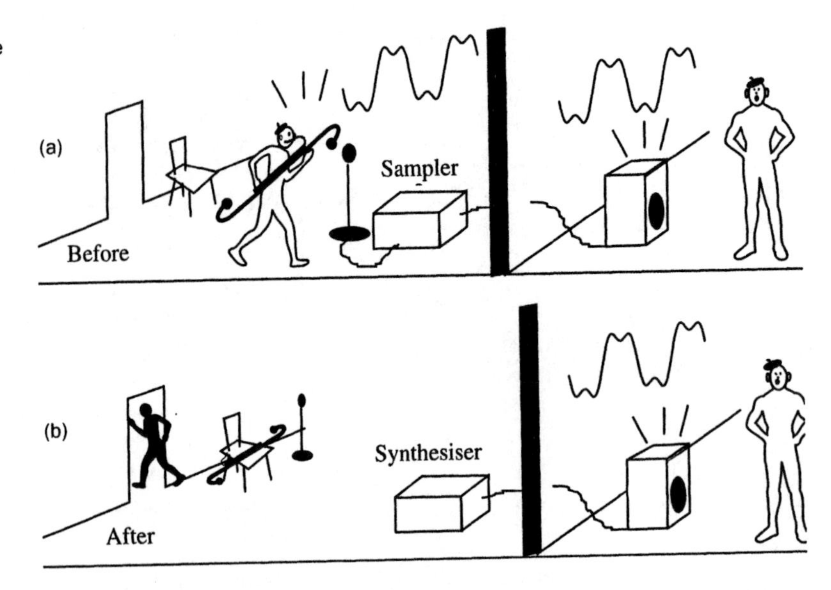

Figure 3.16 (a) and (b) Use of a synthesiser to emulate the operation of a sampler.

structed and played back. In part (b) of Figure 3.16, the player has gone for a tea-break. He has substituted a synthesiser to cover his absence. The question is: what does the synthesiser have to do as the harmophone music changes pitch so that the listener in the other room is unaware that s/he is now listening to a synthesiser?

Figure 3.17 illustrates the *sampler* output produced by the player for two different notes, sampled at a given fixed rate. At the lower pitch, sample points X0 to Xn are output from the sampler. At the higher pitch, sample points Y1 to Yn are output. This is the behaviour which the synthesiser has to emulate, so that the listener is unaware of the difference between the two systems.

In practice, the synthesiser has many samples from one cycle stored in memory. In order to emulate the output of Pitch 1 from the sampler, the synthesiser has to output those samples corresponding to points X0 to Xn. In order to emulate Pitch 2, the sampler must select points corresponding to points Y1 to Yn. Figure 3.18 shows the samples which must be selected to achieve this. It is clear from the diagram that the synthesiser must skip over many more samples in memory in between sample points in order to output the higher pitch. Notice also that the second cycle of the waveforms do not necessarily start with the first sample in memory, since the first sample for the second cycle may be located some way into the cycle, as is the case for Y4 for instance.

We should not worry that we are skipping over samples in order

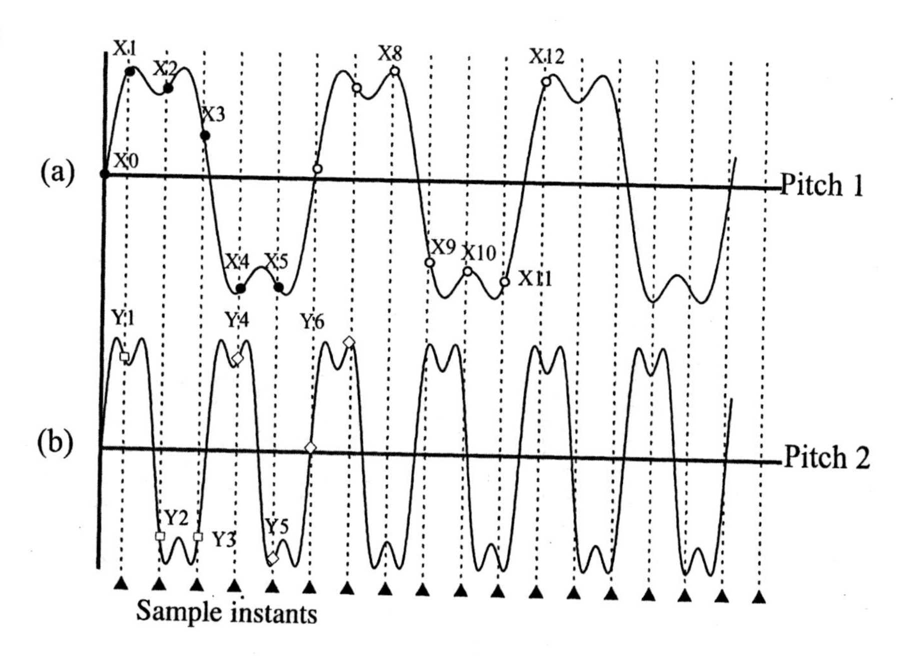

Figure 3.17 (a) and (b) Samples of two different 'harmophone' pitches.

to raise the pitch of the synthesiser output. The sampling theorem allows us to define a waveform by means of discrete samples. It is just a question of how many samples we need per cycle to avoid aliasing. We can therefore miss out many sample points in memory between individual output sample values, up to a point where aliasing sets in.

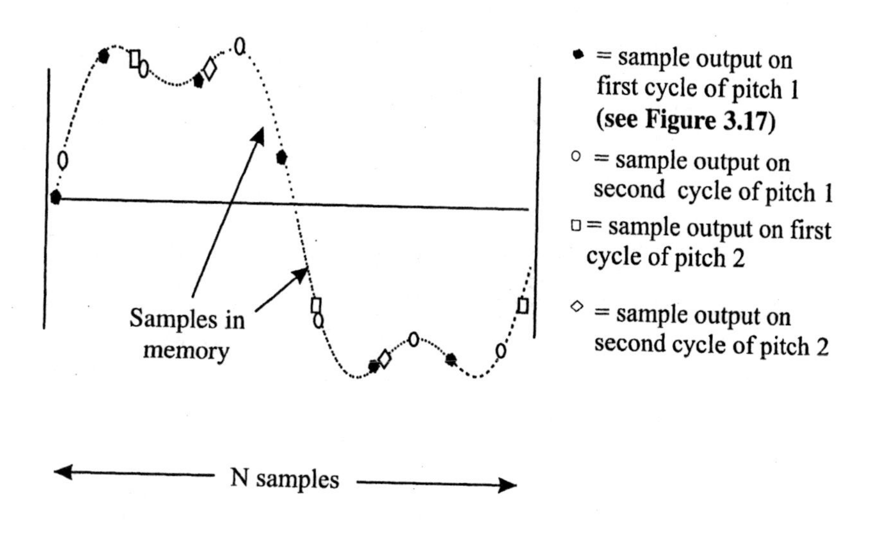

Figure 3.18 Samples which must be chosen to synthesise the waveforms in Figure 3.17.

These aliasing effects will be discussed later. Let us now concentrate on the number of samples which we should skip over in order to achieve a given output pitch. Suppose that there are N samples in memory, as shown in Figure 3.18, and further let us suppose that these are *all* output on every cycle of the waveform at a rate f_s. This means that the interval between each sample is $\tau_s = 1/f_s$. The time taken to output one complete cycle of the output waveform would be $\tau_o = N\tau_s$, or output pitch, $f_o = 1/\tau_o = f_s/N$.

If we now skip over every SI sample in memory, in between each output sample, then we will scan through the memory faster, at a rate inversely proportional to the 'skip interval', SI. For instance if we skip over ten samples every output sample, we will scan through the table in one tenth of the time. Thus, now:

$$\tau_o = N\tau_s / SI, \quad \text{or} \quad f_o = 1/\tau_o = SI \times f_s/N$$

If we want to know what skip interval to use in order to achieve a given output pitch, f_o we can obtain this by rearranging the above equation:

$$SI = N \times f_o/f_s$$

Unfortunately, using this equation it is quite possible to obtain fractional skip interval values. For example, if there are 1000 samples in the wavetable, with a 50 000 samples per second sampling rate, and we want an output pitch of 290 Hz, then SI = (1000 × 290/50 000) = 5.8. What does a skip interval of 5.8 mean? There is a sample at location 5, and another at 6, but there isn't a location 5.8! The ways in which these fractional skip intervals are dealt with give rise to different forms of oscillator. The simplest scheme is just to ignore the fractional part. This is called a *truncating oscillator.* For SI = 5.8, we would choose as first sample, the sample in location number 5, even though the sample in location 6 would probably be nearer to the correct value. A *rounding oscillator* chooses the sample location nearest to the point where the fractional skip interval would fall. The first sample chosen with our SI = 5.8 would therefore be in location 6. The final form of oscillator in common use is the *interpolating oscillator.* For SI = 5.8 this would take 80 per cent of the amplitude of the sample in location 6, and add 20 per cent of the sample amplitude in location 5. It can be shown that this is equivalent to drawing a straight line between the sample amplitudes in location 5 and 6, and taking a point 80 per cent of the way along the line (corresponding to a first sample in 'location' 5.8).

Any inaccuracy incurred through these various methods of approximation appears as a form of noise in the output of the oscillator. The interpolating oscillator is the most accurate, then comes the rounding oscillator, and finally comes the truncating oscillator. Unfortunately, the amount of effort to do the relevant calculation varies directly with the accuracy, so that the interpolating oscillator takes the longest time to calculate an output sample. This means that a synthesiser will support less interpolating oscillators than it would if it used truncating oscillators.

It was stated above that we could step over samples in order to raise the output pitch of our oscillator, as long as we stayed within the limits imposed by the onset of aliasing. What are these limits?

Referring to Figure 3.8(b), we said that aliasing will occur if the sampling rate is less than the twice frequency of the highest frequency component in the baseband. We can put this another way: *there must be at least two samples on every cycle of the highest frequency component in the baseband.* Considering samples of our harmophone signal, as shown in Figure 3.19, which also shows the superposition of components making up the composite – taking samples X1 to Xn, as shown in the figure puts us on the brink of aliasing. This is because the samples we have chosen through our skipping process correspond to taking just *two* samples per cycle of the highest harmonic, as can be seen by examining the lower waveform in the figure. Any attempt to raise the output frequency by skipping over yet more samples in memory would result in our taking *less than two* samples per cycle of the highest harmonic and aliasing would inevitably occur.

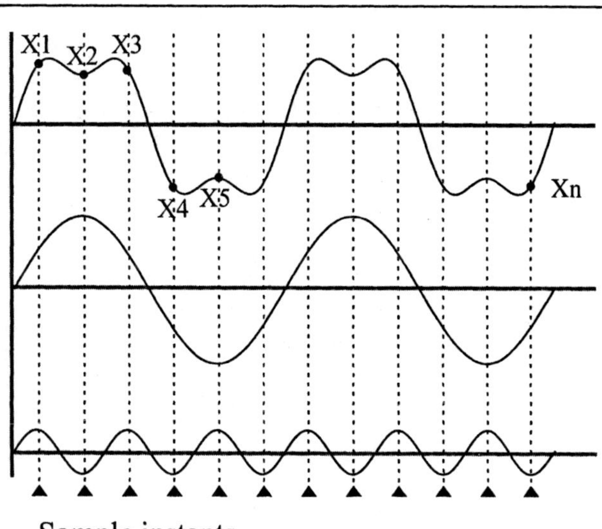

Figure 3.19 Aliasing in sample skip oscillators.

The digital oscillators described in this section could be implemented as pieces of specialised hardware in a synthesiser, or perhaps more typically as pieces of program code (software) running on a computer in a synthesiser. The code (or hardware modules) are often designed so that the output of one oscillator can feed into input of another. The user (or designer) can change the configuration of the network of oscillators to produce different sound effects. This concept of 'plug and play' is not just limited to the use of oscillators. Digital filters, envelope generators, panning units and general mathematical operators can all be written so that the output of one device can drive the input of another. Devices used in this way are often described as *unit generators* (sometimes *opcodes* or *operators*), and they can be integrated into *reconfigurable* sound processing networks to give a very flexible control over a wide variety of sounds. This provides the extensive sound palettes which we associate with synthesisers.

The process of using one oscillator to drive another produces a manipulation of spectral content/timbre known as *modulation*, a topic which we will examine in Chapter 5.

The user is given some appropriate means to define the configuration or interconnectivity of unit generators in order to achieve some sound synthesis task. This interconnection facility might be based on the use of the familiar menus and selector buttons on the front panel of a synthesiser, or it might be provided, in the form of a specialised programming language, or possibly as an extension to an existing language. This 'language' approach to the use of unit generators is described further in Chapter 10, where we describe the use of the MIDAS system which operates with unit generators in this way.

When the configuration of unit generators has been thus defined, the system then repetitively scans down the network of unit generators, transmitting the output of each unit generator to the input of the next in the chain, until the output of the network is reached and the output sample is thus produced. This happens on a sample-by-sample basis. The whole network is thus run as a sequence of 'snapshots', conforming to our model of a sampled audio system.

The use of language-based unit generator structures is likely to become important in synthetic audio for multimedia systems. The amount of information which needs to be transmitted from the programme supplier to the consumer to define a unit generator configuration is minuscule in comparison with the information content (numbers of bytes, bandwidth) of the digital audio itself. It therefore makes good sense to have the sample-by-sample synthesis of sound carried out in the consumer's local unit, rather than transmitting it all over the network (e.g. telephone lines, Internet connections, etc.) connecting the consumer to the transmitting station. It is much more efficient to use this network to download the unit generator configuration, plus any sequencing triggers, to *direct* the configured unit generators as the sound evolves in performance. It is because of the importance of this approach that we have included in Chapter 10 a description of a system (MIDAS) which is capable of operating in this way.

3.6 Summary

In this chapter we have considered the effects of sampling an audio signal. We have seen that this makes the process of converting such a signal into a digital representation possible, through the use of PCM. We have examined the advantages and disadvantages of such a digital description of signals, and have some concept of the reasons why this form of signal representation is so widely used in contemporary audio systems.

We are now ready to consider the *application* of such techniques within commercial music technology and studio equipment.

Suggested further reading

Chamberlin, Hal (1980) *Musical Applications of Microprocessors*, Hayden, ISBN 0-8104-5753-9.

Dodge, Charles and Jerse, Thomas A. (1985) *Computer Music: Synthesis, Composition and Performance*, Schirmer Books, ISBN 0-02-873100-X.

Roads, Curtis (1995) *The Computer Music Tutorial*, MIT Press, Cambridge, Massachusetts, ISBN 0-262-68082-3.

Watkinson, John (1998) *The Art of Digital Audio*, Butterworth–Heinemann, Oxford, ISBN 0-240-51270-7.

Part 3

Music technology systems

This section deals with the structure of contemporary music technology systems, set within the context of the way in which people have used such systems over the years. The MIDI specification is introduced and discussed in Chapter 4. This leads to a description of the essential functional components of modern music technology systems in Chapter 5.

Most of Chapter 4 can be read without prior knowledge, although Section 4.7 will make more sense if read after Part 2 has been studied. Chapter 5 begins with an overview of sound processing, and this is followed by a section that describes signal processing techniques which are commonly found in digital audio and music technology systems. The latter parts of this chapter are a little mathematical, but we have tried to keep this within the bounds of most readers.

4 MIDI: connecting instruments together

Overview

Chapters 2 and 3 have introduced the ways in which sound can be produced in electronic instruments. This gives an indication of the way in which the *voice* of such instruments can be produced. However, this says nothing about the way in which the voice is *controlled*. Any viable instrument needs a performance interface through which a musician can play it. This chapter considers one popular means of achieving this musical task.

The Musical Instrument Digital Interface (MIDI) has been with us since 1983. In this relatively short period of time it has been responsible for a revolution in making music with electronic instruments. Opinion is divided over whether or not it is a good thing, but MIDI has firmly established itself as an integral part of the electronic music industry. Some people claim that it limits musicality, others would not have been able to create music without it.

There are many books and articles that focus specifically on MIDI. Many of these treat it as a topic that stands on its own and whose details can be learnt in isolation. However, it is also important to understand the more general issues of why we should be linking instruments in the first place.

This chapter gives the reader an overview of the issues involved in connecting instruments together through MIDI, looking particularly at the way that MIDI came into existence and has developed over the years.

Topics covered

- The musical background that led to the invention of MIDI.
- A chronicle of MIDI's development.
- Problems and challenges encountered when designing connections between instruments.
- The MIDI specification – its messages and musical implications.
- The format and construction of messages within MIDI.
- Extensions to the MIDI specification.
- Limitations of MIDI.
- The future of instrumental connections.

4.1 Musical background

Consider a church organ. The player of such an instrument has control over many different sounds. Each of the hands (and the feet) can play an independent musical timbre. Therefore the organist is not only an instrumental performer, but a sort of 'conductor' – co-ordinating the timing of several instrumental lines. Before the invention of the organ, the only way of playing many independent musical parts (on different sounds) was to have a group of human performers (each with his or her own instrument) and someone to keep them in time (either a conductor or one of the performers).

Church organists can also customise their own sounds. This is done by selecting combinations of 'stops' which control the airflow to various pipes, each with a distinct sound quality. In addition to this, those organs with several manuals (keyboards) can form *couplings* between the manuals. Thus what is played on one manual is automatically played (via a mechanical network) on another manual. The result of this is a further layering and enrichment of the final sound.

In this context the development of the church organ can be seen as a radical departure from the norm. A single musician can perform and control a suite of customised sounds – all synchronised and co-ordinated. These ideas have blossomed with the

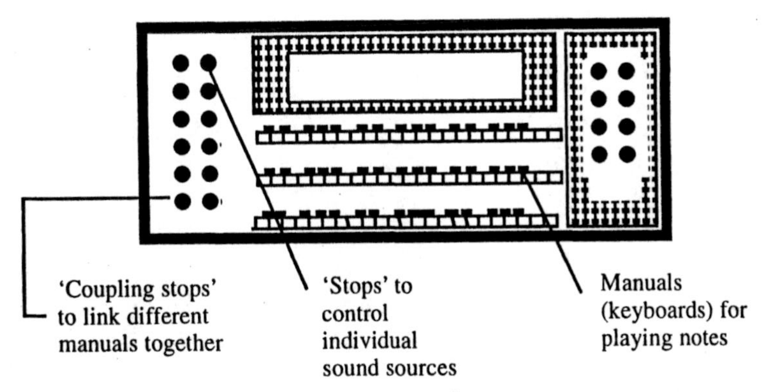

Figure 4.1 A typical church organ console consisting of several keyboard manuals and 'stops' to select the tone quality.

invention of electronic music. Much early electronic music involved the *non*-real-time production of music on multi-track tapes and computers. However, it was the proliferation of *performance-orientated* keyboard synthesisers that launched an interest in discovering just how many sounds an individual player could create and control.

4.2 The environment from which MIDI emerged

In the 1970s keyboard players had increasing access to synthesisers for live performance. The instruments were mostly monophonic (only one note at a time) and, although the sounds were radically different to those available on other instruments, the timbres tended to be very thin and 'flat'. Consequently, players were on the look-out for ways of 'fattening' up the sound. A simple way to do this was to have many synthesisers playing the same thing, or to have many different sounds playing at once. This required having several keyboard players, or a single highly skilled player who developed acrobatic techniques for playing many different keyboards (and foot pedals) at the same time.

It seemed to be a waste of human resources to have to play exactly the same part on two keyboards just because the individual sounds were considered to be too thin. So players collaborated with engineers to find ways of linking keyboards together – rather like the aforementioned coupling stops on the pipe organ. The internal workings of an analogue synthesiser involved sending a voltage from the keyboard mechanism to the electronics which generated the sound. Manufacturers found that these *control voltages* could be sent from one synthesiser to another. One keyboard could therefore control two (or more) sound sources. The control voltages were not entirely standard between manufacturers. Hence in the 1970s players could link

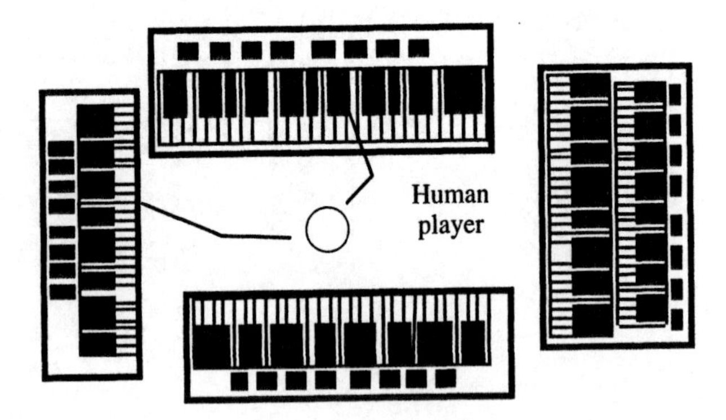

Figure 4.2 A performance set-up consisting of many keyboards but played by a single player. Many of the keyboards can only play one note at a time. This sort of arrangement was very common in the late 1970s and early 1980s.

together only certain monophonic synthesisers, but that was as far as keyboard connections went.

When *polyphonic* synthesisers became available in the late 1970s this method of linking keyboards would no longer work. A polyphonic keyboard allows several notes to be played at once and therefore cannot have a single control voltage to denote 'which single key is pressed'. Instead there is a microprocessor inside the keyboard that scans the keys, and produces an electronic key-code representing each key that is currently pressed. Manufacturers used this facility to produce the first *sequencers*. Sequencers were able to store the key-codes played by the musician, then play them back at varying speeds and with assorted transpositions. However the codes were held secret by each company, so there was no chance of competitors' products being connected to them.

4.3 The development of MIDI

4.3.1 Plans leading to the introduction of MIDI

In 1981 Dave Smith (of the Sequential Circuits company) suggested a radical solution to the problems encountered by players of polyphonic synthesisers. He proposed the invention of a *Universal Synthesiser Interface (USI)* which would allow keyboards and sequencers from different manufacturers to be connected together seamlessly.

This was a unique idea, since until then manufacturers had decided to keep the internal workings of their products secret. The argument went something like this 'If any old sequencer will work with our keyboard, then the customer is not compelled to buy our sequencer and we will lose a potential sale'.

It is therefore most surprising that between the years of 1981 and 1983 several major electronic music companies got together to design such an interface. They risked a potential loss of sales, but hoped that the market would grow due to the increased flexibility of connection between instruments. During this time of discussions there were many negotiations about how complex the interface should be. We shall examine some of these issues in section 4.4.

4.3.2 The launch of MIDI

In 1983, the committee announced its results to the musical world. They had come up with a plan for connecting keyboards and other electronic musical instruments together which had the approval of many of the leading electronic music manufacturers. The plan had been renamed the 'Musical Instrument Digital Interface' and MIDI had arrived to change the musical world.

MIDI consists of a set of standards to which electronic musical devices should conform. At its very basic level it allows keyboards from different manufacturers to be coupled together (as was first demonstrated at the 1983 Frankfurt music fair). It also includes many other features such as time synchronisation between sequencers and drum machines, and data transfer options (for example backing up a synthesiser's memory to a data storage device).

4.3.3 Widespread acceptance of MIDI

By 1985 the impact of MIDI was being felt around the musical world. Most large manufacturers were conforming to the specification for fear of being 'left out'. Customers began to ask 'Has it got MIDI?' before buying a new piece of equipment. *Sound modules* (synthesisers without keyboards) could be controlled from an existing keyboard via MIDI to extend the range of sounds available to a user. The price of synthesisers began to fall, and subsequently more people could afford them.

With the benefit of hindsight we can see that the initial manufacturers were right to take this collaborative gamble. The entire electronic music market has blossomed since the 1983 decision was taken. Products are now interchangeable, cheaper and are sold in much greater numbers.

4.3.4 MIDI as a developing standard

In more recent years the MIDI specification has been expanding. In 1987 the *MIDI time code (MTC)* and *sample dump* standards (described later) were introduced, and customers had yet more reasons for buying and using MIDI compatible equipment.

Prices fell so much during the late 1980s that the role of many recording studios was irrevocably altered by the fact that thousands of people could now compose, produce and record their own music at home. In the 1990s many MIDI products became more portable, to the extent that the 'studio in a briefcase' was now possible.

The introduction of *General MIDI (GM)* has meant that there is a universally agreed set of sounds that all MIDI devices should be capable of playing.

4.4 Designing instrument connections

It is important to consider the issues and compromises that are involved in designing any *standard* for connecting things together. Sometimes the task is relatively easy (imagine trying to connect two keyboard-based sound devices) and at other times it is extremely tricky (imagine a violin-type instrument being connected to a drum-type controller).

The challenges usually emerge as variations on the following questions:

- **How is *pitch* transmitted?** Are there discrete pitches (like a keyboard), or is pitch continuous (as in a violin string)? If we were connecting a piano to a violin would there be a single pitch for an individual note 'event' or should the pitch be constantly changing?
- **How are the different playing techniques reconciled?** A drum consists of a one note playing surface, but it can be hit in a variety of ways to produce different sounds and rhythms. How would you connect this to a violin which has four strings, each of which can play a continuously variable pitch?
- **Which instrument is in charge?** Does the control work both ways? If a 'light-beam instrument' is used to control the sound of an electric violin, can the connection work the other way round ?
- **How complex should the interface be?** Should every function on one instrument be controllable by the other instrument? What if there is no corresponding function? Would you convert it to something else, or just ignore it?

Most serious users of MIDI instruments have criticised the MIDI standard at some point. There are indeed limitations, and these are discussed later. However, it is useful to have an understanding of the issues relating to 'standard connections between instruments' in order to appreciate what was actually achieved by getting several companies to agree to anything in the first place!

In Section 4.5 we begin to discuss the details of what those companies decided. As you read it, you are urged to think about the *implications* of each decision which made up the 1983 MIDI specification.

The MIDI specification is a technical document which describes how devices should conform to a particular standard in order that they can claim to be 'MIDI compatible'. Version 1.0 of this document was completed in August 1983 and consists of two major sections: the **hardware** required, and the **messages** transmitted.

4.5 The MIDI specification – hardware

The MIDI specification states what type of plugs and sockets are to be used for MIDI connections. It also outlines how the electronics should work. The main points of the hardware specification are summarised here. There are three possible sockets on any MIDI-compatible device. They are labelled:

- **MIDI OUT** This sends data to other devices (e.g. a keyboard is played and 'note messages' are sent out from the MIDI OUT socket).
- **MIDI IN** This receives information from other devices. For example a keyboard's OUT is connected – via a MIDI cable – to a sound module's IN, so that it can produce sound on behalf of the keyboard, see Figure 4.3.
- **MIDI THRU** This relays messages arriving at the MIDI IN port, so that further devices can be chained together. (Typically a keyboard's OUT is connected – as above – to a sound module's IN, then the module's THRU is connected to another module's IN. In this way both modules can be driven by the same keyboard.)

MIDI sends data at a fixed speed: 31 250 bits of computer data per second (31.25 Kbaud). This determines the maximum rate at which messages may be sent. It approximates to 500–700 notes per second.

MIDI works by switching current on and off. Five milliAmps

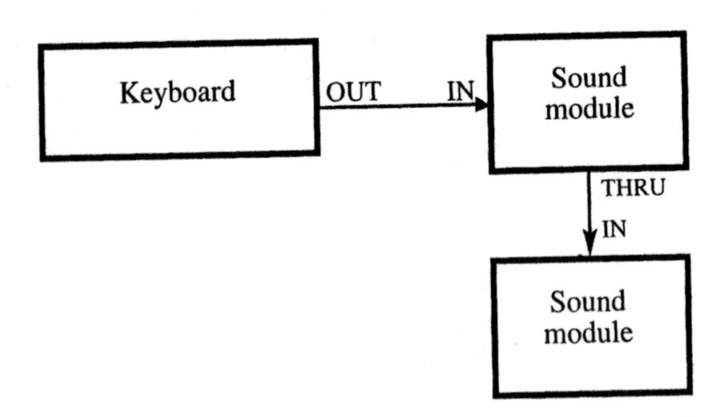

Figure 4.3 MIDI sends messages from the OUT port of a device (such as a keyboard) into the IN port of another (typically a sound module). It can be passed on to other devices by using the THRU port.

(mA) of current flows between two connected devices. An opto-isolator ensures that mains electricity cannot accidentally flow from one damaged piece of equipment to the next and thus destroy all the devices connected together.

Hardware interfacing is dealt with in more detail in Chapter 8.

4.6 The MIDI specification – messages

The MIDI specification not only determines how the hardware should work, but also what messages can be transmitted down the MIDI cables. This section gives an overview of these messages. Later we shall examine them in more detail.

There are two basic sorts of messages – *channel* and *system*. Channel messages are the most common, and they are described below.

4.6.1 Channel messages

MIDI allows 16 different instrumental sounds to be controlled at the same time. The messages for each instrument are kept separate by using *MIDI channels*. Up to 16 channels of information can be sent down one MIDI cable. You may find it helpful to think of each instrumental unit as being like a television set; many channels come across the airwaves together, but each TV can tune into a specific channel and ignore the rest.

There are seven distinct channel *messages*. Each is given a name and carries a particular meaning. As you read about each message remember that MIDI was invented *to connect keyboards together.*

'Note on'

This is the most commonly used MIDI message. It is sent whenever a note is played (usually by pressing a key on a keyboard). Like all channel messages it carries several pieces of information. Each 'note on' message comes with the values of *channel, pitch* and *velocity*.

The *channel* is a number (between 1 and 16) that identifies which of the 16 possible instrumental parts this message is intended for.

The *pitch* represents which key has been pressed. MIDI defines a number for each semitone on a keyboard. Middle C is numbered 60, and so C# is 61, D is 62, an octave above Middle C is 72 (since a musical octave consists of 12 semitones), and an octave below Middle C is 48. MIDI pitch values are in the range 0 to 127.

The *velocity* is a number (0 to 127) which represents how fast the key has been pressed. A fast key-press is usually interpreted as being 'full of energy' and therefore triggers sounds quite loudly. A slow key-press is interpreted as being a soft sound.

'Note off'

This message turns off a particular note. It too has values of *channel, pitch,* and *velocity*. Channel and pitch are exactly the same as for 'note on', and velocity refers to the speed that the key is *released*.

The 'note off' message is used much less often than 'note on'. This may seem surprising at first, but there are not many keyboards which bother to measure the 'release velocity' of a note. This has meant that manufacturers tend to turn notes off by *sending another 'note on' message with a velocity of zero*. Therefore notes can be started and stopped by streams of 'note on' messages. We will see later how this can enable slightly faster transmission of notes.

'Pitch-bend'

Most keyboards have a pitch-bend control in the form of a wheel or a sideways-moving handle. Keyboard players use it to bend (or 'inflect') the pitch of all the notes currently being played. It generally returns to its central position (no bend) when the player lets go. MIDI encodes the movement of such a wheel into the pitch-bend message. Its data consists of *channel* (1–16) and *pitch-bend value* (0–127). A value of 0 represents 'fully bent downwards in pitch', 127 is 'fully bent upwards in pitch' and 64 is the central (no bend) position. Pitch-bend can also be sent at a much higher *resolution* with values from 0 to 16 383.

A problem with the pitch-bend message is that the amount of *actual* note bending that occurs is determined by the settings on the receiving device. For example a maximum value of 127 might bend the note by a semitone *or* an octave (depending on the current setting on the sound module).

Another problem is that *all the notes on the specified channel* are affected. Whilst this is representative of how two *keyboards* can be satisfactorily connected together, it renders this message awkward to use for affecting individual notes such as might be required by a violin or guitar. Since pitch-bend alters all the notes on one channel, so each note (or each 'string' of the guitar) must have its own channel, yet there are only 16 channels altogether. If we are not careful, much of MIDI's entire capacity can be taken up with representing a single instrument.

'Program change'

This message is used for selecting different patches (or 'programs') on a piece of electronic music equipment. On a synthesiser this usually changes the pre-set sound. Again it carries the values of *channel* (1–16) and *program-change number* (0–127). When MIDI was invented in 1983 it was thought unlikely that keyboards would have more than 128 sounds. Within a few years technology had advanced so rapidly that synthesisers regularly had banks of several hundred sounds. Later changes to the MIDI specification (known as 'General MIDI' or 'GM') have attempted to remedy this problem (see section 4.8.3).

Table 4.1 Summary of MIDI channel messages

Message type	Channel	Data required
Note on	1–16	Pitch (0–127) Velocity (0–127)
Note off	1–16	Pitch (0–127) Release velocity (0–127)
Pitch-bend	1–16	Bend-value (0–127) (sometimes 0–16383)
Program change	1–16	Program-change value (0–127)
Aftertouch	1–16	Aftertouch value (0–127)
Polyphonic aftertouch	1–16	Pitch (0–127) Aftertouch value (0–127)
Control change	1–16	Controller type (0–127) Value (0–127)

'Aftertouch'

Some synthesisers have a sensor beneath the keyboard which can detect pressure. Thus a keyboard player can add 'aftertouch' by pressing down (usually quite hard) on the keys after notes have been played. This can be used, for example, to swell the volume or to make the sound brighter. The MIDI encoded message consists of the usual *channel* (1–16) and an *aftertouch value* (0–127) – where 0 means 'no pressure', and 127 is 'maximum'. As with Pitch-bend, this affects all the notes on the specified channel, and it is up to the receiving device to decide how it uses this data.

'Polyphonic aftertouch'

A few expensive keyboards allow *individual notes* to have their own aftertouch response. The MIDI-encoded form of this feature has the values *channel* (1–16), *pitch* (0–127) and *aftertouch value* (0–127). The pitch is encoded in exactly the same way as for the note-on message. Relatively few keyboards have this feature and therefore this message tends not to be used very often.

'Control change'

This last channel message is perhaps the most complex because it represents a **set** of messages. The original MIDI design team presumed (very wisely) that there would be lots of different controllers (pedals, sliders and wheels) in the future which would be useful to send over MIDI. Hence this message sends the following data: *channel* (1–16), *controller type* (0–127) and *controller value* (0–127). The controller type determines what sort of message is being sent (for example if it is a 'sustain pedal' or a 'modulation wheel' message). Various types of message have been standardised over recent years. A selection of these can be seen in Table 4.2.

Table 4.2 Examples of MIDI control change messages

Controller type	Message purpose	Values
1	Modulation wheel	0 (no modulation) ... 127 (max.)
2	Breath controller	0 (no breath) ... 127 (max.)
7	Master volume	0 (silent) ... 127 (max. volume)
10	Stereo panning	0 (left) ... 64 (central)... 127 (right)
64	Sustain pedal	0 (off) ... 127 (sustain on)
91	Reverb level	0 (no reverb) ... 127 (max. reverb)

For example, consider the following message:

Control change: channel (1), type (64), value (127)

Referring to Table 4.2 we can see that this message indicates that the sustain pedal (controller type 64) is 'on' (127 = sustain on) and will thus sustain any notes on channel 1.

As another example, consider the message:

Control change: channel (2), type (10), value (64)

This means that notes on channel 2 should *pan* to the centre of the stereo field.

4.6.2 System messages

System messages are sent to *all devices* in a MIDI system – they are not limited to a particular 'channel'. System messages allow a more varied range of data to be sent. This makes them generally more difficult to understand than channel messages, so we will deal with them in a later section. For now, all you need to know is that there are three basic types of system message:

- **Real-time:** Allows devices to synchronise together (for example a drum machine and a sequencer can start and stop at the same time and follow the same tempo).
- **Common:** Allows devices to agree on some common musical issues (such as 'are we in tune?' or 'which song are we playing?').
- **Exclusive:** Sends data exclusively to one device type. Manufacturers use 'system exclusive' messages for sending data to particular makes of synthesiser or other device. These messages are also used for 'add-ons' to the MIDI specification. It is this ability to expand and develop which has enabled MIDI to remain a viable standard – even though the technology has changed dramatically since it was invented.

4.7 Format and construction of MIDI messages

As engineers and designers we often need to understand the internal workings of the systems we are using. This section introduces the concepts of *how* MIDI messages are formed from the fundamental computing 'bit' and 'byte' building blocks. As a *user*, you may simply wish to accept the definitions of MIDI messages that are outlined above and skip to Section 4.8. As a *designer* of systems which use and interact with MIDI, you will benefit greatly from knowing how MIDI works 'on the inside'.

4.7.1 Number systems

In the following sections we need to refer to the numerical rep-

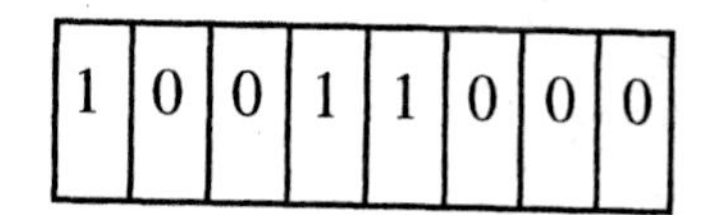

Figure 4.4 A typical MIDI byte (made up of 8 'bits').

resentations of how computers store information. If the terms *binary, bit, byte, decimal* and *hexadecimal* are unfamiliar to you, you might wish to refer to Chapters 6 and 7 for a more detailed description.

4.7.2 The MIDI language

In the English language the 26 letters of the alphabet are put together in different combinations to make many tens of thousands of words – each of which has a meaning. The MIDI language also consists of 'letters' and 'words'. The MIDI equivalent of a letter is the *byte* (a fundamental computing building block consisting of 8 binary digits – or *bits*): see Figure 4.4.

MIDI *messages* are the equivalent of words. Just as English words are made up from a series of letters, so MIDI words (messages) are made up from a series of MIDI letters (bytes).The leftmost bit of each byte (known as the most significant bit or MSB) is very special. Looking at the example in Figure 4.5, we can see that the first byte has an MSB of 1, whereas the other two have an MSB of 0. A byte which begins with a 1 is known as a *status byte*, and a byte which begins with a 0 is known as a *data byte*.

MIDI messages always begin with a status byte. The status byte determines the *type* of message (note-on, note-off, pitch-bend, etc.). It will be helpful at this point to examine the way in which the status byte is made up from its component bits.

The top bit (MSB) is always 1 in a status byte. The next three bits determine the message *type*. Three bits can give eight combinations of numbers (000, 001, 010, 011, 100, 101, 110 and 111), and that is why there are eight basic MIDI messages. The first seven of these messages (summarised in Table 4.1) work on a specific MIDI channel, and this channel is encoded into the remaining four bits. Four bits give sixteen possible numbers (from 0000 through to 1111), and that is why there are sixteen MIDI channels available (see Figure 4.6).

1	0	0	1	1	1	0	0

0	1	0	1	1	0	1	0

0	0	1	0	1	0	1	1

Figure 4.5 A typical MIDI message (word) – made up of three bytes (letters).

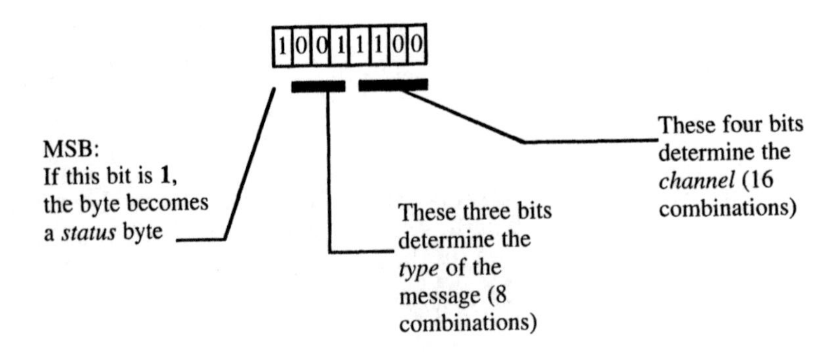

Figure 4.6 A MIDI status byte, and the meaning of its component 'bits'.

Referring to our example MIDI message in Figure 4.5, let us now consider the two bytes which begin with a 0. They are called *data* bytes because they carry the information (or data) that belongs to a particular message type. As we saw in Section 4.6.1 each 'note on' message has three values associated with it. The *channel* value is encoded as part of the status byte. The *pitch* and *velocity* values are each stored in their own data byte. The first data byte always refers to the pitch, and the second to the velocity.

The top bit of a data byte (MSB) is always 0 in order to distinguish it from a status byte. This leaves seven bits that can be used for storing data. Seven bits can represent the numbers from 0000000 to 1111111 (i.e. 0 to 127), a total of 128 combinations. This is why most MIDI values tend to be coded as numbers between 0 and 127.

Some people like to think of the status byte as the engine of a goods train which is pulling behind it a series of carriages that represent the data bytes. This analogy turns out to be quite useful as you *always* need an engine for the train to move, but you can have a variable number of carriages connected to it, depending on what goods are being delivered. This is equivalent to saying that MIDI always requires a status byte to be sent, but that different numbers of data bytes can follow behind, depending on the type of message being transmitted.

4.7.3 Hexadecimal representation

The structure of MIDI messages is best understood by representing them in binary (as in Figures 4.4 and 4.5). However, unlike computers, humans do not find it easy to recognise large numbers of 0s and 1s as anything meaningful, so a different method of representation is sought when writing computer programs. Hexadecimal notation involves the digits 0 1 2 3 4 5 6 7 8 9 A B C D E F, i.e. 16 combinations. Since four bits of binary (from

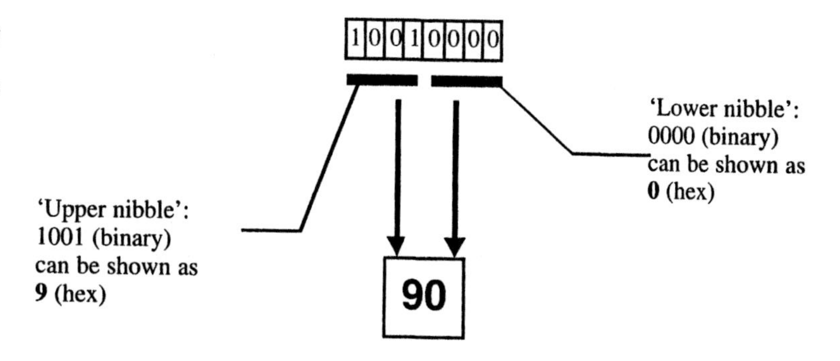

Figure 4.7 A status byte can be thought of as consisting of two parts. Each part can be represented by a hex digit. So 10010000 can be written as 90.

0000 to 1111) also have 16 combinations, each four bits of binary can be replaced by a single hexadecimal digit. Hence a MIDI *byte* (8 bits) can be shown by *two* hex digits (see Figure 4.7).

This helps greatly with identifying status bytes as the first digit shows the type of status byte. Consider the status byte for a note-on message on channel 1.

The upper four bits (referred to as the 'upper nibble', i.e. 'half a byte') determines the *type* of message, i.e. note-on. The bits 1001 can be replaced by the hex digit 9. The lower four bits (0000) determine the *channel*. So, the hex form of 10010000 is 90 (you should say 'nine zero', not 'ninety').

Note that MIDI users generally refer to channels 0 to 15 as 'channels 1 to 16'. So in the example in Figure 4.7 the message is said to be on 'channel 1'.

Table 4.3 shows the binary and hex representations for the eight major types of status byte.

Table 4.3 MIDI status bytes and associated data bytes

Upper Nibble of Status Byte			**Status Byte is followed by**	
Binary	**Hex**	**Message Type**	**Data Byte 1**	**Data Byte 2**
1000	8	Note off	Pitch	Velocity
1001	9	Note on	Pitch	Velocity
1010	A	Poly aftertouch	Pitch	Value
1011	B	Control change	Type	Value
1100	C	Program change	Voice number	–
1101	D	Aftertouch	Valye	–
1110	E	Pitch bend	Value (fine tune)	Value
1111	F	SYSTEM	Variable number of data bytes	

Examples 4.1 and 4.2 take MIDI messages in hexadecimal and analyse their meaning by referring to Table 4.3.

Example 4.1: 90 47 40 The first byte (90) is the status byte which can be seen to be a note-on message on 'channel one'. As it is a note-on message, the second byte will be pitch data. 47 in hex corresponds to $(4 \times 16) + 7 = 71$ in decimal; the B above Middle C. The third byte will be the velocity data. 40 (hex) is $4 \times 16 = 64$ (decimal) i.e. halfway between silent and full volume.

Exercise: Try to work out what tune is played by the following set of MIDI messages. (Hint: Section 4.6.1 explains how the MIDI pitch numbers work.)

90 40 40, 90 40 00, 90 40 47, 90 40 00, 90 42 57, 90 42 00, 90 40 41, 90 40 40, 90 45 71, 90 45 00, 90 44 67, 90 44 00.

Example 4.2: B3 07 7F The first byte (B3) is the status byte. Table 4.3 shows this to be a 'control' message on 'channel 4'. Therefore the second byte is the type of control message; 07 is 'volume' (see Table 4.2). The third byte is the value of the controller; 7F is maximum. This message thus represents a MIDI volume control set to maximum on channel 4.

4.7.4 System messages

System messages are special in that they are not 'performance' messages for one channel. A system message has a status byte that begins with an F (this is usually signified as Fx, where x simply means 'any hex digit can go here'). We have seen that in *channel* messages the lower nibble (4 bits) of the status byte is used to represent which of the 16 possible channels the message is intended for. With a *system* message there is **no channel**, and so the lower nibble of the status byte is used to denote what *type* of system message it is (see Table 4.4). You will notice that not all 16 possible system types are yet defined.

Section 4.6.2 outlined the three basic categories of system message – real time, common and exclusive.

- *System real time messages:* These are *single status bytes* which can be sent *at any time*. You may find it helpful to think of a train engine without any carriages! They can appear anywhere within the MIDI message stream (even in the middle

Table 4.4 Types of system message

Status Byte		Message Type	Followed by Data Bytes
Binary	**Hex**		
1111 0000	F0	System exclusive	Various number
1111 0002	F1	Quarter frame message	1
1111 0010	F2	Song position pointer	2
1111 0011	F3	Song select	1
1111 0110	F6	Tuning request	0
1111 0111	F7	End-of-exchange (EOX)	0
1111 1000	F8	Midi clock	0
1111 1010	FA	Start	0
1111 1011	FB	Continue	0
1111 1100	FC	Stop	0
1111 1110	FE	Active sensing	0
1111 1111	FF	System reset	0

of other messages!) since they have a high priority. Drum machines and sequencers can synchronise together by a combination of the 'start', 'stop' and 'midi clock' messages. Their ability to appear in the middle of any other message makes these bytes a common source of errors for designers of systems which receive MIDI. A good system should be able to recognise 'real time' bytes *as they arrive* at the MIDI IN port, deal with them immediately, then continue decoding the message in progress.

- *System common messages:* These commands are used to select which song is playing or to point out the position within a song. They are used in combination with 'real time' messages to act as a sort of 'conductor' for the two devices. A typical example would be to synchronise the song positions and playback speed of a sequencer and a tape deck (fitted with a tape-to-MIDI synchronisation box).
- *System exclusive messages:* System exclusive messages are used to transmit exclusive (private) data to a particular type of device (for a example a specific make of synthesiser). The detailed format of these messages is left to the individual manufacturer, but the general format is shown in Figure 4.8.

Each MIDI manufacturer has an individual 'ID' (identification) code which is the second byte in the message. There then follows a stream of data bytes (defined by the manufacturer) which is used to transmit such data as synthesiser parameters or memory 'dumps'. The stream (which can be any number of data bytes long) must be terminated with the end-of-exclusive (EOX) command byte (F7).

Some of the manufacturer ID numbers have been reserved for

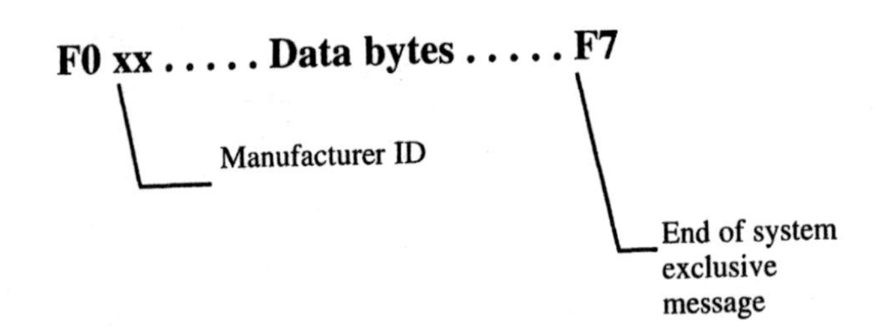

Figure 4.8 The general format of a system exclusive message.

special use. It is now possible to *extend* the MIDI specification by using one of these IDs (see Section 4.7.6).

4.7.5 Running status

The original MIDI specification includes some 'small print' which allows more messages per second to be sent down a MIDI line. Consider the following two notes (two note-ons, two note-offs):

90 40 7F   90 43 7C   80 40 41   80 43 21

The note-off status bytes (80) can be replaced by note-on messages with a velocity of **zero**:

90 40 7F   90 43 7C   90 40 00   90 43 00

Since all the status bytes are now the same, the running status principle allows us to send *only the first one*:

90 40 7F   43 7C   40 00   43 00

Running status means that a device only needs to send a status *byte when it is different from the last one*. In this way, up to 33 per cent more data can be sent down a single MIDI line.

Example 4.3 The example tune (shown earlier) might look like this with the inclusion of real-time bytes, and the use of running status. See if you can decode it:

90 40 40 FE 40 00 40 F8 47 40 F8 00 42 57 42 F8 F8 00
40 41 F8 40 00 F8 45 FE 71 45 F8 00 44 F8 67 44 00 F8

Messages encoded in 'running status' are more complex to decode, as the receiving system must constantly check to see if a new status byte has been sent. This situation is often compounded by the presence of 'real time' bytes in the middle of the message!

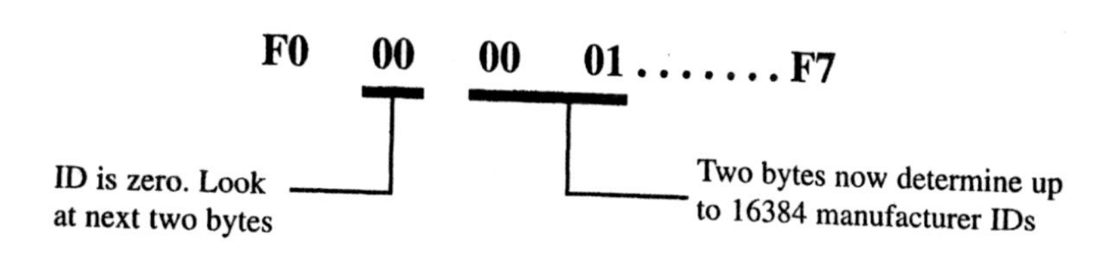

Figure 4.9 How MIDI has coped with more than 128 manufacturers.

4.7.6 Extended sys-ex messages

From Figure 4.8 we see that the second piece of data in a system exclusive message is the 'manufacturer ID'. This would lead us to believe that there are only 128 possible manufacturers of MIDI equipment. Actually, even within the first few years of MIDI's existence there were many more than this. To cope with the possibility of 'running out of manufacturer IDs' a new code was added. Nowadays if the ID is *zero* we are to expect *two extra bytes* which are used to represent 16 384 (128 × 128) possible manufacturers. This is shown in Figure 4.9.

The idea of using a special number in the 'manufacturer ID' was quickly adopted as a way of expanding the entire MIDI specification. Three other values of the ID were reserved for special use. These are:

- 7D *non-commercial*: People can develop their own system exclusive codes as long as they do not commercially release a product. For this to happen they must obtain an official manufacturer ID.
- 7E *non-real time*: New messages that can be processed 'at leisure' by the receiving device.
- 7F *real time*: New messages that must be acted on immediately.

The latter two of these messages are sometimes called *universal SysEx* messages, because they are no longer intended to apply to one device only. Even though the original specification said that system messages are those which are not allocated a 'channel', these new universal message IDs are followed by a *SysEx channel* data byte which gives up to 127 destinations for these messages (and one option for 'everything to receive this message').

4.8 Extensions to the original MIDI specification

As MIDI has developed through the years it has been able to 'move with the times' because of the way that certain of the original messages were left open for future expansion. This section explains what expansion has occurred, but does not go into the details of the new messages. Since MIDI is still developing, it is more appropriate to check the latest information on the Internet, or from the International MIDI Association.

4.8.1 MIDI time code

MIDI time code (MTC) is one of the features which is coded within these new messages. MTC is the method for allowing devices to communicate the *time elapsed since the start of the piece.* MIDI clock messages (Section 4.7.4) are related to the current *tempo*, and this can vary throughout a piece of music, making communication between devices cumbersome. MTC uses a representation known as SMPTE. SMPTE code is an international standard for film and television timing (named after its founders, the Society of Motion Picture and Television Engineers). Figure 4.10 shows how the time is coded.

The messages that are coded within MTC allow radically different devices (e.g. a video recorder and a MIDI sequencer) to stay in time with each other, and to move to the same place within a recording. The basic 'clock tick' for MTC is not one of the large universal SysEx messages, instead it has been elevated to being one of the basic system messages. It is called a 'quarter frame message' (status byte F1) and is listed in Table 4.4.

4.8.2 Sample dump

Another major use for the universal SysEx messages was the sample dump standard (SDS). This was developed in the 1980s as a way of transferring *audio sample* data from one sampler to another across MIDI. Since MIDI was never designed to transmit the large quantities of data required for digital audio signals, this transfer process is very slow. As more computer systems are *integrating* the processing of sound data and MIDI data within a single piece of software, the requirement for the sample dump standard will diminish.

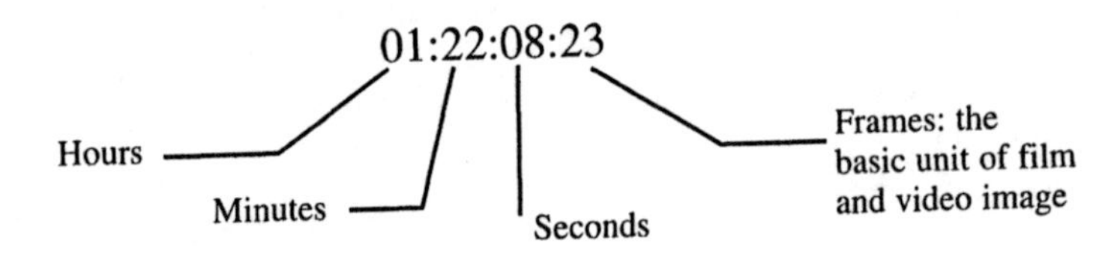

Figure 4.10 The SMPTE timing standard.

4.8.3 General MIDI (GM)

Several criticisms were aimed at the original MIDI specification, particularly concerning the lack of standardisation of sounds.

Since MIDI was designed to transmit *key-press* and *control* information (and thus not 'audio') there are many situations where what is played on one keyboard sounds completely different on another. While musicians can always use this difference to their advantage there were times when people could not guarantee the correspondence between MIDI 'program change' messages and the sounds they wanted.

General MIDI is a way of standardising the sounds (across all GM compatible devices) for a given 'program change' number. It also defines a basic level of compatibility between devices. For example it specifies that channel 10 is the 'drum-track', and states which percussion sounds should correspond to which MIDI note numbers. GM also expects the device to react polyphonically on all 16 channels, and to have a polyphony of at least 24 voices.

Modern devices often have several 'banks' of sounds. GM ensures that the first bank of 128 sounds will always be standardised (e.g. Program 1 will be an 'acoustic piano' sound). This set of 128 sounds includes many imitations of orchestral instruments, rock band instruments and sound effects, arranged into groups. The *sounds* themselves are still left up to the manufacturer – only the *names* of the sounds are defined.

Whilst General MIDI offers a basic set of assumptions for the mass market (for example a computer game programmer can send a MIDI message to 'program change' 125 and confidently expect to hear a 'telephone ring'), many are worried that this is the *complete opposite* of what creative synthesis should be about. They fear the 'worst case scenario' of manufacturers deciding that it is only financially viable to create mass-market GM modules which allow no individual synthesis.

Perhaps the 'best scenario' is a large number of devices which offer a basic level of compatibility, yet allow infinite creative sculpting of sound by individuals.

4.8.4 Standard MIDI files (SMF)

A 'file' is a package of data that is typically stored on a computer disk. Most computer programs save their data in files. A standard MIDI file (SMF) is a file which stores *sequences* of MIDI data. Whilst a MIDI cable transmits messages in real time, a SMF *stores* that data as a series of message events, each with its own *timestamp*. SMFs can be stored on computer disk (and thus moved between computers), or sent across a computer network.

SMFs are used for transferring MIDI 'songs' from one sequencer to another (even on different types of computer). When played back on General MIDI sound devices, these files can be used to distribute musical (note-based) material to a variety of sound devices and computers. Many companies sell disks of 'backing-tracks' (to famous songs or pieces of classical music) which can be run on any device or sequencer that can read SMFs and is compatible with General MIDI. People working in remote locations can send an SMF by disk (or electronic mail, or over the Internet) and thus work on pieces of music together, within the limitations of MIDI and General MIDI.

4.9 Limitations of MIDI

Much has been written about the limitations of the MIDI specification. It is important to be aware of the deficiencies of any specification, but let us begin by recognising MIDI's achievements. MIDI is the result of an international group of competitive companies negotiating with each other to reach a common standard. It has promoted the growth of the modern music technology industry and has provided a low-cost method of linking radically different pieces of musical electronic equipment together. The range and variety of musical devices that are now available are due, in no small part, to MIDI. There are people involved in music who could not otherwise have been (due to physical disability, or lack of access to equipment). In short, MIDI has transformed the electronic musical world.

However, it is important to remember that MIDI is based on 1980s' technology and was originally intended purely for keyboards, so there are some important limitations.

MIDI does not support *sound*. Although it was never MIDI's intention to specify sound, MIDI has tended to be used as a musical 'cure-all'. However, it is a *control* protocol intended for connecting keyboards. Therefore MIDI messages do not lend themselves for specifying creative synthesis, let alone sound waveforms (the sample dump standard is too slow to qualify as a real-time sound control medium).

MIDI is built around *keyboard pitch*. MIDI happily transmits messages generated by a typical keyboard. However, for other types of instrument (for example a MIDI clarinet) the restriction of playing *discrete semitone* pitches and 'bending' these with pitch-bend messages is considerable. An even greater problem exists for players of such instruments. MIDI uses a single note-on *velocity* to define the dynamic response of the note, whereas an acoustic player *continuously shapes* the volume, timbre and pitch of the note, as it is played.

MIDI has a limited *bandwidth*. MIDI transmits a 3-byte note message in one millisecond. This means that it can cope with a maximum of 500 notes (each consisting of a note-on and a note-off) per second (or about 750 notes, if 'running status' is used). This is more than adequate for connecting a few keyboards together, but other data can 'clog' up the MIDI bandwidth considerably. Any type of *continuous controller* message (such as pitch-bend, aftertouch, filter cut-off frequency, etc.) produces vast amounts of data. When this is multiplied by the 16 available MIDI channels, the data rates required are just not possible within the MIDI bandwidth. MIDI's 16 channels are also now considered a limitation. MIDI time code also generates a significantly large amount of data to slow other things down. It is therefore sent down its own dedicated MIDI cable wherever this is possible.

In the past it was often the custom to blame MIDI for problems actually caused by the sluggish response of a keyboard or a module's internal processor. A designer should take care to distinguish between the two effects – one is a MIDI-based transmission delay (due to MIDI's limited bandwidth), the other is simply a slow sound-producing device.

4.10 The future of instrumental connections

MIDI is just one way of connecting instruments together – it happens to be a very popular, internationally agreed way. Apart from the simple keyboard interconnections (which it was originally designed to cope with) MIDI will seem increasingly outdated with the ever-increasing speed of multimedia home computers and entertainment systems. There are various things that can be done to build upon what MIDI has achieved and to extend the range of musical control that is possible. Some of these approaches are described below:

- *Providing multiple MIDI ports*: Several MIDI cables can be used to carry data. Many computer interfaces are available nowadays to support individually addressable MIDI ports. To some degree this allows more channels, less congestion and retains compatibility with existing MIDI devices.
- *Adjusting the MIDI spec*: The addition of new messages to the MIDI specification has allowed MIDI to change with the times. This trend could continue with new types of message, but this does nothing to help the speed (and in fact tends to

slow it down as the new messages typically contain more bytes of data).

- *Incorporating MIDI into networking protocols*: Since connecting computers together (via Ethernet networks for instance) became a relatively cheap task, there have been various attempts to code MIDI as a subset of a faster protocol. A network connection can be made to carry audio, video *and* several channels of MIDI information down a single cable. It still needs MIDI-compatible plugs for the equipment at the far end (so there is a data bottle-neck in and out of each MIDI device), but a single cable (or optical fibre link) can now carry information to many MIDI devices.
- *Replacing MIDI with a high-speed multiple media protocol*: The above methods all involve keeping MIDI at its current data rate. At some point a new standard of audio-visual instrumental connection will be proposed. Audio data and sound synthesis parameters at high data rates will be incorporated along with a more flexible definition of pitch. Many of today's inter-computer connections could be adapted for this purpose. The single biggest obstacle to the introduction of a new standard is the 'inertia' caused by the current industry. Think how many companies, users and educational establishments around the world have invested in MIDI hardware and software. It will be difficult for a new standard to emerge that would satisfy the millions of current MIDI users that their investments so far have not been wasted.

4.11 Summary

MIDI is an internationally agreed way of connecting electronic musical instruments together. It consists of a standardised description of electronic hardware and digital messages. The messages are coded versions of human keyboard gestures, such as playing notes and moving the pitch-bend lever. However, there are ways of expanding the original specification by the use of variable-length system messages. MIDI equipment and use are extremely widespread, but the technology is ageing and a new standard will probably be required.

Suggested further reading

The web-site for this book (http://www.york.ac.uk/inst/mustech/dspmm.htm) contains links to related sites on the Internet covering these topics. Other useful books include:

Rothstein, J. (1995) *MIDI: A comprehensive introduction*, A-R Editions, Madison, ISBN 0895793091.
Rumsey, F. (1994) *MIDI systems and control*, Butterworth–Heinemann, Oxford, ISBN 0-240-51370-3.
Russ, M. and Rumsey, F. (1996) *Sound synthesis and sampling*, Butterworth–Heinemann, Oxford, ISBN 0-240-51429-7.

5 The structure of common music technology systems

Overview

This chapter takes a look at what goes on inside the type of electronic music devices found in a typical high-street music shop – samplers, synthesisers, sequencers and effects units. The complex functionality of these devices is broken down into a set of more easily understood functional components.

The sound generation techniques used in these devices are described and the reader is directed to other parts of the book as appropriate. An introduction to effects units is given, and their internal signal processing techniques are described. Where necessary we explain the mathematical methods that are required to produce these audio processing operations.

The chapter is divided into three major topics. The first (Section 5.1) gives an overview of the kinds of MIDI device available commercially. The second (Sections 5.2 and 5.3) gives an *overview* of the sound synthesis and processing techniques used in such devices. The third (Sections 5.4 and 5.5) gives a more detailed description of the 'inner workings' of these synthesis and processing techniques. This latter topic may be skipped on first reading, especially for those readers who do not require a detailed understanding.

Topics covered

- Types of electronic music equipment.
- Functional categories of common systems.
- Methods of sound generation.
- Effects units.

5.1 Types of electronic music equipment

Manufacturers of music technology equipment market a bewildering range of products. Each product has its own name (and often a number) in order to appeal to its potential purchasers. The names for very similar devices are often wildly different. For example you might be surprised to discover that a 'polyphonic performance synthesiser' could contain exactly the same technology as an 'intelligent digital piano'. There are numerous other examples of ways in which operationally equivalent technology can be dressed up in different packages and given different names.

The purpose of this section is to try to 'decode' the mystifying array of names into a few simple functional components. In other words we will define a set of basic features that can be found in different devices.

Several of these categories refer to 'MIDI'. MIDI is described in detail in Chapter 4, but for our purposes here we need only to think of it as standard method of linking electronic musical instruments together (by passing digital messages between them). The categories are listed below, and are then described in more detail in the following sections.

1 MIDI control device
2 MIDI sound generator
3 MIDI message record/edit/playback
4 MIDI message processor
5 Sound record/edit/playback
6 Sound processor

5.1.1 Category 1: MIDI control device

A MIDI control device is a piece of hardware that converts human gestures (actions) into MIDI messages (see Figure 5.1).

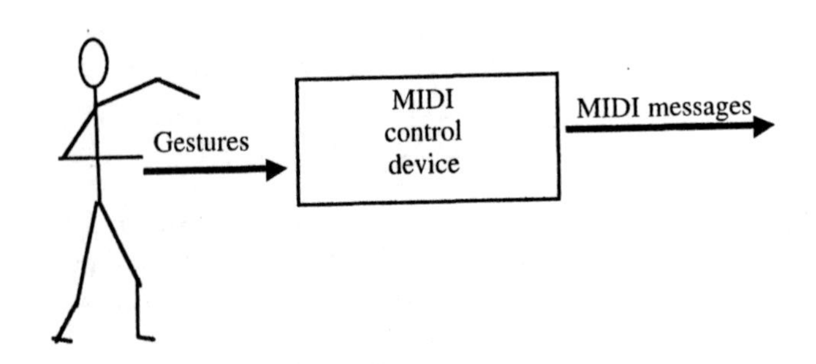

Figure 5.1 MIDI control devices turn human gestures into MIDI information. They can take many different forms, such as keyboards and drum pads.

The most common type of control device is the keyboard, but there are many other types such as drum-pads, guitar interfaces, 'organ-type' pedal boards and wind controllers.

Be aware that 'MIDI control devices' are often called 'MIDI controllers' – a name that is also used for the MIDI *messages* which transmit control change information (see section 4.6.1).

In their purest form MIDI control devices make no sound at all. 'Master' keyboards (sometimes called 'mother' keyboards) rely on the use of externally connected MIDI modules to make sound.

A desktop computer that generates MIDI messages can be considered to be a type of MIDI control device too. Remember that underneath the casing of any digital synthesiser is a computer in disguise – usually in the form of a dedicated microprocessor and lots of pre-programmed memory.

5.1.2 Category 2: MIDI sound generator

A MIDI sound generator receives MIDI messages and uses them to control the creation of sound. The most popular form of such a device is the MIDI sound module. These contain the hardware and software required to synthesise sound, but do not contain a control interface. They rely on being coupled to a MIDI control device (see Figure 5.2). Such devices are sometimes intended for general use and they are given names such as 'multi-timbral sound module' or 'sound expander box'.

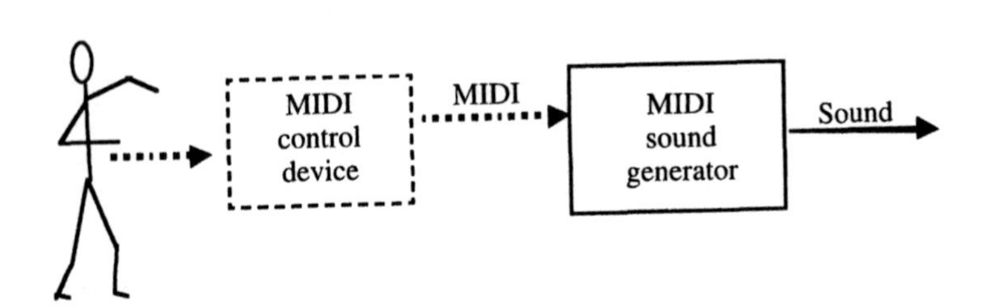

Figure 5.2 A MIDI sound generator receives MIDI messages from a MIDI control device in order to play sound.

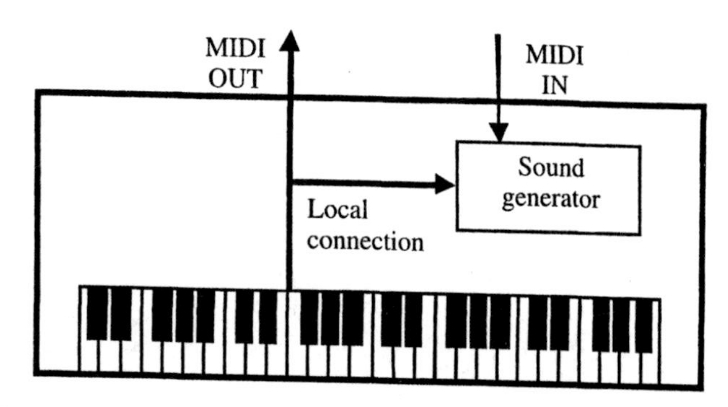

Figure 5.3 The internal connection inside a typical synthesiser. The keyboard is connected to the MIDI OUT *and* to the internal sound module.

Others are aimed at a specialist market and have specific types of sound, for example a 'percussion module' or an 'orchestral sound bank'.

Manufacturers make an assortment of products that contain *both* Categories 1 and 2. In other words they provide a 'control device' and 'sound generator' in the same box. A typical example is a keyboard that (when its keys are pressed) generates MIDI messages *and* its own sound. A variety of names is given to such devices, for example 'multi-timbral synthesiser', 'programmable synth', or 'digital piano'. While it is true to say that each device has its own special characteristics, the basic functional model is the same.

Figure 5.3 shows the internal connections in such a device. The keyboard is played by the user to generate note events. These events are passed to *both* the MIDI OUT socket and to the internal sound generator. Sometimes it is useful to use the keyboard to send *only* MIDI messages. To do this the local connection to the sound generator is switched off. This is usually referred to as 'LOCAL OFF'. Any MIDI messages coming *into* the device are routed directly to the sound generator, and it acts just as if it were an independent MIDI sound module.

5.1.3 Category 3: MIDI message record/edit/playback

This category caters for devices that store MIDI messages, allow the user to edit them, then play them out at a later time (see Figure 5.4).

These systems are typically called *sequencers* due to the way in which they allow the user to put together successions (or 'sequences') of notes. Once the notes have been entered (typically from a MIDI keyboard) they can be edited, added to, deleted, and played back (over MIDI) at a variety of speeds on various instruments. Again, various alternative names appear

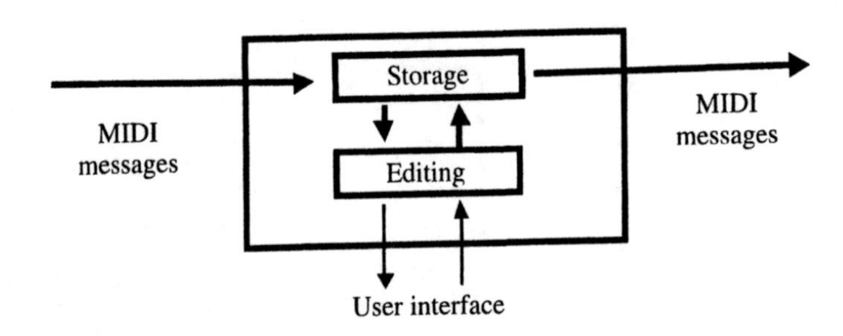

Figure 5.4 The set of functions needed to record, edit and playback MIDI messages.

for products that have basically the same abilities; 'micro-composer', 'portable sequencer' and 'rhythm unit'.

Musicians find this type of functionality useful as the musical structure (the notes and rhythms) can be manipulated *independently* of the final sound. The piece is only heard when the MIDI messages are converted to sound by playing them through a MIDI sound generator.

Some devices combine the functions of Categories 2 and 3. In other words they allow the manipulation of MIDI information, but also can play it back as sound, using an internal sound module (see Figure 5.5).

Names given to this type of product include 'MIDI player', or 'studio production system'. However, if the sound module is dedicated to percussion sounds they are called 'drum machines' or 'rhythm composers'.

Many manufacturers produce equipment that incorporates the functionality of categories 1, 2 and 3. Typically this is a keyboard, with its own internal sound generator and sequencer (see Figure 5.6). They are generally called 'workstations'.

However an equally valid combination of Categories 1, 2 and 3 would be a drum-pad interface that created its own

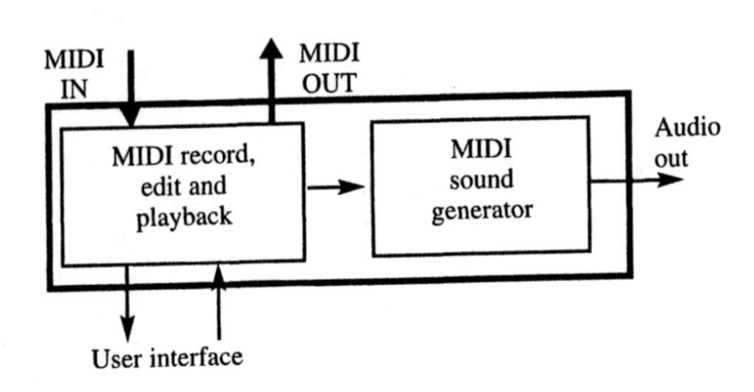

Figure 5.5 A device that combines MIDI recording with sound generation.

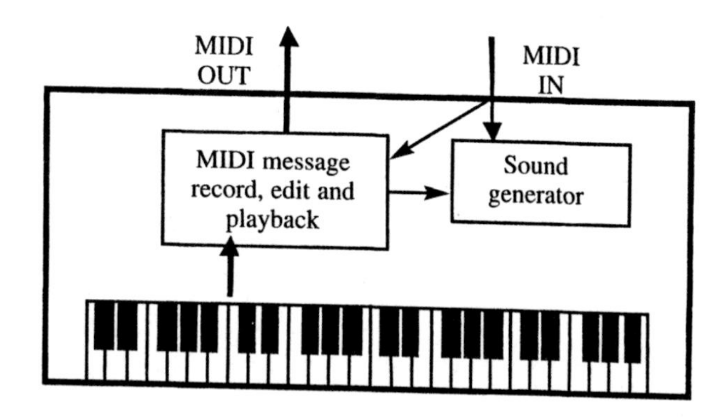

Figure 5.6 Schematic of a keyboard 'workstation'. MIDI messages can be taken from the keyboard and/or the MIDI IN. Messages can be sent to the sound generator from the keyboard, the MIDI IN, or the MIDI playback area.

sound and could record and play back what the drummer performs.

5.1.4 Category 4: MIDI message processor

Some devices process MIDI information in real time. Their purpose is to transform MIDI messages as they are played (according to commands set up in advance by the user). A 'MIDI processor' box could be used to change the MIDI notes being played on an external keyboard. For example, the notes could be sent out on different channels (and thus to different instruments) according to how hard the keys were pressed.

In order to use these devices effectively, the user must understand something of the structure of MIDI messages, and must be able to think about how one message type could be transformed into another.

Some MIDI processors are called 'MIDI patchers' or 'MIDI mixers'. Their job is to *route* MIDI information around a complex set-up of MIDI equipment, such as may be found in a modern studio. They allow the studio connections to be under the control of a computer, rather than relying on the manual 'patching' of leads from one piece of equipment to another.

5.1.5 Category 5: Sound record/edit/playback

Devices in this category allow sound to be stored in digital form and played back. Some equipment allows you to record your own new sound, others will only play back pre-recorded pieces. More advanced equipment allows the sound to be edited before it is played back. Figure 5.7 shows some of the options available.

A 'MIDI sample player' is a device that can play back samples when triggered by a MIDI message. By contrast a 'MIDI sampler' can record new sounds from the outside world, and can often

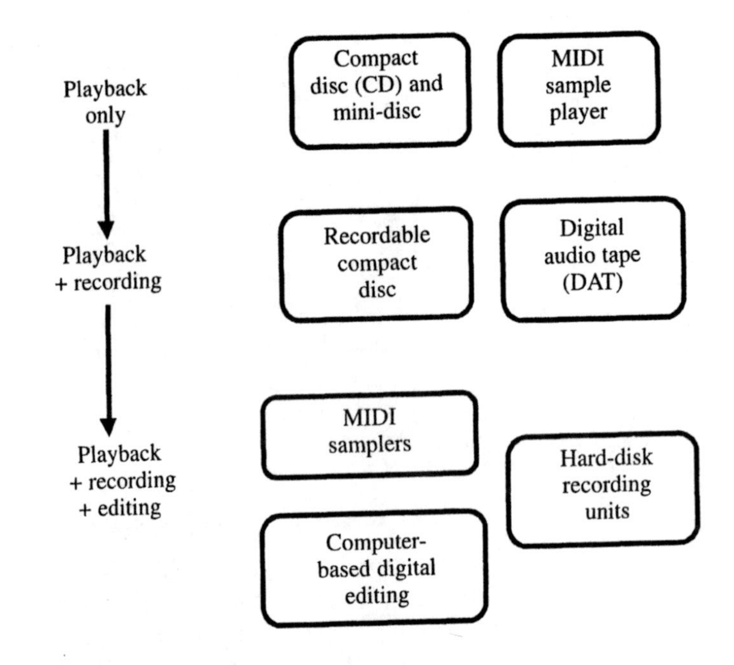

Figure 5.7 A selection of devices that work with digital sound. Towards the top of the diagram only pre-recorded data can be played back. Further down the diagram sound can be recorded and later edited.

allow some editing of the stored sounds. Hard-disk recording units permit a greater duration of sound to be recorded, but they cannot generally be triggered 'live' by a MIDI keyboard. Computer-based digital audio editing offers extensive screen-based editing of sound samples, as well as editing of the structure of the musical piece being constructed.

5.1.6 Category 6: Sound processor

The final category concerns the manipulation of digital audio signals in order to change sound or to add 'effects'. If the sound can be processed 'live' (i.e. the processed sound is output as it is played in) then it is a *real-time* sound processor, and is usually referred to as an 'effects unit'. Section 5.3 considers these devices in more detail.

If a system is running a sound processing algorithm that requires more computer power than is available for it to run in real-time, then the processed sound must be stored. Sound is taken in from a 'soundfile' and then the processed sound is written bit-by-bit to an output soundfile.

5.2 Methods of digital sound generation

This section gives an overview and summary of the most common ways that sound is generated within digital sound systems. Many of these techniques are dealt with in some detail later in the chapter. We have divided them into three main categories:

1 Synthesis by oscillator
2 Synthesis by complex modelling
3 Sampling

5.2.1 Synthesis by oscillator

An 'oscillator' is an entity that produces a periodically fluctuating signal. In other words an oscillator generates a varying output that is regularly repeated. It has three basic attributes:

- The 'shape' or trajectory of the output. This is known as the *waveform*.
- The maximum deviation of the waveform from 'rest' – the *amplitude*.
- The number of times per second that the waveform is repeated – known as the *frequency*.

The oscillator is the basic unit of synthesis for many systems. Oscillators can be built from certain electrical circuits to produce varying voltages. This is the basis for analogue synthesis, where electrical signals are manipulated and then converted into sound. In digital systems, however, oscillators are constructed from computer code. The code repeatedly outputs the 'waveform' by reading it continuously from computer memory at the required frequency.

A single oscillator on its own produces a single tone. The waveform is often a sinewave – a very simple form of oscillation. By itself the sound is very pure, yet piercing, not at all a pleasant tone to listen to. So in order to make an acceptable sound we must combine the tones of the oscillators. The following sections describe the most common ways of combining oscillators.

Additive synthesis

When the outputs of several oscillators are added together to produce a composite sound, this is called *additive synthesis*. The complexity of the composite sound is determined by how many oscillators are added together, and what their individual waveforms, frequencies and amplitudes are. Most additive synthesis systems use sinewaves since it is possible to reconstruct any sound by using the right combination of sinewaves.

When a series of sinewave oscillators are added together, whose frequencies are all exact multiples of the first (known as *harmon-*

ics), then the resulting sound is 'pulled together' by our brain into one complex tone. This is the basis of creating rich sounds by using a combination of simple tones.

Sounds in the real world are generally very complex, and so a large number of oscillators are needed to recreate them. Additive synthesis is considered to be a very flexible way of creating sound, but one in which a lot of computing effort needs to be applied in order to make sounds of acceptable quality.

Subtractive synthesis

An alternative to additive synthesis is *subtractive synthesis*. Instead of building up a complex sound by adding together a large number of sinewave components, we can start from a complex signal and *subtract* harmonics to make it simpler. The act of removing certain spectral components from a signal is called *filtering*.

Musicians often feel that subtractive synthesis produces a very expressive control of sound. This may be because our own vocal mechanism uses a form of filtering to make different sounds. For example if you make an 'eee' sound, then gradually purse your lips until you produce an 'ooo' sound you have just performed a form of subtractive synthesis. The 'eee' is very rich in harmonics, and these are removed as you close your mouth to produce a softer and less harmonically complex 'ooo'. We even give vocal names to subtractive filtering effects. Consider the guitarist's 'wah-wah' pedal which (when pedalled down and up) lets through a variable number of harmonics to produce its characteristic sound.

Frequency modulation (FM)

When the output of one oscillator is used to change the frequency of another oscillator, this is known as frequency modulation. Imagine a pure tone (say of 440 Hz, the pitch that an oboe plays for an orchestra to tune up to). Now imagine slowly varying the frequency of that tone up and down. If it goes up and down by a small amount (a few times per second) it will sound like the 'vibrato' that musicians often add to their voice or instruments when holding a steady tone. If, on the other hand, the frequency were to sweep wildly up and down it might begin to sound like an ambulance or police siren. Both of these effects are due to frequency modulation.

However, a strange thing happens when the speed of the variation becomes quite fast. We cease to hear it as a 'vibrato' or 'siren' sound – but instead the *timbral quality* of the tone changes dramatically. As we continue to change the speed and the depth of the modulation we hear the sound go through a continuous series of tonal changes (some harsh tones, some buzzy, others noisy).

So FM uses a small number of oscillators to produce complex tones. However there is no straightforward perceptual relationship between the modulation values and the timbre produced, and hence it is difficult to use FM synthesis to create a specific sound that you might require.

Amplitude modulation (AM)

If an oscillator's *amplitude* is varied by the output of a second oscillator, we call this *amplitude modulation*. Again, try to imagine a tone of 440 Hz, but this time its amplitude is slowly varied. You would hear tone getting repeatedly louder and softer. Some guitarists call this effect 'tremolo', and it is a feature found on some guitar amplifiers.

When the speed and depth of the modulation are increased, the sound is heard to become increasingly complex. The effect is similar to, but rather less dramatic than, its frequency modulation equivalent.

5.2.2 Synthesis by complex modelling

An increasingly popular category of synthesis uses mathematical models of real-world objects as the basis for sound generation. Objects such as springs, air molecules vibrating within an enclosed tube, and material surfaces are described inside the computer as mathematical equations. This is often known as 'physical modelling'.

In order to generate sound in a real physical object it must be hit or scraped or blown. This is known as *excitation*, where energy is injected into a system so that its component parts vibrate. At some point the energy is dissipated as sound waves. These two concepts must also be modelled within the computer.

The computer stores an equation for every component in the system and for any connections between components. The values corresponding to energy excitation are fed into each equation. All the equations are then run in turn, with energy being passed from one component to another until it reaches a point of dissipation. At that point the fluctuations in energy are converted into sound by calculating the sum of the energies at a particular point which corresponds to an imaginary microphone.

Physical models may consist of tens of thousands of equations, and thousands of connections. It is therefore no surprise that, in order for the whole model to be run fast enough to generate sound in real-time, it needs an enormous amount of computing power.

The great benefit of physical modelling is that it is capable of producing realistic, rich and dynamic sounds of great complexity. As computer systems become faster in the future this method of synthesis will grow in importance and is likely to take over as the dominant type of sound generation.

5.2.3 Sampling

The former two methods of sound generation are based around the creation of sound from equations (either repetitive or complex). Sound *sampling*, on the other hand, is concerned with recording sound from the real world and then using that sound as the source for generating new sounds.

Sound is recorded into a computer using a microphone together with a device that turns analogue sound signals into a digital form. The stream of numbers that represents the sound is stored in a block of computer memory. If the sound is simply played back out (by converting the numbers back into sound at the same rate that it was recorded) the original sound is heard.

However 'samplers' generally allow the numbers to be played out at different speeds. If the numbers are played out faster than they are recorded then the sound is heard at a higher pitch. This will not generally sound like the original instrument playing at these different pitches. A voice sample played back faster than it was recorded will sound 'squeaky' and rather like a cartoon character. MIDI samplers can receive MIDI note messages and replay samples, in real time, at the appropriate pitch. Most complex sampling equipment allows the user to record or import many different samples and have them playing back at the same time.

A common method of sound generation is to *combine* sampling and synthesis. The human ear is very sensitive to the first fraction of a second of any sound. It uses this part of the sound to work out what instrument is playing. Therefore many systems always use a short sound sample of a real instrument at the start of a note, before completing the note with a less-critical synthesised sound.

5.3 Methods of digital sound processing

Digital effects processing units are widely used by keyboard players, guitarists and studio engineers. Their aim is to modify the incoming sound in some way for musical or sonic effect. This section describes in outline the main forms of sound processing that are found in effects units. Many such devices can run several of these effects at the same time, and these are often called 'multi-effect units'. Greater detail is presented in Section 5.5.

5.3.1 Reverberation

Sound that is produced naturally in a room or a concert hall bounces around the room (and its contents) in a complex way. This effect is called *reverberation*. In everyday life we tend not to notice it, except when it is exaggerated in an extremely large hall or a cathedral, where the sound seems to carry on for quite a while after we have stopped generating it. However, we tend to notice its *absence* in many electronically generated sounds, or sounds that have been recorded in a small room. We say that it sounds 'dead' or 'dry', and it is not usually a pleasant listening experience.

Artificial reverberation (or 'reverb') is one of the most commonly used forms of sound processing. The main characteristics of reverb that can be controlled by the user are:

- The *depth* – how much reverb is applied to the signal.
- The *decay time* – how long the reverb takes to die away.
- The *algorithm* – the actual equation for the reverb process that controls how the effect is created. Usually the different algorithms are given individual names, such as 'Hall 1', 'Small room' or 'Reverb plate'.
- The *mix* – the balance between the dry signal and the reverberant signal. Most effect units have this control as it allows the user to control the level of effect from 'subtle' to 'gross'.

5.3.2 Delay

If a 'copy' is made of a sound, then that copy is played back a short while later, it will sound like an echo. In real life situations an echo is caused by sound travelling some distance to a point of reflection and then coming back (some time afterward) to the listener. The echo is usually much quieter than the original signal. The act of copying a sound and then playing it back after a pause is called *delay*. It is a commonly used effect, particularly by guitarists. It is often used to provide a rhythmic effect whereby the *delay time* is set to be related to the tempo of the music. This can give the effect of doubling or trebling the speed of notes played. When the delay time is particularly short it can create a 'ricochet' effect that sounds like the early sound reflections in a small room. Extremely short delay times make it sound like there is more than one instrument playing. This is one of the aspects of the *chorus* effect.

5.3.3 Chorus

Chorusing is a way of enriching a sound signal so that a single instrument is made to sound like a group or 'chorus' of instruments, all playing the same musical line. This can be done in a variety of ways. A popular method is to process the input sound through several delay effects, all with slightly different delay times. Another method is to make several copies of the sound and to shift each pitch (very slightly) so that they are all different.

The main aim is to create the illusion that there are several instruments playing.

5.3.4 Phasing and flanging

If a sound signal is put through a delay, and the *delay time is continuously changed*, then a variety of sonic effects is heard. These are given various names, but the most popular are *phasing* and *flanging*. They are both widely used by guitarists to enrich the sound signal in a dynamic way. Flanging creates a 'sweeping' or 'swishing' movement, whereas phasing tends to produce a gentle undulation in the sound.

5.4 Digital sound generation techniques

In this section, we examine the way in which signal processing techniques described earlier, such as sampling and unit generator synthesis, are used in producing the sound output from common music technology systems. The reader is therefore strongly advised to read the first part of this chapter and Chapters 2 and 3 before proceeding with this section.

5.4.1 Sound synthesisers

The wavetable oscillator described in Section 3.5 forms the basis of many synthesisers in current usage. In Section 3.5, we described the concept of a *unit generator*. This is an audio pro-

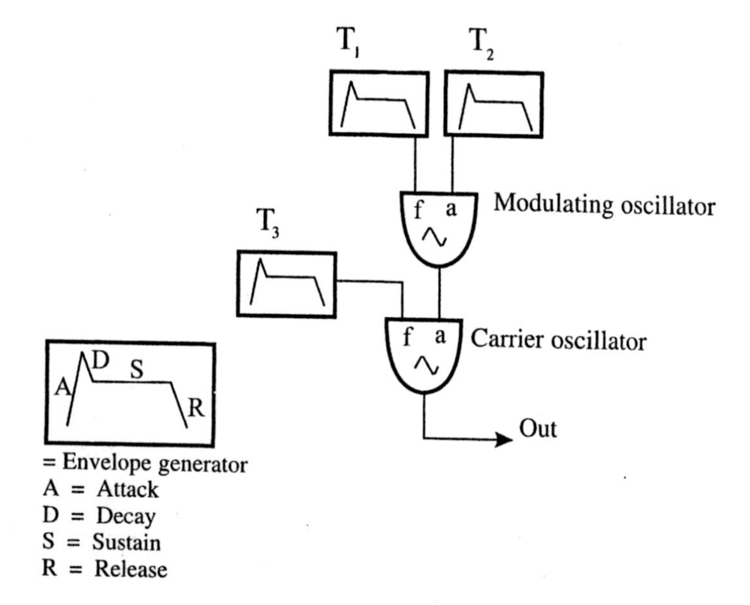

Figure 5.8 Ring modulation using oscillator unit generators.

cessing unit such as an oscillator, a sound transformation device or an algorithm, constructed in a form which allows networks of such units to be put together in different configurations. Even relatively simple configurations of such units can achieve quite radical timbral transformation and synthesis of sound, as we shall see.

Amplitude modulation

Consider the configuration shown in Figure 5.8. This contains two oscillator unit generators, the output of the first (the modulating oscillator) driving the *amplitude* input of the second (the carrier oscillator). This is a form of ring modulation, as we will now show.

The amplitude control input of an oscillator unit generator controls the oscillator's output by *multiplying* output samples selected from the wavetable by the number passed in through the amplitude input. This process of multiplication is easy to carry out, since the unit generators are in reality small pieces of computer program code, and multiplication is available as a primitive operation in most computers.

It was shown in Section 3.1 that the process of multiplying two sinusoids results in DSBSC amplitude modulation, with the spectrum shown in Figure 3.3. If the signal supplied from the modulating oscillator were not a simple sinusoid, but was in fact a composite signal consisting of a number of harmonics (this would be easy to arrange by simply filling the modulating oscillator's wavetable with a composite), then the spectrum at the output would be as shown in Figure 3.4.

It was also shown in Section 3.1 that increasing the frequency of the carrier oscillator produces a *lateral* shift of the sidebands up the frequency axis in Figure 3.4. This lateral frequency shift produces radical changes in the harmonic structure of the sound being manipulated by DSBSC amplitude modulation, and this in turn has a radical impact on the sound produced.

To see how the harmonic structure is affected, imagine that we start with a baseband in the modulating oscillator consisting of a fundamental plus third harmonic, output by the modulating oscillator at 100 and 300 Hz respectively. Suppose this was then sent to the amplitude input of the carrier oscillator, whose frequency input was set to 800 Hz. Looking at Figure 3.4, you will see that the output spectrum would then contain components at 500 Hz, 700 Hz (lower sideband), 900Hz and 1100 Hz (upper sideband). The harmonic signal input to the carrier results in an output where *none* of the components are harmonically related. This is indeed a radical harmonic manipulation. Of course, choosing the frequencies of the two oscillators carefully *could* make the output components harmonically related, and shifting the frequencies of the oscillators could make the outputs move in and out of harmonicity, as might be required by the sound manipulation process. The *process of modulation* has enabled the wavetable oscillator to produce potentially many inharmonic components, and only at the cost of a couple of oscillators!

Envelope generators

The envelope generators shown in Figure 5.8 can be used to assist in timbral manipulation with DSBSC. An envelope generator is a unit generator which is rather similar to an oscillator, in the sense that it outputs a sequence of values from a stored internal data table. However, unlike an oscillator, instead of 'wrapping around' when it gets to the end of the table, to form a continuously evolving waveform, the envelope generator typically runs through its table once (usually at a fairly slow rate) and then stops. Its function, once triggered, is to provide a time-varying flow of parameters to an input within a configuration of unit generators, so that the behaviour of the network can be shaped over the duration of a musical event.

A typical application of an envelope generator might be to shape the output amplitude of a network of oscillators, perhaps in response to a 'note-on' MIDI event. In this case, once triggered, the sound would build up during the attack phase (see Figure

5.8), reach a peak, and then fall back during the decay phase to a 'sustain' level, which might be held constant at the output of the envelope generator whilst the performer holds a key down on a MIDI keyboard. When the key is released the output of the envelope generator gradually decays away to zero, through the release phase. The envelope generator in this application therefore defines the amplitude of a note event synthesised by the network of unit generators. However, there is nothing to say that the envelope generator should be constrained to control only amplitude. It could be used to sweep the cut-off frequency of a filter unit generator for instance, so that the spectrum (timbre) of a sound is dynamically controlled during a musical event. This is an important facility, since one of the criticisms often levelled at electronic sound is that it is static and therefore dull, so dynamic control in this form is valuable.

In Figure 5.8, envelope generators are used to modify dynamically the operating frequencies of the DSBSC. T_1 controls the output frequency of the modulating oscillator via its frequency input, and will define the fundamental frequency of the modulating oscillator. This will therefore control the *width* of the sidebands of the DSBSC spectrum, so that this will fluctuate during a musical event (review Figure 3.4). T_3 on the other hand controls amount of translation of the DSBSC sidebands along the frequency axis, and as described above, this radically affects the harmonic relationships within the signal.

Frequency modulation

If the output of the modulating oscillator was connected to the *frequency* input of the carrier oscillator, then a different form of modulation, *frequency modulation (FM)*, would result. Here, the *frequency* of the carrier oscillator will be proportional to the *amplitude* of the output of the modulating oscillator. Let us assume for the moment that the amplitude of the carrier oscillator is held constant. Figure 5.9(a) illustrates the time domain output of the FM signal which would result from this connection.

K_1 in Figure 5.9(c) ensures a positive output frequency, even at the most negative swing of frequency f_1.

Figure 5.9(b) is a schematic representation of the frequency spectrum of the FM signal. Within the spectrum envelope there are many individual spectrum components. It is not possible to relate directly the output spectrum of FM to the spectrum of the modulating oscillator, as it was in DSBSC/ring modulation. All we can say is that the trivially simple connection of two oscillators in this configuration has produced a complex output spec-

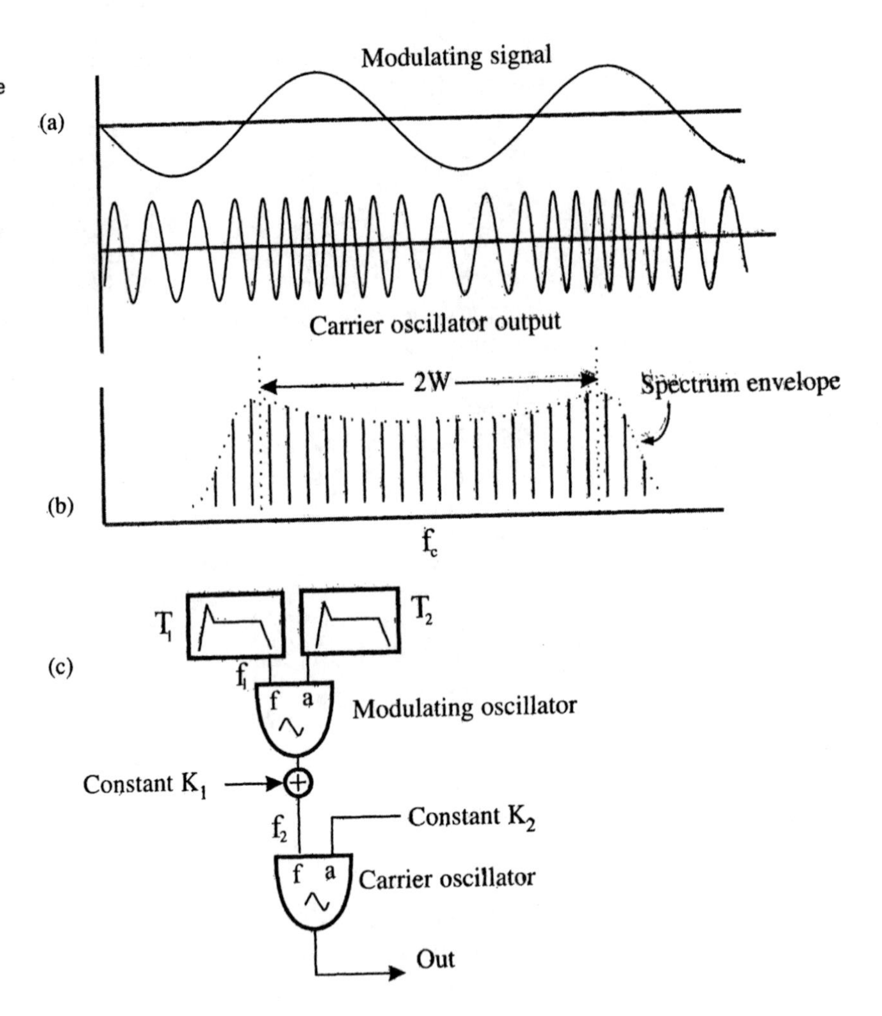

Figure 5.9 A frequency modulated signal and the unit generator configuration which produces it; (a) time domain; (b) frequency domain; (c) configuration of unit generators for frequency modulation (FM).

trum, with a correspondingly complex timbre. It is not even possible in general to say how the spectrum will be affected by changing parameters in the synthesis network, for instance by increasing the amplitude of the output of the modulating oscillator. This will increase the frequency swing of the output of the carrier oscillator away from its quiescent value. This will increase parameter W in Figure 5.9(b), and hence the overall bandwidth of the spectrum. Changing constant K_1 will move the spectrum along the frequency axis, and hence further modify the timbre since it controls the unmodulated frequency of the carrier, shown as f_c in the figure. In practice, envelope generators may be used to impart a dynamic character to the sound produced in a similar way to the DSBSC spectra described above.

It is quite difficult to predict the timbral effect of changing any of the synthesis parameters such as W or f_c in FM. This is because there is in general, a highly non-linear relationship between the numerical values of these parameters and the spectrum produced. Nevertheless, it is possible to proceed by a process of auditioning the output produced by a particular parameter set, and characterising the sound produced by some evocative label, such as 'metal violin' or 'moonscape grains'. In practice, a combination of amplitude modulation and frequency modulation is used in the design of 'electroacoustic instruments', further making the timbral design more flexible although increasingly complex.

Other forms of synthesis

We see from the discussion above that modulation of various forms is capable of generating complex spectra relatively cheaply in terms of the numbers of unit generators used. To some extent, this is bought at the cost of predictable behaviour. Nevertheless, these kinds of configurations (or 'patches') of synthesis units are widely used in commercial synthesisers. They are also used in more experimental synthesis applications, perhaps using the Csound language or the MIDAS system. The popularity of this approach arises because of the relative simplicity (and therefore computational cheapness) of these modulation based approaches for a given spectral complexity.

We could always synthesise a given spectrum simply by taking as many oscillator unit generators as there are frequency components, and simply adding their outputs together, perhaps using envelope generators to provide a dynamic character. This approach, known as *additive synthesis*, has the almost the reverse characteristics of the modulation approach: it is very controllable, since we have direct control over the frequency and amplitude of each partial produced by a given oscillator. Nevertheless, we would need very many oscillators to synthesise a spectrum of any complexity, and this may make the computational cost prohibitive.

An alternative approach is to start with a noise-like signal which has a wealth of spectral components, and to remove unwanted components using a bank of filters. This *subtractive synthesis* scheme is also directly controllable, but the cost of the numbers of filters may be unacceptably high in any given application.

5.4.2 Samplers

The idea of a stored wavetable is also central to the operation of the sampling synthesiser, or 'sampler'. However, in this case, the samples from a live sound are placed into digital memory, rather than the pre-calculated synthetic waveforms used in the synthesis techniques described in Section 5.4.1. The musician is provided with a microphone to record sound. This is sampled, PCM encoded and then the PCM code groups are stored in digital memory. The wavetable synthesis techniques described in Chapter 3 can then be used to re-synthesise the sound. Often the sampler is controlled via MIDI to control pitch and other performance parameters.

Although this may be regarded by the naïve as a worrying attempt to capture instrumental or orchestral sound, and 'put it in a box' in order to make orchestras redundant, the process is technically difficult to achieve effectively for the reasons described below. Orchestras have a role to play for the foreseeable future!

During playback of the samples, the wavetable memory is treated as consisting of several segments, corresponding in many ways to the ADSR envelopes described in Section 5.4.1. Let us suppose that the sampler is to be played using a keyboard. When the musician strikes a key, the sampler plays through the stored attack and decay transient. If the sound is to be a sustained note, for instance that of an organ, then while the musician holds the key down, the sampler will *loop* through a predefined set of samples in memory usually located within the 'sustain' segment of the stored waveform. This will produce successive cycles of the output signal, exactly as described for the wavetable oscillator in Chapter 3. When the key is released, the sampler leaves the loop, and plays through the decay transient, before ending the sample playback, and thus completing the note. This scheme allows a sampler to use a realistic envelope for the sound synthesis. This is important, since envelope parameters such as the attack and decay transients are important psychoacoustical cues in the cognition of sound.

The musician is usually given the means to select the positions of the 'loop points', or the boundaries in memory of the looped waveform, so that the sonic material making up the sustained sound can be selected.

Pitch is controlled using the sample skip method described in Chapter 3 for the wavetable oscillator, where the pitch needs to rise. If the sound is to be played back at a lower pitch than that recorded, then sample interpolation is used to provide additional synthetic output samples lying between those actually recorded into memory.

The difficulties in this scheme arise in the way a sampler treats the spectra of the recorded signals. Since the signal which will be played is a 'snapshot' of one particular note, then the sampler produces exact replicas of this snapshot on playback. This sounds unnatural, since no live musician can produce such a uniform sound. The pitch shifting techniques described above replicate the harmonic structure of the recorded note(s), albeit at a different pitches since the waveshape is retained, even though the pitch is increased. This replicated harmonic structure is also very unnatural. In real instruments, sonic parameters such as spectra and envelope parameters change quite significantly throughout the compass of notes available, and if samplers are to do a reasonable job of replicating the instrumental sound, then this behaviour must also be replicated. For this reason, even a modest sampler would provide the means to use different note samples every 3–4 semitones within the compass of the sound.

Even more difficult is the problem posed by inharmonic components within the frequency spectrum. As discussed in Chapter 2, any system (e.g. a sampler) which replicates identical waveforms in successive cycles will not be capable of dealing adequately with inharmonic frequency components in the sound. Since real instruments contain some elements of inharmonic sound in the general case, this will cause difficulties for the sampler.

The difficulty emerges in the selection of suitable loop points. These are chosen so that the waveform segment played out within the looped section contains an exact integer number of cycles of the fundamental. By definition, this means that the waveform segment will also contain an exact number of cycles of any harmonics present. The sample can then be played out end-on-end to give a continuous replication of successive cycles without any waveform discontinuities, since the end of the looped segment corresponds exactly to the beginning of the segment (since it contains an integer number of cycles).

Consider now the fate of inharmonic components. By definition, if the waveform segment contains an exact integer number of cycles of harmonic components, it must contain a *non-integer* number of cycles of the inharmonic components. The last cycle of any such inharmonic component in the segment will be incomplete. When this incomplete cycle is looped, there will be a regular discontinuity in the waveform produced, and this will produce a rough-sounding electronic artefact. This will be particularly true if the looped segment is short, containing a small number of cycles. The discontinuity will therefore appear at a relatively high rate on playback, causing a persistent 'buzzing' noise.

These problems are ameliorated (they cannot be removed completely) by taking the loop segment over many cycles of the recorded waveform. This means that the repetition rate of the segment will be quite low, so that the discontinuities occur relatively infrequently, perhaps every second or so. This will make them less invasive, producing a 'click' every second in this case. The effect of the click is often further reduced by joining the end of the loop back to the beginning under a cross-fade (the end of the loop is faded out whilst the beginning of the loop is faded in). These techniques control the problem of inharmonic components to within acceptable limits, for many applications.

5.5 Digital sound processing techniques

We now turn to a study of the internal architecture of a typical effects unit, so that we may gain some understanding of the operation of the signal processing techniques used. Many of the techniques described in this section build on concepts described in Chapters 2 and 3, and the reader who wishes to proceed further is strongly advised to read those chapters first. The (simplified) layout of a typical unit is shown in Figure 5.10.

As described earlier, the purpose of an effects unit is to process an input (analogue) sound with a variety of sound processing techniques, in order to transform the input 'dry' sound into a 'wet' sound which has different acoustic properties. The actual output sound is usually a mix of the direct dry sound and the processed wet sound.

The processing is usually carried out in digital form, often in computer based algorithms, although sometimes direct hardware implementations of these digital 'algorithms' are used to enhance speed of operation. This digital processing requires that the sound is sampled and PCM encoded before being passed on to the processing chain, with the output samples reconstituted back into an analogue signal in the normal way for a PCM system.

The heart of an effects unit is some form of internal 'sound interchange bus', which allows the processing units to share and exchange transformed sound data in a variety of configurations of the sound processing chain. So for instance, Figure 5.10(b) shows a specific configuration of the sound processing chain which might be implemented with the generalised architecture of Figure 5.10(a).

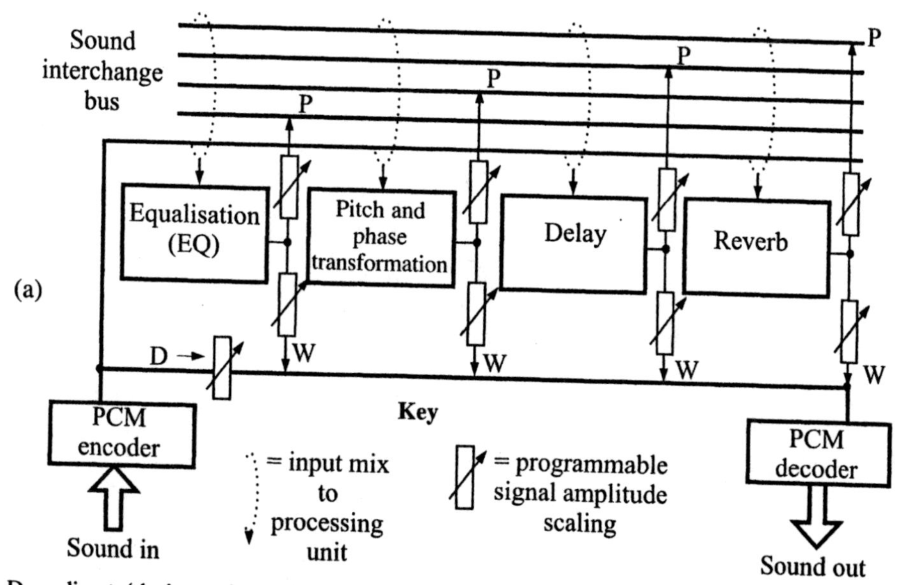

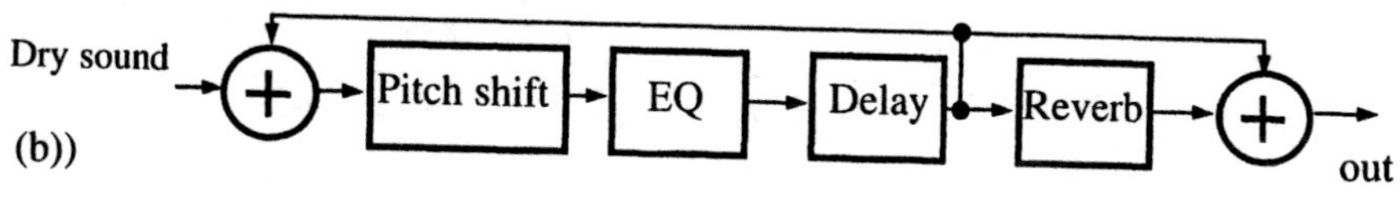

Figure 5.10 (a) Internal structure of a typical effects unit; (b) Example of an effects processing chain based on the generalised architecture in (a).

The functions of the major transformation units in Figure 5.10(a) are described in the subsections below, however, they are all mainly based on a variety of uses of memory delay lines. In such a delay line, PCM samples are shunted into digital memory, to be retrieved later and mixed in with other signals. An understanding of the effects of *delay* on signal characteristics is therefore central to this section. This is especially true of the 'EQ' (equalisation) section, where a variety of *digital filters* are used to manipulate the spectrum (and hence timbre) of sound. Since digital filtering employs delay and mixing, it therefore follows that any process which uses these techniques (e.g. reverberation) must, of necessity, exhibit some implicit filtering characteristics. It is important to be aware of these implicit properties, so the section starts with a careful look at the relation between delay and filtering within the context of digital filters.

5.5.1 Equalisation and digital filters

As stated above, equalisation consists of the application of a variety of different kinds of filters to a sound in order to manipulate the timbre, or to emphasise (or de-emphasise) different parts of the sound spectrum. The filters used include all of the forms we have met before (lowpass, highpass, bandpass and bandstop), but they are usually implemented as computer algorithms, in various different, but related forms. This subsection introduces the techniques used, and shows how they are based on the mechanisms of scaling and delaying of signals. As stated above, an understanding of the frequency selective behaviours of these primitive operations is important, since they appear in many of the other effects, not just in equalisation.

Phase shifts and filtering

In analogue systems, filters are made from physical electronic components such as capacitors, resistors and amplifiers. It is not at all obvious at first sight how a computer program, or some comparable digital process can achieve an equivalent effect on signal spectra. The key to understanding the way in which digital systems do this lies in understanding the effect of *relative phase differences* on resultant amplitude, when two signals of the same frequency (but different phase) are superposed or added together. This is illustrated in Figures 5.11 and 5.12.

It was explained in Chapter 2 (Section 2.2) that a phase shift in a signal is equivalent to shifting a waveform laterally along the time axis. It was also explained that phase is measured in degrees or equivalently in radians, and that a phase shift of 360° (or 2π radians) corresponds to changing the phase of the signal by one complete cycle. In order to understand the nature of the output signal in Figure 5.11, we need to know the phase shift imposed on a signal of a given frequency caused by the fixed time delay τ.

Now frequency is defined as the number of cycles (of phase 360°

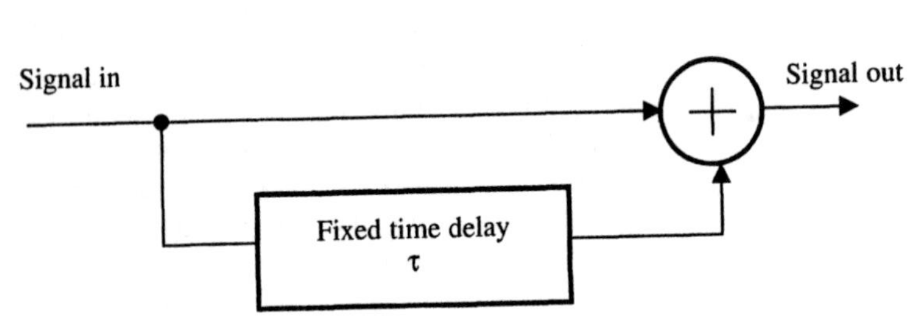

Figure 5.11 Superposing a signal with a delayed version of itself.

or 2π radians) produced by a signal in one second. There are, in fact, two measures of frequency. The first we have already met: the number of cycles per second, usually given the symbol f and measured in Hz. The other measure is known as the *angular frequency,* usually given the symbol ω and describes the numbers of *radians* produced by the signal in one second. Since each cycle consists of 2π radians, the relationship between the number of cycles produced per second (f) and the number radians produced per second (ω) is given by:

$$\omega = 2\pi f \qquad (5.1)$$

If ω is the number of radians produced by the signal *per unit time,* then the phase shift (delay) produced in our fixed time delay of τ seconds is $\omega\tau$ radians. Notice that given a fixed time delay τ, this phase shift is *proportional to frequency* (ω). This is so important to our understanding of digital filters that we shall say it a different way: if signals of varying frequencies are applied to the system in Figure 5.11, then we will be superposing signals at the output with corresponding phase shifts which vary with frequency.

What is the effect on the superposed output signal *amplitude* of these varying phase shifts? This is shown in Figure 5.12.

The time interval imposed on the delayed signal is the *same* in each case, no matter what the input frequency: it is in fact one half of the period of the highest frequency signal in the example shown in Figure 5.12. You could confirm the results shown in the figure yourself with a piece of graph paper and some patience: it is just a process of adding together (superposing) the two input waveforms to produce the output. (In Figure 5.12 all of the outputs have been scaled by a factor of 0.5 to make them comparable in amplitude to the inputs, but this does not affect the generality of the argument.) Notice that the *amplitude* of the output signal reduces as the phase difference, $\omega\tau$ radians increases with input signal frequency ω. Changing amplitude with frequency is a property of a filter: our fixed time delay system shown in Figure 5.11 is in fact a simple (lowpass) filter!

It is quite difficult to make a fixed (frequency invariant) time delay in analogue electronics, but in digital audio systems it is astoundingly simple: we simply store our PCM samples (binary numbers) in a digital memory such as RAM, and let them lie fallow until we are ready to use them at some later time. The length of the delay will not, of course, vary with frequency. Such digital memories are described in Chapters 6 and 7 where we describe the operation of computer systems. This mechanism

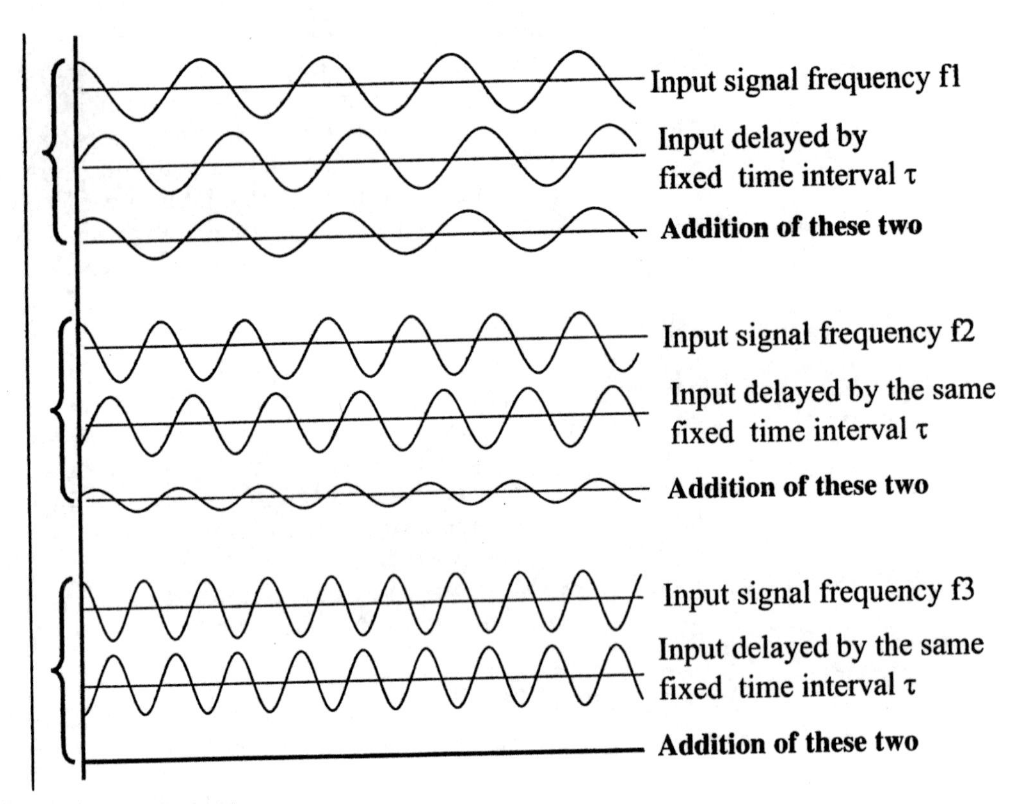

Figure 5.12 The effects of the fixed time delay in Figure 5.11 on the output amplitudes formed from the superposition of different input signal frequencies.

gives us the basis by which, for instance, a computer program can implement a filtering process.

Representing phase shifts with phasors

Whilst our method of superposition has showed us that a fixed time delay can implement a filtering process, it would be quite difficult for us to predict the effect of having several such delays in operation, perhaps additionally scaling the amplitudes of the delayed signals. Also, what would be the effect of adding in delayed versions of the output? All of these things would presumably have some more or less radical effect on the filtering action of the system. We could ultimately get to the answer through superposition (although dealing with delayed versions of the output might tax the brain somewhat). However, the practical difficulty of adding together wiggly lines representing signals has defeated us again, just as it did in Chapter 2, where we were considering the nature of complex signals. Again, we need a more convenient way of representing the information. In Chapter 2, when we hit this difficulty, we introduced the notion of the *frequency domain*, where we used a single line to represent

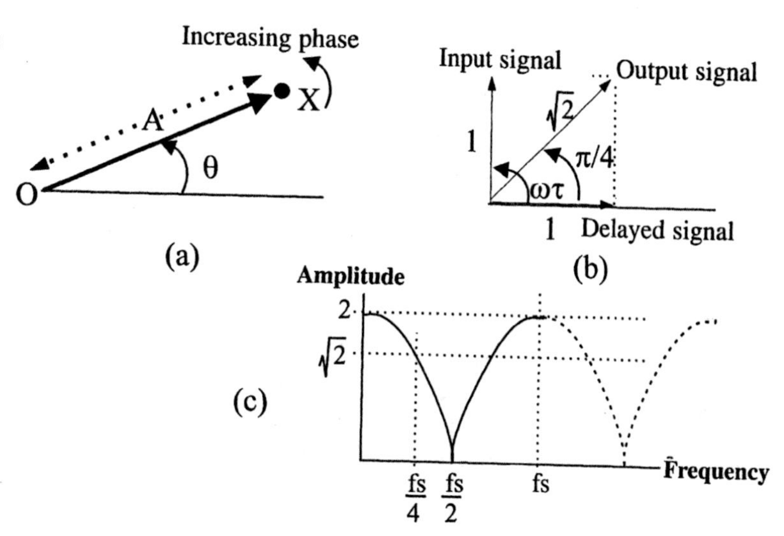

Figure 5.13 (a), (b) and (c) The phasor and frequency response representation of the output of the fixed delay superposition system shown in Figure 5.11.

a whole sinusoid. Can we look to this single line representation to inspire us in finding a more convenient representation of phase, amplitude (and perhaps frequency) of a sinusoid?

Indeed we can. Figure 5.13(a) shows the concept of a *directed line*, where the length of the line (A) represents the amplitude of a sinusoid, and the angle θ measured with respect to some phase origin (conventionally taken as the horizontal line) represents the phase of the sinusoid. Other names for such a directed line include the term 'vector' and more commonly in the world of signal theory, the term 'phasor'.

We can even represent frequency in a certain way with such a diagram. If we recognise that a sinusoid is a signal whose phase is increasing proportionally with time and frequency, i.e. $\theta = \omega\tau$, then a sinusoidal signal would be represented in Figure 5.13(a) by rotating the phasor counter-clockwise, with angular velocity ω. It is even possible to unify the phasor representation with the frequency spectrum diagram such as that shown Figure 2.6. If we imagine looking *along* the line of the frequency axis, and further imagine that each line in the frequency spectrum is actually a phasor whose frequency is given by its position on the frequency axis, then a rotation *about* that axis would represent the phase of the frequency component. Figure 5.13(a) is a representation of what we would see looking at one frequency component, along the line of the frequency axis.

Figure 5.13(b) is a phasor diagram of our fixed time delay (τ) system shown in Figure 5.11. All of the phasor signals shown in the diagram are rotating counter clockwise, at angular rate ω. This simply tells us that the signals involved are all sinusoids. Of more interest is the fact that the phasor diagram shows the delayed signal lagging behind the input signal by a phase angle $\omega\tau$. The amplitude and phase of the superposed output is therefore found by adding the input signal and delayed signal vectorially, in diagrams such as Figure 5.13(b). Note that the angle between the input signal and the delayed signal varies with frequency, being proportional to the frequency of the signals, ω (since τ is fixed). If the frequency, ω, of our signals is zero (0 Hz or 'DC'), then both phasors for the input signal and the delayed signal line up horizontally pointing to the right, and the resultant amplitude is 2 (assuming that the input and delayed input signal have amplitude 1). If the frequency ω increases so that the angle $\omega\tau = 180°$ (π rads), then the 'input' phasor will lie horizontally pointing to the left in relation to the delayed signal (horizontally to the right), so the overall resultant output amplitude will be zero.

Suppose that we say that the time delay τ involved is equal to the interval, τ_s, between samples in this sampled digital audio system. That is, suppose we delay our samples by one sample interval before adding them to the input samples. Then at the frequency, when $\omega\tau_s = \pi$, we have:

$$\omega = \pi/\tau_s = \pi f_s \text{ (remembering that frequency, } f = 1/\text{period, } \tau)$$

remembering also that $\omega = 2\pi f$ (Equation 5.1), we have:

$$2\pi f = \pi f_s, \text{ or } f = f_s/2 \qquad (5.2)$$

That is, the phase shift in the sample length fixed time delay, τ_s, is π rads (180°) for an input frequency at half the sampling rate. At intermediate frequencies, between 0Hz and $f_s/2$, we have to add the phasors vectorially, as shown in Figure 5.13(b). The example shown in the figure is the case for $\omega\tau_s = \pi/2$, or $f = f_s/4$. In this case the output amplitude is $\sqrt{2}$ and the output phase is $\pi/4$. Taking all of this information into account, we arrive at the amplitude frequency response shown in Figure 5.13(c). We have just analysed our first digital filter! It is easy to conceive of a computer program, or even a bit of digital hardware which would carry out the operation described here (adding a signal to a delayed version of itself). The amplitude clearly varies with frequency, and this variation comes about because we are using a fixed time delay (1 sample period in this case), which amounts to a *phase* delay proportional to frequency.

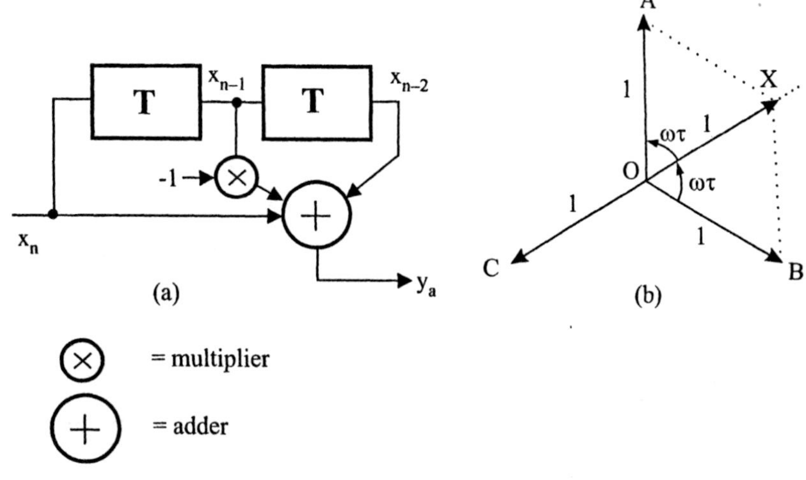

Figure 5.14 (a) and (b) A multi-delay element digital filter and its phasor diagram.

Multi-delay element filters

Let us now apply our new-found tool to a slightly more complicated digital filter structure such as that shown in Figure 5.14. The nomenclature in Figure 5.14 requires a little explanation. A box labelled 'T' represents a 1 sample delay. x_n represents a signal sample at time 'n' (the nth sample), and therefore x_{n-1} represents a delayed version of the input sample (i.e. the previous sample produced by the first delay element, T). x_{n-2} represents the input sample from two sample instants ago and y_n is the output sample at time 'n', that is to say, the current output sample.

Figure 5.14(a) represents a computer algorithm (program) which adds the current input sample to the previous input sample (scaled by a factor –1), and to the input sample obtained two sample instants ago, to produce the current output sample.

Figure 5.14(b) shows the phasor diagram for this filter. Phasor OA is the input signal and phasor OB is the input signal delayed by two sample instants, which amounts to a phase delay of $2\omega\tau_s$. Phasor OC is the input delayed by one sample instant (which would make it lie along the line OX), but multiplied by a constant, –1. Multiplying a sinusoid by –1 amounts to phase-shifting it by 180°, as the third set of waveforms in Figure 5.12 shows. This 180° phase shift therefore moves the OX phasor round, so that it lies along line OC. Notice that the phasors all have amplitude 1 (relative to the input amplitude x_n), as specified by the filter structure in Figure 5.14(a).

What can we say about the amplitude response of the filter? For samples of an input signal of frequency of 0 Hz (DC), $\omega = 0$, and hence $\omega\tau_s = 0$. Phasors OA and OB both lie along the line OX, to give a total amplitude of 2. This is offset by the phasor OC (amplitude 1) 'pulling' in the opposite direction, to give a net amplitude of 1 acting in direction OX. At some frequency, ω_o, the net amplitude of phasors OA and OB acting along OX will be 1, and this will be exactly cancelled out by phasor OC, to leave a total amplitude of zero. This is the situation actually illustrated in Figure 5.14(b). What is the frequency ω_o? A little elementary geometry will show that when this frequency occurs, $\omega_o\tau_s = \pi/3$. Using the same kind of analysis which led to Equation 5.2, we find that this corresponds to a frequency $f_s/6$. At frequencies above this, the phasors OA and OB swing around clockwise and counter-clockwise respectively towards OC, with steadily increasing total output amplitude until at a frequency $\omega\tau_s = \pi$ ($f_s/2$) we have a maximum output amplitude of 3.

We have just analysed our second digital filter. However, we are running into trouble. Although our phasor representation is conceptually accurate, and will always work in principle, it is actually quite hard to visualise what is going on, even in our simple filter in Figure 5.14(a). We still have not considered the effect of scaling any of the signals so that their amplitudes are different to the value 1, and if we were to add in many more delayed versions of the input signal, we would have many more phasors in our phasor diagram, and its behaviour would become impossibly difficult to visualise. We need a more powerful and compact model than our simple phasor representation.

An algebraic model for phasors

Let us see if an algebraic approach will help. Is it possible to find an algebraic model for our phasor? Returning to Figure 5.13(a) (reproduced as Figure 5.15(a)), the directed line or phasor is really describing the point 'X' in the so-called phasor space. Any other model which adequately (i.e. uniquely) describes X would

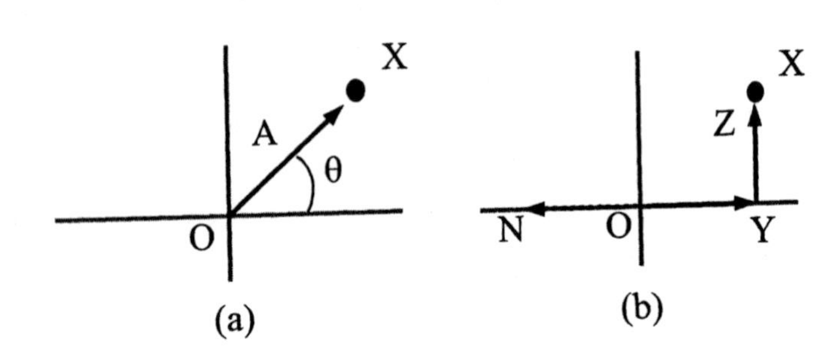

Figure 5.15 (a) and (b) The resolved components of point X in the phasor space.

do as well. For instance a pair of directed lines such as Y and Z shown in Figure 5.15(b). *Taken together*, this *pair* of lines points the way to X just as unambiguously as the line OX in Figure 5.15(a).

Y and Z are sometimes called the 'resolved components' of the phasor OX. Alternatively Y is called the 'in phase' component of the point X, and Z the 'quadrature' component. They form the basis of a mathematical model. Any point in the phasor space (i.e. any signal of a given magnitude and phase) can be represented by such a pair of resolved components. The magnitudes of Y and Z can be related to A and θ by the simple trigonometric relations:

$$Y = A\cos\theta \quad \text{and} \quad Z = A\sin\theta \qquad (5.3)$$

We can treat the pair of lines Y and Z as one *compound* entity representing X by *labelling* the quadrature component in some way, and then taking the pair of lines together. Suppose we prefix Z with the label 'j'. Then the point X could be described by the compound statement:

$$X = A\cos\theta + j\,A\sin\theta = A(\cos\theta + j\sin\theta) \qquad (5.4)$$

Here 'j' simply labels Z as the quadrature component, and reminds us that since Z has a 90° phase shift with respect to Y, we cannot simply add Y to Z arithmetically to define X (for instance). This is because the phase of Z is significant, and must not be ignored. This notation is now very convenient. We can use the normal processes of algebra for example, and rely on our 'j' label to carry phase information through our calculations and analysis.

Complex numbers

Many readers will be struck by the similarity between our resolved components phasor model and the way in which so-called *complex numbers* are represented. Most people first become aware of complex numbers through the solution of the time honoured equation:

$$\left[\frac{-b \pm \sqrt{b^2 - 4ac}}{2a}\right] \qquad (5.5)$$

to find the roots of the quadratic polynomial $ax^2 + bx + c$. A complex number arises when the term 4ac is greater than b^2, leading to a negative number under the square root sign. This state of affairs could easily happen with certain given values of a, b and c, and causes a degree of difficulty in ordinary arithmetic, since no real number, multiplied by itself, can give a negative result. We cannot find the square root of a negative number in ordinary arithmetic. This is dealt with in mathematics by writing the solutions to Equation 5.5 under these conditions as:

$$r_1, r_2 = d \pm if \qquad (5.6)$$

where i is a *label* for $\sqrt{-1}$, and r_1, r_2 are the required roots. The form d + if (or d – if) is known as a 'complex number'. The need for a label reminds us of our phasor model, and indeed, the two separate parts of the complex number are often represented diagrammatically in a form identical to Figure 5.15(b). The justification that this form is appropriate for a complex number, and that our mathematical form for a phasor (Equation 5.4) can therefore be regarded as a complex number will be given shortly. It is extremely valuable if this is the case, since there is a well-developed branch of mathematics, the algebra of complex numbers, which can then be used to represent and analyse our signals!

The equivalence of 'j' and 'i'

For the moment, let us assume that our label 'j' amounts to the same thing as the label 'i' ($\sqrt{-1}$). Returning to Equation 5.4, this can be written in a more compact form:

$$X = A\,e^{j\theta}, \qquad (5.7)$$

since by a standard identity

$$\cos\theta + j\sin\theta = e^{j\theta} \qquad (5.8)$$

The form for X described in Equation 5.7 is often written in the form $A\angle\theta$. Identity (5.8), known as Euler's equation, can be quite easily proved with a bit of simple differentiation:

Let $P(\theta) = \cos\theta + j\sin\theta$

Then:

$$\frac{dP(\theta)}{d\theta} = -\sin\theta + j\cos\theta$$

given that $j^2 = -1$ (by definition)

$$\frac{dP(\theta)}{d\theta} = -\sin\theta + j\cos\theta = j(\cos\theta + j\sin\theta) = jP(\theta)$$

The only function which satisfies $dP\theta/d\theta = jP(\theta)$is the function $P(\theta) = e^{j\theta}$. Hence $\cos\theta + j\sin\theta = e^{j\theta}$, as described in Equation 5.8.

We can now show that our quadrature phasor label 'j' and the complex number operator 'i' amount to the same thing by observing that if we treat i as a phasor we get a consistent behaviour with our understanding that $i = \sqrt{-1}$, as we will now show.

For instance if i ($= \cos \pi/2 + i\sin \pi/2 = e^{i\pi/2}$) really can be represented by a phasor such as Z of unit length in Figure 5.15(b), then does i^2 give a consistent representation of –1 on our phasor diagram? If it does, then we are justified in treating i as a phasor such as Z.

Now $i^2 = e^{i\pi/2} . e^{i\pi/2} = e^{i\pi}$ (adding exponents) = –1 (substituting π for θ in Equation 5.8). Now if $e^{i\pi/2}$ is a quadrature phasor (i.e. with phase angle 90^0 or $\pi/2$ rads), then to be consistent, $e^{i\pi}$ would be a phasor with phase π rads, i.e. located on the horizontal axis, pointing to the left of O in Figure 5.15(b). This is shown as the dotted phasor ON in this figure. If phasor OY corresponds to the value +1, then ON must correspond to value –1. In other words, treating i (= $e^{i\pi/2}$) as a phasor such as Z is consistent with the mathematical properties we expect for i (such as i^2). The same can be shown to be true for i^3 (= –i) and i^4 (= +1) – try it! Our quadrature label j is therefore one and the same thing as our complex number operator i and we are justified in using the mathematics of complex numbers to represent our phasors.

The use of 'j' and 'z' to represent phase and delay

Of particular importance to us will be the ability to use the exponential form of Euler's equation to represent a phasor. Explicitly, a phasor of magnitude A and phase θ can be represented as $Ae^{j\theta}$. Whenever we see an expression of this form, we are entitled to interpret anything after the 'j' in the exponent as *a phase shift*, and any real number such as A as an amplitude.

An important specific example of this exponential representation of a phasor is the signal output of one of our unit sample delay units, 'T' in Figure 5.14(a), such as x_{n-1}. If the sampled signal S is of frequency ω, then the output of the delay unit could be described as $Se^{-j\omega\tau s}$. The expression $e^{j\omega\tau s}$ is used so frequently that it is usually given its own symbol: z. The expression z^{-1} ($=e^{-j\omega\tau s}$) therefore describes the phase shift imposed on a signal of frequency ω by a fixed time delay,τ_s.

How does all of this mathematical spade-work help us with the understanding and analysis of digital filters? Returning to the filter shown in Figure 5.14(a), we can algebraically represent each output sample, y_n as:

$$y_n = x_n - x_{n-1} + x_{n-2} \qquad (5.9)$$

Where (for instance) x_{n-2} means an input sample delayed by two sample intervals. If we now stop speaking about individual samples, y_n , x_n ,and speak instead of a whole *stream* of samples, making up *signals* X and Y, then equation (5.9) generalises to:

$$Y = (1 - z^{-1} + z^{-2})X \qquad (5.10)$$

If you are not sure about this, try a 'stream' consisting of two input samples, x_n, and x_{n+1}, together with their corresponding outputs, y_n and y_{n-1}. You will find that Equation 5.10 holds for the whole stream X.

The expression in brackets is called the *transfer function* of the filter. This is because it describes the changes imparted on the input signal X, as the filter transforms (or 'transfers') the input, X to the output Y. It is also called the z transform of the filter. It is clearly a function of z, which in turn is a function of frequency, ω. Therefore, in general we must expect magnitude and phase of the transfer function to change with frequency. These changes, when imposed on the input signal, X, constitute *the* filtering action of the algorithm described in Figure 5.14(a). Do we have any convenient ways of visualising how the magnitude (and phase for that matter) of the transfer function vary with frequency? The answer is yes, and this will lead us to a powerful way of representing the behaviour of such systems, but we need to do a bit more work to get there.

The zeros of the transfer function

We will be working later with the roots of the transfer function. For this purpose, it will be helpful to have the polynomials in the transfer function as positive powers of z. Let us convert the equation to this form by factoring out z^2. The transfer function then becomes:

$$\left[\frac{z^2 - z + 1}{z^2}\right] \tag{5.11}$$

We can now apply our Equation 5.5 for finding the roots of the quadratic in the numerator of Equation 5.11, to give a pair of roots:

$$\zeta_1, \zeta_2 = 0.5 \pm j\sqrt{3}/2 \tag{5.12}$$

we can therefore rewrite Equation 5.11 as:

$$\left[\frac{(z - \zeta_1)(z - \zeta_2)}{z^2}\right] \tag{5.13}$$

with the values of ζ given in Equation 5.12. Roots of the *numerator* of the transfer function (such as ζ here) are called *zeros* of the transfer function, since when z takes the value of one of these roots (e.g. $z = \zeta_1$), the value of the whole transfer function has a value of zero.

Finding the magnitude frequency response of the filter therefore comes down to working out how the magnitude of factors such as $(z - \zeta)$ vary with frequency. Since we have two such factors we simply multiply the two frequency-dependent behaviours together, to give the overall frequency-dependent behaviour of the magnitude of the numerator. What about the magnitude of the denominator? Since z is the phasor $e^{j\omega\tau s}$ (by definition), the magnitude term (A in Equation 5.7) has value 1. We have two such terms (z^2) and they therefore do not affect the overall amplitude response in this case. (The phase is a different matter.)

The unit circle

Can we find a convenient (e.g. graphical) way of representing the frequency dependent behaviour of factors such as $(z - \zeta)$? Well, z is a phasor of amplitude 1, as already noted, with phase angle $\omega\tau s$ (i.e. its phase is proportional to frequency). ζ in Equation 5.13 is, in general, a complex number (i.e. a phasor) whose value is found by the application of Equation 5.5. The expression $(z - \zeta)$ is therefore a phasor (vector) *subtraction* of these two phasors. This is illustrated in Figure 5.16.

In the figure, the phasor z is the line OZ, with its phase angle $\omega\tau_s$ clearly shown. The phasor ζ is the line $O\zeta_1$. Its phase angle is ϕ. Given the values for the roots in Equation 5.12, a little geometry shows that this angle amounts to $\tan^{-1}\sqrt{3}$, which is equal to $\pi/3$ radians. Subtracting phasor ζ from phasor z is equivalent to *adding* phasor z to phasor $(-\zeta)$. We form phasor $(-\zeta)$ by taking the negative values of the two components of ζ and plotting these as the components of the phasor $(-\zeta)$. This gives the line $O,-\zeta_1$ in Figure 5.16. Adding the phasors z and $-\zeta_1$ therefore gives the line OX in the figure. Notice that a line (phasor) of equivalent amplitude and phase is given by the line $\zeta_1 Z$. *This can be formed by simply drawing a line from the position of the root, ζ_1, to the end of the phasor z*. Therefore, if we wish to know how the magnitude of a

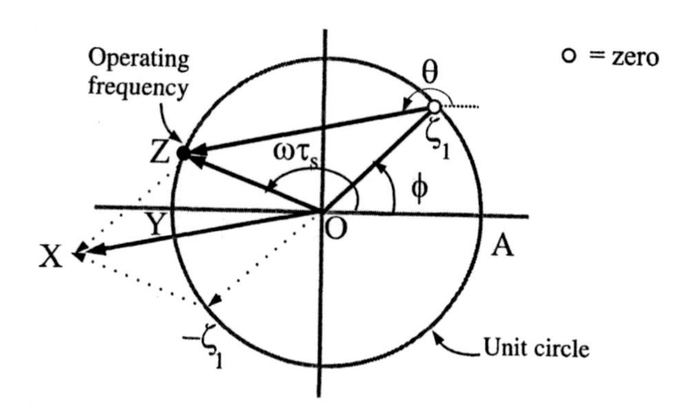

Figure 5.16 The formation of the phasor difference $(z - \zeta)$.

factor $(z - \zeta)$ varies with frequency, we draw a line from the position of the zero to the end of the phasor z, and take the length of the line $\zeta_1 Z$ for various frequencies, ω. (The phase response of $(z - \zeta)$ could be measured with varying ω as well; this is the angle θ in Figure 5.16).

Changing the input frequency ω means rotating the phasor z counter-clockwise, since its phase angle, $\omega\tau_s$, is proportional to frequency. The magnitude of phasor z is 1 (unity, since $z = e^{j\omega\tau s}$), so the tip of this phasor describes a circle of radius 1, as the frequency increases. This is known as the *unit circle.* The whole diagram, including the unit circle, plots of the positions of zeros (and also things known as poles, which we will meet later), and any constructions based on them is known as the *z plane.* When the z phasor gets half-way around the unit circle, to Y in Figure 5.16, then $\omega\tau_s = \pi$, and this corresponds to a frequency $f = f_s/2$, as Equation 5.2 shows. Similarly, with further increasing frequency, z will come back to A on the unit circle, and this corresponds to $f = f_s$. Any further increase in frequency involves another transit around the unit circle, beyond A. More of this later.

Frequency response of multiple delay element filters

Let us return to our task of finding the frequency response of the algorithm (filter) shown in Figure 5.14(a), and described by Equation 5.13. We can use the unit circle to find the magnitude response of the factorised transfer function, as we will now show. The process is illustrated in Figure 5.17.

We select a number of frequencies, ω_1, ω_2, ω_3 at which we are interested in measuring the frequency response. We plot their positions on the unit circle, scaling these on the basis that one half rotation corresponds to the Nyquist frequency, fs/2. We then plot the positions of the roots of the transfer function (in this case zeros) on the z plane, in relation to the unit circle. In this case,

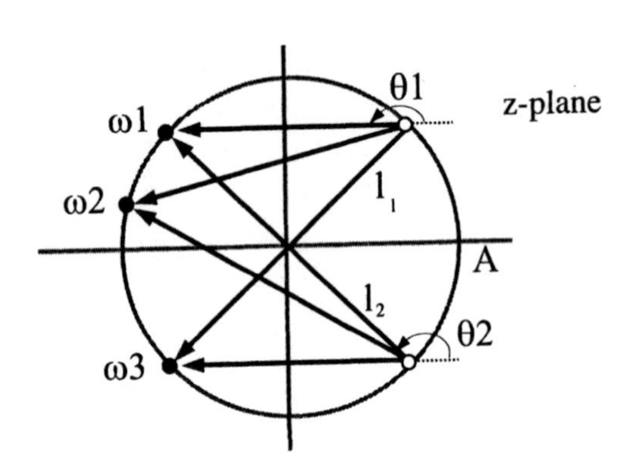

Figure 5.17 The z-plane diagram for $(z - \zeta_1)(z - \zeta_2)$.

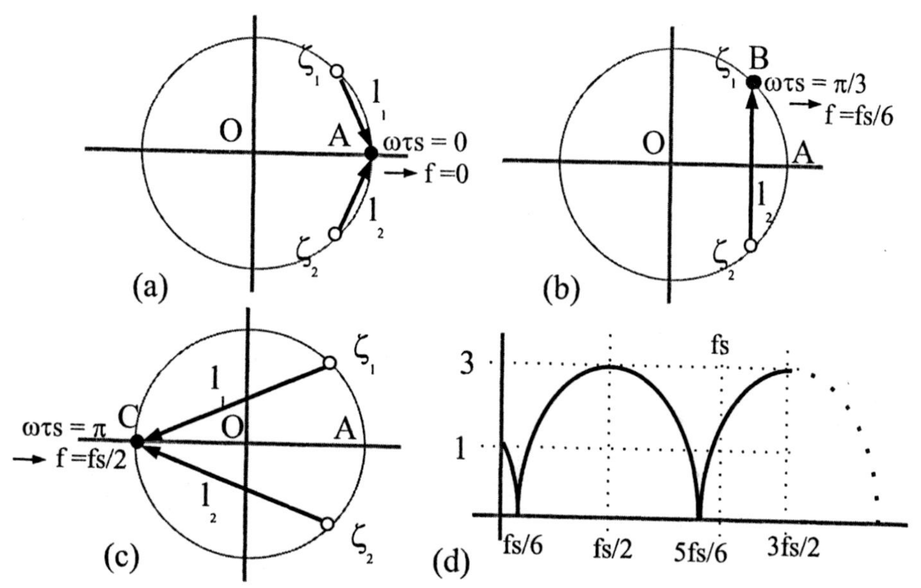

Figure 5.18 Unit circle plots, and corresponding magnitude plot for a 2-zero system.

they happen to lie on the unit circle, although this will not always be the case. We then plot the lines l_1, l_2 from these zero positions to the frequencies of interest and measure/calculate their lengths. The overall magnitude response for the filter (i.e. magnitude of Equation 5.13) can be found by *multiplying* the lengths of l_1, l_2 for a range of frequencies. The z^2 term in the denominator has no effect on the magnitude because of the unity amplitude of this term (see 'The zeros of the transfer function', above). If we plot the results we obtain against frequency, we have our magnitude response!

Taking spot frequencies, we can see that in Figure 5.18(a) at frequency 0 Hz, l_1 and l_2 should be drawn to point A. Given that the angle $AO\zeta_1$ is $\pi/3$, and the lines OA and $O\zeta_1$ are of length 1 (being the radii of the unit circle), simple geometry tells us that each of the phasor magnitudes l_1 and l_2 are of length 1. Therefore the magnitude of the transfer function at 0 Hz is also 1. This accords exactly with the conclusions we drew from the behaviour of the phasor diagram, Figure 5.14. What happens at the other cardinal frequencies which we analysed in conjunction with this figure? Let us take a frequency fs/6 where $\omega\tau_s = \pi/3$. This is shown as point B in Figure 5.18(b), and it corresponds exactly to the position of one of the zeros. l_1 is therefore a line drawn from ζ_1 to itself, and correspondingly has length 0. The overall magnitude response of the transfer function at this frequency is therefore 0, again confirming the result predicted by

Figure 5.14. The final frequency tested in conjunction with this figure was at fs/2. The z-plane construction for this is shown in Figure 5.18(c), where point C corresponds to this frequency. Measurement and/or geometry shows that the lengths of the lines l_1, l_2 are now of magnitude $\sqrt{3}$, so that the overall response at this frequency is 3, again a result confirmed by Figure 5.14. Figure 5.18(d) summarises these results in a frequency response diagram.

As an additional exercise, you might like to draw the phasor diagram and z-plane construction for a filter which subtracts samples delayed by two sample instants from the input samples, to form the output samples. You should find that both methods (the phasor diagram and the z-plane) give the same results, namely a magnitude response of 0 at 0 Hz and fs/2, with magnitude 2 at frequencies fs/4 and 3fs/4. This affirms that the two approaches are consistent with each other – two different means of visualising the same phenomena.

The analysis presented so far has been mainly concerned with the magnitude response of a digital filter. For a phase response, we *add* the angles such as θ_1 and θ_2 in Figure 5.17. This is because they appear directly in the exponent after the 'j', and we add exponents when we multiply exponentials. Similarly we *subtract* the phases of any terms appearing in the denominator of the transfer function, since dividing by such numbers implies subtracting the exponents. If we plot the results against frequency, we have the phase response. In Equation 5.13 we should take account of the phases of the terms z.z (z^2) in this way, if we wish to know the phase response for the filter in Figure 5.14.

It will be clear from our unit circle construction that the frequency response will be replicated for frequencies beyond the sampling frequency (point A in Figure 5.17). This should not alarm or surprise us. We are working with sampled (digital audio) systems, and we know from our analysis of the sampling theorem that we get replications of signal spectra above the sampling rate, f_s (see Figures 3.7 and 3.8 for instance). The replication produced by the z-plane diagram is consistent with this.

We now have a very powerful analytical and visualisation tool at our disposal. There is no particular problem about scaling the outputs of the delay taps in Figure 5.14, i.e. we are not limited to multiplying them by ±1. If we multiply them by other values, this simply changes the coefficients in the filter Equation 5.9, and hence the numbers we plug into Equation 5.5 to find the roots. This will change the position of the zeros on the z-plane, but the method outlined above for finding the frequency response will still work with complete generality.

Neither should we be concerned about analysing filters with more delay taps than the two shown in Figure 5.14. If we had more taps, then we would have terms z^3, z^4 etc., in the filter equation, perhaps scaled by associated coefficients. It clearly becomes difficult to use the phasor diagram to visualise the frequency response, since there are many more phasors involved. Instead, we should proceed in the way outlined above – convert the transfer function into a power series (polynomial) in powers of z, and then find the roots, so that the frequency response can be found from the zero plot on the z-plane. The only problem is that we do not necessarily have a convenient equation such as 5.5 to find the roots of these higher order polynomials. Again, we should not concern ourselves about this unduly. These days, there are plenty of computer based packages around to factorise polynomials. An example is the 'roots' function in the Matlab program. It is therefore common practice simply to use these as a tool to find the roots, and we do not really need to worry ourselves how they work, no more than we should worry about the way in which our calculator finds a square root.

IIR and FIR filters

From the discussion given in this section, it should now be clear that adding time-delayed *input* samples into a mix of samples gives a frequency-dependent (i.e. filtering) behaviour. What happens if we mix in delayed versions of the current *output* samples? It turns out that this gives a whole new dimension in the construction of digital filters which we can now analyse, given our more refined visualisation tools. Consider the algorithm structure shown in Figure 5.19(a).

If we input a single 'pulse', as shown in Figure 5.19(b), (this is known as an 'impulse' in the jargon), then ignoring the effect of the second order delay for the moment (e.g. by setting $b = 0$), the output sequence of samples would be as illustrated in the figure. One impulse gives rise to a whole series of output samples, which in theory could go on forever! This gives rise to the name infinite impulse response (IIR) for this kind of filter. Its behaviour is explained by the fact that we are *feeding back* scaled output samples to the input. As shown in the figure, depending on the value of scale-factor 'a' – whether it is less or greater than 1, (remember $b = 0$ temporarily), the output impulse response can gradually decay away to 0, or it could grow steadily, without bound! The output 'blows up', or becomes unstable for values of $a > 1$. IIR filters are potentially dangerous as far as the signals

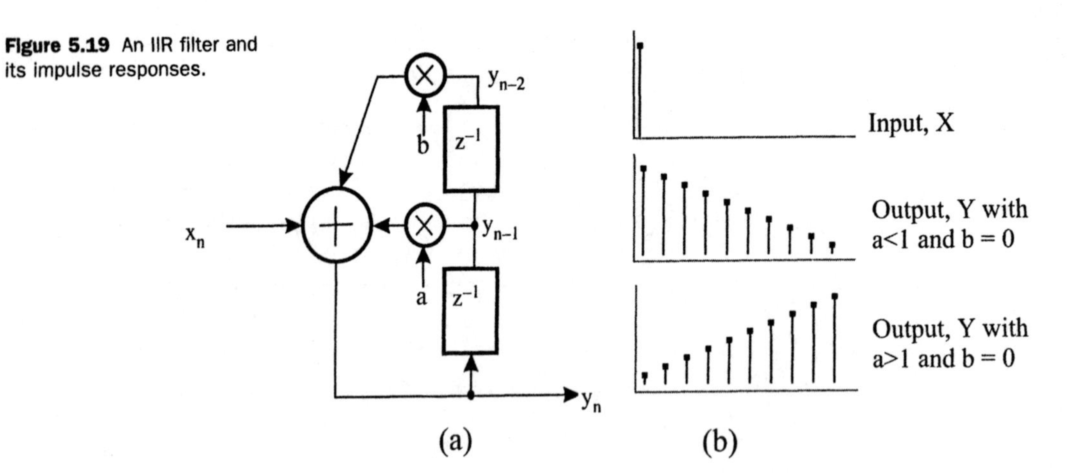

Figure 5.19 An IIR filter and its impulse responses.

being processed by them are concerned, since an unstable filter will swamp the signal samples with this ever-increasing impulse response.

By contrast, the 'feed-forward' filters which we have been looking at previously cannot exhibit this unbound impulse response, since a single input impulse can only produce as many samples in the output response as there are delay taps in the filter. There will always be a finite number of these in practice. These kinds of filters are therefore called finite impulse response (FIR) filters.

Let us return to our IIR filter in Figure 5.19. If we have a stream of samples forming an input signal, X, how can we visualise the output Y? We would be in severe difficulty if we could only use our method of addition of sinusoids, or our phasor diagram representation (Figure 5.14(b)). Whilst this approach is still *conceptually* valid, the difficulty arises since we are dealing with a potentially infinite number of signals to be added together. This is because of the configuration of the output delay feedback path; there is something of the very first output signal sample in *every subsequent* output sample. Very fortunately, our z-plane representation comes to our aid like a knight in shining armour, as we will now show.

Using the values shown in Figure 5.19(a), output y_n is given by:

$$y_n = x_n + ay_{n-1} + by_{n-2} \quad (5.14)$$

Alternatively, thinking about output *signals* X and Y (that is, streams of samples) $Y = X + az^{-1}Y + bz^{-2}Y$ or, rearranging:

$$Y(1 - a\,z^{-1} - bz^{-2}) = X \quad (5.15)$$

This bears a striking resemblance to Equation 5.10, except that the roles of X and Y have been interchanged, and we have used general values for the scaling factors, a and b, rather than the specific values ±1 associated with Figure 5.14(a). Equation 5.15 can be written as:

$$Y = \frac{1}{(1 - az^{-1} - bz^{-2})} X \tag{5.16}$$

This is a transfer function of the same essential form as Equation 5.10, except that the polynomial in z now appears in the *denominator* (whereas previously it had appeared in the numerator). Also, the scaling factors in the filter algorithm (a,b here) appear with a negative sign, whereas in Equation 5.10, for the feed-forward IIR filter, they would not.

Otherwise, all of the analysis devised above, including the use of roots with the unit circle applies exactly as described above. The roots of the denominator of the transfer function (p), arising from feed-back terms in the filter, are known as *poles*. The filter structure shown in Figure 5.19(a) is therefore known as a two-pole filter. Equation (5.16) can therefore be re-written as:

$$Y = \frac{z^2}{(z - p_1)(z - p_2)} X \tag{5.17}$$

where p_1, p_2 (the poles) are the roots of the denominator. If we want to plot the frequency response (magnitude and phase) of our two-pole filter, we place the positions of the poles in the z-plane, measure or calculate the lengths of the lines l_1 and l_2, and *divide* by these lengths in the transfer function. We divide because the factors (z–p) appear in the denominator in this case. For a similar reason, for the phase response of the poles, we *subtract* the phase angles ϕ. See Figure 5.20 for the graphical construction. Our z-plane construction has got us around the problem of having to add infinite numbers of phasors!

Notice that the poles in Figure 5.20 are mirror-images of each other. They are *complex conjugates* of each other in the jargon. The factorisation of the polynomial will always yield poles in conjugate pairs, where the coefficients of the polynomial (a,b here) are real (i.e. not complex) numbers. A similar thing holds true for pairs of conjugate zeros.

An interesting behaviour emerges for poles which are located near to the unit circle. Since we divide the transfer function magnitude by the lengths of the phasors such as l_1 and l_2, the closer

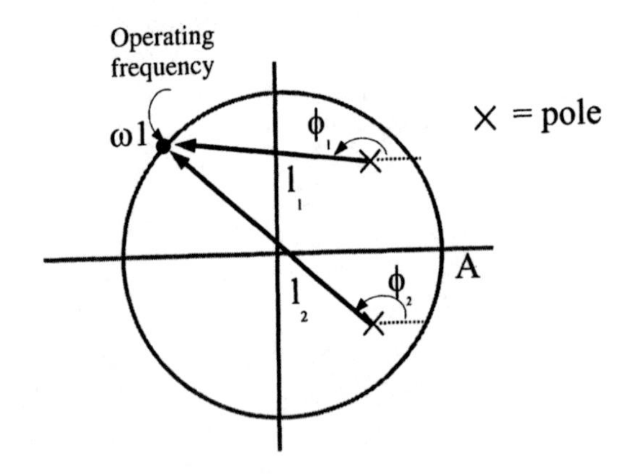

Figure 5.20 Z-plane plot for a two-pole filter.

a pole lies to the unit circle, the greater its contribution to the magnitude response. Dividing a function by a small number ($l_{1,2}$ in this case) makes the overall magnitude of the function larger. In the limit case (the pole lying on the unit circle), we divide the transfer function by 0 at the frequency corresponding to the position of the pole. This makes the gain of the system *infinite* at this frequency. This would make the system unstable. However, for poles lying near, but not on the unit circle, we would obtain a peak (or *resonance*) at or near the corresponding frequency. This behaviour makes the IIR a useful basis for band-pass filters.

Cascaded IIR and FIR filters

Let us now examine the transfer function of a filter which contains both poles and zeros. For the moment, let us imagine that we have a cascade of a FIR and IIR filter, as shown in Figure 5.21. The two addition units can be coalesced into one unit, as indicated in the bracketed part in Figure 5.21(b). We choose second order filters for the illustration (two-element delay lines), but the argument can be generalised to any order.

The overall transfer function for the cascade, Y/X, is given by the products of the individual transfer functions, W/X and Y/W, since:

$$\frac{Y}{X} = \frac{Y}{W} \quad \frac{W}{X}$$

Following the analysis given earlier in this section,

$$\frac{W}{Y} = (1 + a_1 z^{-1} + a_2 z^{-2}) \text{ and } \frac{Y}{W} = \frac{1}{1 - b_1 z^{-1} - b_2 z^{-3}}$$

and

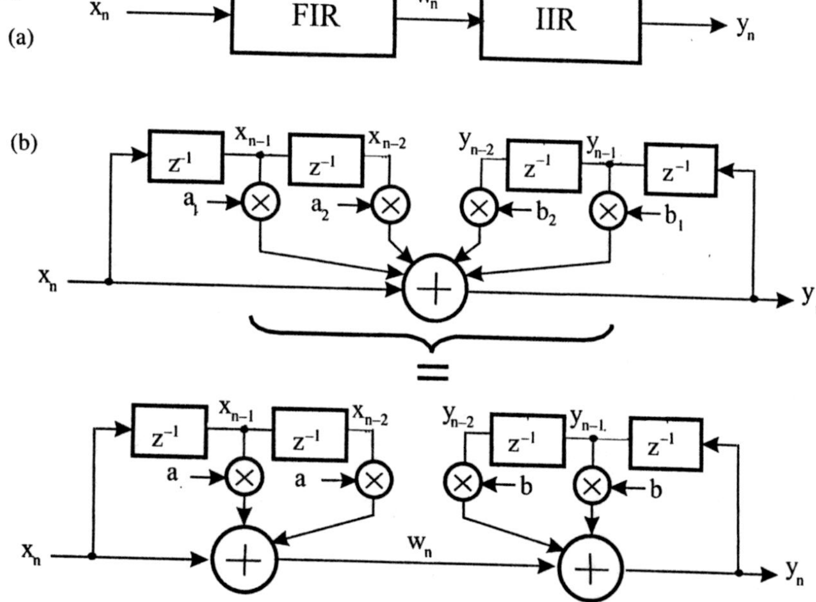

Figure 5.21 Cascaded FIR/IIR filter, (a) in outline, (b) in more detail.

Hence:

$$\frac{Y}{X} = \frac{z^2 + a_1 z + a_2}{z^2 - b_1 z - b_2} \tag{5.18}$$

This general form for the filter structure shown in Figure 5.21 allows us to realise the filter coefficients (the scaling factors a_n, b_n in Figure 5.21). If we have a particular configuration of poles and zeros, corresponding to a specific frequency response, we can write the transfer function out as products of factors $(z - \zeta)$ in the numerator (for zeros) and $(z - p)$ in the denominator (for the poles). We can then multiply out the factors to give us numerator and denominator polynomials in z, in the form of Equation 5.18. We can then equate the coefficients of the polynomials we get with the values of a_n and b_n in Equation 5.18 (taking account of the negative signs for the 'b's), to give us the numbers which we should plug into the multipliers in Figure 5.21(b). This will give us a filter which will have the frequency response specified at the outset i.e. this is a way of *synthesising* a filter.

The process can be worked in reverse to *analyse* a filter whose coefficients are given. The coefficients from Figure 5.21(b) are put into Equation 5.18. The numerator and denominator are both factorised to give the roots, and hence the positions of the poles and

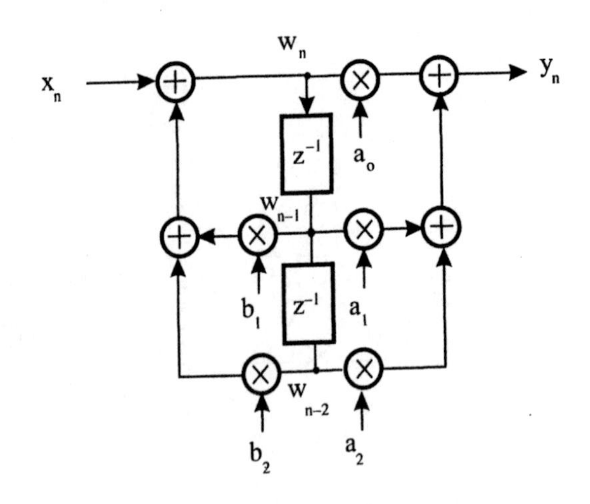

Figure 5.22 Generalised second order FIR/IIR filter.

zeros on the z-plane. The z-plane construction will then give us the frequency response.

Generalised form for FIR/IIR filters

Figure 5.22 shows another, particularly elegant generalised form for second order FIR/IIR filters. The operation of this filter can perhaps be best understood by considering the signal W (made up of samples w_n) at the head of the delay line:

$$W = X + (b_1z^{-1} + b_2z^{-2})W$$

Rearranging, this becomes:

$$W = X/(1 - b_1z^{-1} - b_2z^{-2}).$$

Now:

$$Y = (a_0 + a_1z^{-1} + a_2z^{-2})\, W$$

Substituting for W from the equation above and rearranging, we get:

$$\frac{Y}{X} = \frac{a_0 + a_1z^{-1} + a_2z^{-2}}{1 - b_1z^{-1} - b_2z^{-2}} \qquad (5.19)$$

a form which leads directly to an equation of the form of Equation 5.18. In other words, the generalised structure in Figure 5.22 is equivalent in operation to that of Figure 5.21. We can make higher order filters by cascading several of these second order 'sections' together. You will often see filter structures based

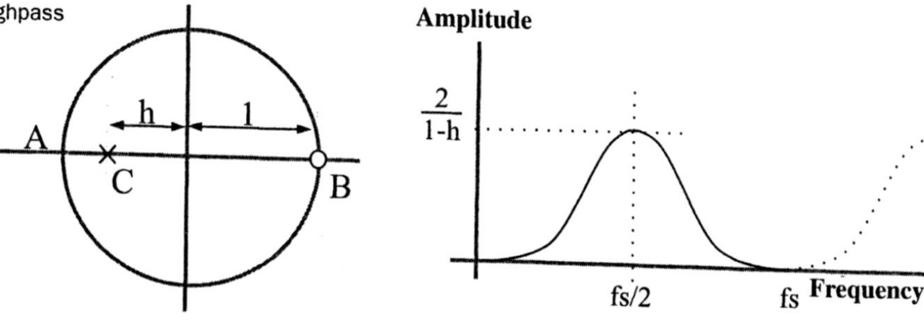

Figure 5.23 Pole-zero plot and corresponding highpass response.

on Figure 5.22. An example is the *all-pass* filter which we will use in reverberation and phasing units later in this section.

Example: a high-pass filter

We complete this section on digital filters with an example concerning the realisation of a highpass filter. The pole-zero diagram of Figure 5.23(a) would yield the amplitude response shown in Figure 5.23(b). The zero at f = 0 Hz and fs (the sampling rate) ensures the amplitude response is 0 at these frequencies. At the Nyquist frequency (half the sampling rate) the magnitude of the zero factor is at a maximum (stretching all the way across the unit circle from B to A), whereas the pole factor at –h is at its minimum, stretching from C to A. This gives the response curve in part (b) of the figure. In practice, in any sampled system, we are only interested in frequencies up to the Nyquist frequency, so this amplitude response has the desired high-pass characteristic.

Forming the transfer function from the numerical values for the zero and pole given in the figure:

$$\frac{Y}{X} = \frac{z-1}{z-(-h)} = \frac{(1-z^{-1})}{(1+hz^{-1})} \tag{5.20}$$

Comparing this with Equation 5.19, and interpreting the coefficients in the light of the filter section shown in Figure 5.22, we have b2 = a2 =0. a0 = 1; a1 = –1; b1 = –h. This filter can therefore be realised using the structure shown in Figure 5.24(a). The alternative form based on Figure 5.21(b) is shown in Figure 5.24(b).

It is interesting to note that the magnitude response peak at fs/2 arises from the location of the pole on the horizontal axis of the z-plane diagram (Figure 5.23(a)). If we had wished to construct a *bandpass* filter, with centre frequency somewhere between 0 Hz and fs/2, we would have had to locate a pole near the unit circle

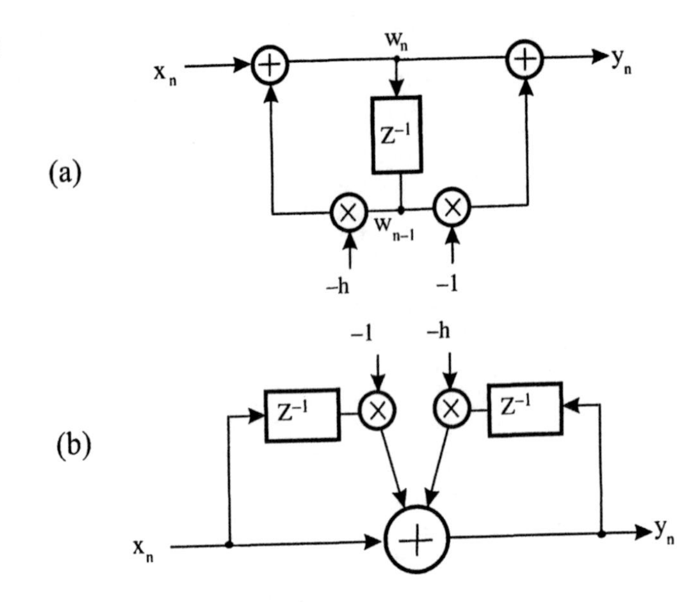

Figure 5.24 (a) and (b) Digital filter realisations of the transfer function given in Equation 5.20.

at a point corresponding to the centre frequency such as ω_0, as indicated by the point ζ1 in Figure 5.25(a). This pole must have a complex conjugate at ζ2, since they must appear in complex conjugates pairs. We are therefore dealing with a second-order filter in this case. Its form is shown in Figure 5.25(b). Its derivation is left as an exercise for the reader.

We have now seen all of the important filter types implemented as digital filter, and this gives us a good basis for the EQ section of an effects unit. Since they are in reality different forms of (computer) algorithm, it is relatively easy to change some of the operating parameters of the algorithms, to change the cut-off frequencies of the filters for instance. This makes EQ based on such devices very flexible in use. Adaptive filters used in this way form the basis of the so called 'parametric equaliser'.

5.5.2 Delay, phase and pitch transformation

Having considered the operation of the EQ units in the effects unit block diagram, Figure 5.10, let us now consider the remaining units in the effects processing chain. These are pitch shifting, delay and reverberation. A number of techniques used in effects units fall into the delay and pitch shifting categories. Amongst them are the following effects: flanging, phasing, chorusing and pitch shifting.

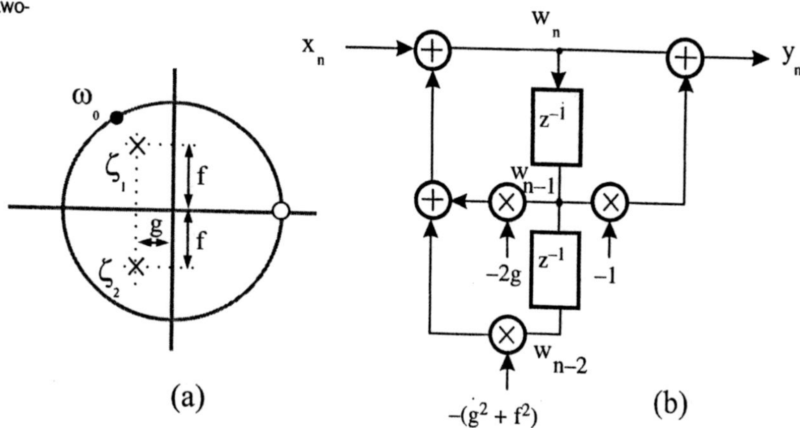

Figure 5.25 (a) and (b) A two-pole bandpass filter.

Flanging

The idea in flanging is to mix a signal with a time delayed version of itself. In some designs, delayed output samples are mixed in with input samples, to form the next generation of output samples. In other designs, input samples are delayed, and then mixed with subsequent generations of inputs, to form the current generation of output sample. This, of course, is just another way of describing the operation of IIR and FIR filters, described in the previous subsection, and the block diagram of a flanging unit would essentially be the same as that of these filters. See Figures 5.11 and 5.14(a) for the case of the FIR for instance.

Flangers are often provided with a control over the amplitude of the delayed signal before it is added into the mix (we called this a *scaling factor* in our study of digital filters), and also a control over the length of the delay. The delay unit may be many samples in extent, and thus by selecting various taps on the delay line, it is possible to vary this delay, perhaps under the control of an external source such as a low frequency oscillator (LFO).

The filter structure implemented by the flanger has the notch filter frequency response associated with many of the filters which we have already analysed. See the response in Figure 5.13(c) associated with the 'flanger', Figure 5.11 for instance.

Changing the delay parameter (e.g. with a low frequency oscillator) results in sweeping the nulls, or notches in the response up and down the frequency axis, and this is often described as a 'churning' or 'whooshing' sound. The scaling factor controls the depth of these notches.

Phasing

'Phasing' effects are closely related to flanging, but are subtly different. The churning effect of the flanger may be avoided to some extent by using *all-pass* filters. These are filters which have a flat amplitude response but have a defined phase response. The filtering action is thus primarily manifest in this phase response. The churning effect is ameliorated by the flat amplitude response, although there will still be notches caused by the addition of the phase shifted 'wet' signal with the dry input. The notches, though, are controlled by the number and phase shifting characteristics of the filters present. This is in contrast to the regular notched characteristic produced by FIR/IIR filters and common forms of flangers.

The configuration of one common form all-pass filter is shown in Figure 5.26. It is a special case of the filter shown in Figure 5.22, with the following values: a0 = –k, a1 = 1, b1 = k, a2 = b2 =0. Substituting these into the transfer function (Equation 5.19), and simplifying will show that the magnitude of the transfer function is constant, and does not vary with frequency – it is 'all-pass'.

Chorusing

Chorusing attempts to simulate the effect of many voices in an sound ensemble (i.e. a 'chorus') by mixing in the dry sound with versions of itself which are slightly delayed and detuned. This produces a fine emulation of the gentlemen's chorus of the Puddletown Choral Society(!) – with apologies. The pitch shifting used for the detuning can take a variety of forms. One is based on the wavetable sample skip technique described in Chapter 3, although interpolation ('guessing' the sample values which would lie between samples in the wavetable, if only they were present) is often used to give the small increases and decreases in pitch required for chorusing.

Another technique uses the 'harmoniser' approach, where input samples and output samples are read in/out at slightly different

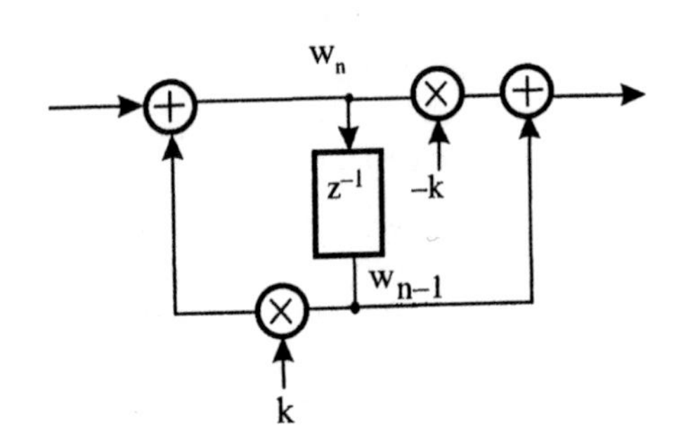

Figure 5.26 An all-pass filter.

rates, to give an apparent shift in pitch. If one side (reading or writing) runs ahead of the other, then samples must be repeated or skipped, as appropriate. If the sample output reading process must skip to another place in the sample stream in order to 'catch up', then this is done under a cross-fade to minimise the interference.

Time granulation is yet another technique used for pitch shifting. The sound is split up into time segments of 50 ms or so, and sound signals are generated by 'bolting together' varying numbers of these segments, or grains. To shift the pitch down, whilst keeping the duration of the sound fixed, the sample playback rate is reduced, and the number of grains reduced, so that the system takes the same time to play through the whole sound. The converse process is used to raise the pitch.

Delay lines are also used to mix in grossly delayed versions of the dry signal to produce various forms of echo effect. These can be used to build up dense layers of interrelated sounds, and also to impart various rhythmic figures within the echoes. The delay is varied by picking off different taps in the delay line. If this selection of the delay tap is placed under the control of a time varying process such as a low frequency oscillator, then some modification of pitch can also be imparted to the sound, since frequency is synonymous with the *rate of change* of phase.

5.5.3 Reverberation

Reverberation in a natural acoustic space is caused when sound waves emanating from the source are reflected from the many surfaces surrounding the space, so that multiple reflections arrive at the listener all with slightly different time delays. The differing time delays are caused by the slightly different path lengths to and from the different surfaces. Reverberation ('reverb') units are provided in effects processors so that the

Figure 5.27 An elementary reverberant structure.

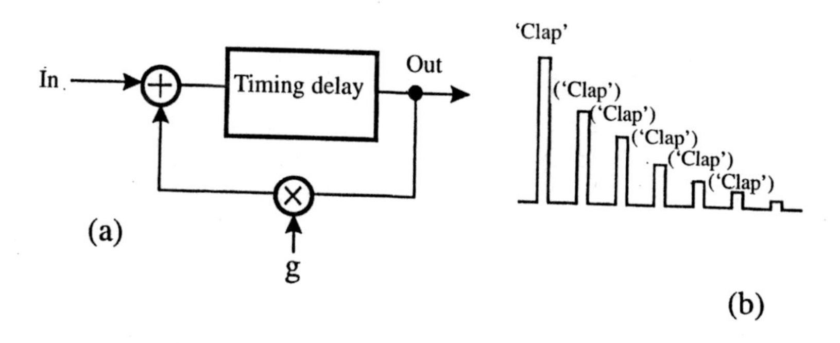

reverberation present in the natural sound can be modified, and often augmented.

Figure 5.27 represents a simple-minded approach to the electronic simulation of the reverberant behaviour described above.

If you imagine standing in a large hall and clapping your hands once ('clap'), then you can imagine that you would hear multiple reflections of your initial hand clap (clap), (clap), (clap), (clap)...., diminishing in intensity with time. Actually, you would hear reflections from many sources, so this is probably an over-simplified model of what we would hear, but it is at least a starting-point. By injecting an initial acoustical stimulus, and measuring the response, you have actually measured the *impulse response* of the space. The intention in Figure 5.27 is to come up with a structure which has a similar impulse response to the acoustic space. The parameter g would typically have a value <1, chosen to make the sound die away at the appropriate rate.

We have seen these kinds of structure and impulse response before. Figure 5.27 is really a single delay tap (i.e. first order) version of the second order IIR filter shown in Figure 5.19(a), and its impulse response in Figure 5.19(b) corresponds to that shown in Figure 5.27(b). Not surprisingly then, the elementary reverb structure will also have the filtering characteristics associated with a first order IIR filter. The frequency response of this filter may modify or *colour* the sound in an unintentional way. We therefore need to develop our ideas beyond this elementary approach, firstly to model the multiple reflections with different path lengths we would get from the many surfaces in a real space, and also to manipulate the filtering characteristics so that we can control the colouration of sound.

One approach used is to run multiple units of the form shown in Figure 5.27 in parallel, each with a different time delay factor and scaling value, g. This would simulate the path delays and refection properties of different surfaces, and also have the effect of smearing out the frequency colouration by spreading the cut-off frequencies of the filters. For a given input impulse, the number of 'reflections' in the response would be equal to the sum of the individual impulse responses, since the filters are operated in a parallel connection.

Another approach used is to make a series connection (or cascade) of all-pass filters. As described in Section 5.5.2, an all-pass filter has a largely flat amplitude response, and so its colouration of sound will be reduced accordingly. The structure of an all-pass unit is shown in Figure 5.26. The cascade connection means that each pulse in the impulse response of any one filter generates many such pulses in the next filter in the chain. The total number of output pulses from the cascade per input impulse is therefore equal to the *product* of the numbers of pulses in the individual impulse responses. This provides the potential for very dense reverberant effects to be produced.

In practice, various 'algorithms' each consisting of different configurations of all-pass and IIR structures are provided in reverb units. The reverberation characteristics of these different structures are then labelled with descriptions such as 'bright hall', 'plate' and so on.

5.6 Summary

This chapter has presented the structure of the sound synthesis and transformation devices found in common use in the studio and concert hall. A classification of MIDI devices has been presented, and the way in which they treat sound described. This description includes common synthesis techniques such as amplitude and frequency modulation, as well as the more common 'effects' such as EQ, reverb, flanging and chorusing. It was found that these effects all depend on similar techniques to do with signal delay. Because of this common root, a detailed analysis of the behaviour of signal delay in filters and related structures was presented.

Suggested further reading

Denbigh, Philip (1998) *System Analysis and Signal Processing*, Addison-Wesley, ISBN 0-201-17860-5.

Dodge, Charles and Jerse, Thomas A. (1985) *Computer Music: Synthesis, Composition and Performance*, Schirmer Books, ISBN 0-02-873100-X.

Lynn, Paul A. and Fuerst, Wolfgang (1989) *Introductory Digital Signal Processing*, John Wiley and Sons, ISBN 0-471-91564-5.

Roads, Curtis (1995) *The Computer Music Tutorial*, MIT Press, Cambridge, Massachusetts, ISBN 0-262-68082-3.

Steiglitz, Ken (1996) *A Digital Signal Processing Primer, with Applications to Digital Audio and Computer Music*, Addison-Wesley, ISBN 0-8053-1641-1.

Watkinson, John (1998) *The Art of Digital Audio*, Butterworth–Heinemann, Oxford, ISBN 0-240-51270-7.

Part 4

Computer fundamentals

This section describes the computer systems which are used to realise the sound processing techniques described in Parts 2 and 3. It serves as an introduction to computer fundamentals for those who have not studied computers before, and who wish to understand the way in which they are used in digital audio and music technology systems. These chapters concentrate on the issues common to all conventional processors, and do not focus on any one specific device, so that the material will remain relevant as the technology changes.

Chapter 6 introduces the elementary logic gates that are fundamental to computer operations. Chapter 7 considers the structure of a typical computer system – its hardware and software. Chapter 8 then describes how the computer interacts with the digital audio world by considering the topic of computer interfacing. The chapter concludes with examples of computer based sound processing devices which illustrate the concepts dealt with earlier in the section. These include digital signal processors (DSPs), PC sound cards and the digital operation of the MIDI interface.

In places we have used shading in the margin alongside sections of the text which can be read as an overview of more technical material, or alternatively, which may be used as an adequate summary of the essential material for those not wishing to descend into the detail on a first reading.

6 An introduction to digital logic

Overview

We now have an understanding of the way in which digital audio signals are represented, manipulated and stored. It is therefore appropriate to gain an understanding of the digital machines, particularly the computer, which carry out this processing.

In this chapter we will look at the nature of the information handled within computer systems, and at the circuits which deal with, and store this information. These circuits will then form the basis of our understanding of the structure of computer systems, which will be introduced in Chapters 7 and 8.

6.1 Elementary logic and binary systems

All information in conventional computers is encoded into the *binary* number system. That is to say that all information and numbers to be processed in the computer are represented by signals which assume one of two possible states. We are very familiar with such binary signals in everyday life. For example a switch takes one of two states – on or off. A light bulb is another example where the states are on or off. In most computer systems the binary states are represented by voltages 'high' (about 5 volts) and 'low' (about 0 volts). These states are often labelled '1' and '0' respectively.

Binary systems also have a long history in logical discourse and argument, where the states are called 'true' and 'false'. Thus the statement that the 'earth is spherical' would be assigned the binary state 'true'. In fact a whole algebra based on these states, known as Boolean algebra, was developed for the analysis of logical debate during the early nineteenth century. Thus the use of binary systems in computers has its roots in a theory developed to serve this logical analysis, and this is why the circuitry used in a computer is still often referred to as 'binary logic'.

There are two main classes of such binary logic: *combinatorial* (sometimes also called *combinational*) and *sequential* logic. In combinatorial logic, the outcome of the system depends *only* on the combination of the inputs present, not on the order (or sequence) of the application of those inputs. In sequential logic, the outcome of the system depends also on the order of the application of the inputs.

6.2 Combinatorial logic and logic gates

We are very familiar with the operation of combinatorial logic in everyday life, albeit under a different name. Consider the case of the interior light of a car. It must come on (assume a logic 1 state) when any door is opened (assumes a logic 1 state), *or* when the interior light switch is operated. It does not matter in terms of the outcome which door is opened first, or when the interior switch is operated; the result is always the same: if any one of these events occurs, the light comes on. To take another example: a central heating boiler must come on when a time switch operates, indicating that it is the correct time of the day, *and* a thermostat operates, indicating that the rooms in the house are cold, *and* the master switch is set, indicating that the owner wants the system to work. Again, it does not matter in what order the individual switches were set; only that the appropriate *combination* of switches (all set) is present. The operation of these combinatorial systems can be described in tabular form, as shown in Figure 6.1.

A truth table representation, as shown in Figure 6.1, is simply a listing of all possible combinations of inputs (usually represented in an ascending binary sequence), together with the output corresponding to each input combination. An input switch in the 'ON' state, or an active output state (light or boiler

Figure 6.1 Truth table representation of combinatorial systems. (a) Truth table for car interior light; light (X) is true if switches A **or** B **or** C are true. Boolean expression: X = A + B + C; (b) Truth table for boiler control; boiler (X) is true if switches A **and** B **and** C are true. Boolean expression: X = A . B . C.

(a)

Inputs			Output
A	B	C	X
0	0	0	0
0	0	1	1
0	1	0	1
0	1	1	1
1	0	0	1
1	0	1	1
1	1	0	1
1	1	1	1

(b)

Inputs			Output
A	B	C	X
0	0	0	0
0	0	1	0
0	1	0	0
0	1	1	0
1	0	0	0
1	0	1	0
1	1	0	0
1	1	1	1

ON) is indicated with a '1'. The expression which would be used in the Boolean algebraic function equivalent to the operation of these systems is also shown for completeness. Figure 6.1(a) is the truth table for a three-input OR function, Figure 6.1(b) is that for a three-input AND function.

AND and OR are two of the three fundamental logical operators found in binary logic circuits. The third is the INVERT function, described in Figure 6.2. These three operators (AND, OR, INVERT) are fundamental in the sense that any logic circuit, no matter how complex, can be broken down into suitable combinations of AND, OR and INVERT operators. This is as true of the most complex digital computer, as it is for the most simple logic circuit, representing the operation of a car interior light for example.

Notice that there are eight entries (rows) in the tables in Figure 6.1, there being eight (i.e. 2^3) combinations of the three inputs A, B, C. (Three inputs; each with 2 states = $2 \times 2 \times 2 = 2^3$ separate combinations.) If these combinatorial systems had eight inputs, which would not be unusual by modern standards, there would be 2^8, or 256 entries in the table! Clearly, although the truth table is capable of *precise* description of the operation of combinatorial systems, it is hardly *concise*. If you were designing a combinatorial circuit, imagine trying to juxtapose the operation of two eight-input circuits, each with its own 256 entry table, and then recall that this is still a trivial circuit by modern standards. A more *concise*, as well as precise representation is required.

The *logic gate* representation of the logical operators satisfies this requirement. A logic gate is a small electronic circuit, usually fabricated within an integrated circuit, which implements a particular logic function (or truth table) such as AND, OR, INVERT, working with voltage levels, to represent the two binary states.

Figure 6.2 The fundamental logic gates. (a) INVERT gate, its truth table and Boolean expression (the bar above A means invert); (b) Three-input OR gate; (c) Three-input AND gate.

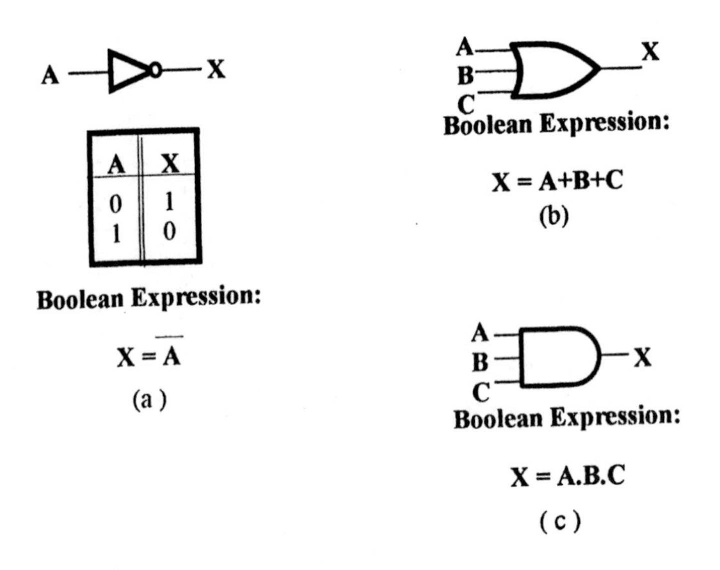

Thus there are AND gates, OR gates and INVERT gates (sometimes called NOT gates), as illustrated in Figure 6.2 to enable us to use the fundamental logical operators in digital circuits. The INVERT gate simply outputs the opposite logic state to that presented to its input. The symbols shown in the figure are used in logic circuit diagrams, allowing complex logical operations to be described concisely. The *functional description* of the operation of these gates, described in Section 6.3 below, defines the way in which these gates transmit or propagate binary logic signals (1s and 0s), and this provides an alternative description of the gates' operation in a convenient, concise and precise form.

Although the gates shown in Figure 6.2 represent the fundamental logical operators as they are found in logic circuits (perhaps with different numbers of inputs to the AND and OR gates), they are not the only kinds of gate found in common use. Three other kinds as illustrated in Figure 6.3, NAND, NOR and XOR, are also commonly found. This time, two input gates are shown, although gates with more than two inputs are available.

The operation of a NOR gate is functionally equivalent to connecting an INVERT gate to the output of an OR gate, hence the name NOR (*N*ot *OR*). This can be confirmed by comparing the truth table for NOR (Figure 6.3(a)) with the equivalent table for OR. The output column of the one is the logical inverse of the other. The 'bubble' on the output of the NOR gate reminds us that the NOR gate gives us the logically inverted operation of the OR gate. A NAND gate (*N*ot *AND*) is similarly related to the AND gate. Note that connecting an INVERT gate to the output

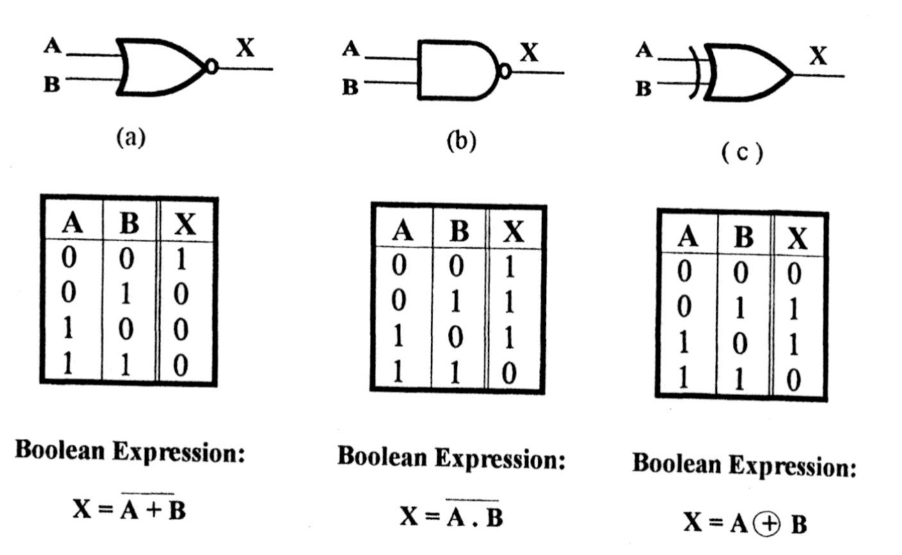

A	B	X
0	0	1
0	1	0
1	0	0
1	1	0

Boolean Expression:

$X = \overline{A + B}$

A	B	X
0	0	1
0	1	1
1	0	1
1	1	0

Boolean Expression:

$X = \overline{A . B}$

A	B	X
0	0	0
0	1	1
1	0	1
1	1	0

Boolean Expression:

$X = A \oplus B$

Figure 6.3 Commonly used logic gates. (a) Two-input NOR gate; (b) Two-input NAND gate; (c) Exclusive OR gate.

of a NOR gate gives a functionality which reverts to that of an OR gate, and similarly a NAND followed by INVERT gives AND.

Up to this point we have assumed that 'things start to happen' in a circuit when we have a logic 1 representing, for instance, a switch being 'ON'. This is described as an 'active logic 1', or that the system uses 'positive logic'. However, this choice of a '1' to *label* the active state is purely arbitrary. There are plenty of systems where things start to happen when a '0' state arises. Think of what happens when your car runs out of petrol ('0' = not full). These are called 'active 0' systems, or 'negative logic' systems.

The graphical form of the symbols shown describes very precisely the *operation* of the gates, as they propagate logic signals (1s and 0s) from their inputs to their outputs. In particular, a bubble on an output *or an input* indicates that the expected *active logic signal level* is a 0. The absence of a bubble indicates that the active logic level is a 1. Thus in the case of the NOR gate, the symbol shown in Figure 6.3(a) indicates that an active 1 on input A or B (the OR function is specified by the shape of the symbol) generates an *active 0* on the output of the gate. Only in the case of no input being active (both 0s), is the output in its *inactive* state, i.e. the output assumes a logic 1. This can be confirmed by looking at the truth table for NOR, Figure 6.3(a). However, note that once the rationale behind the format of the symbol is understood, it is no longer necessary to remember the truth table – the symbol provides its own functional description. This kind of functional description deals with the problem of circuits with large numbers of inputs, and correspondingly large truth tables.

You can tell what is going to happen (the *functionality*) just by looking at the symbol.

A similar interpretation of the NAND gate is available. The *symbol* says that the output is *active* (0) only if inputs A *and* B are *active* 1s. If either A or B is *inactive* (i.e. 0), then the output assumes its *inactive* state (1).

This kind of reasoning gives an additional useful way of interpreting the operation of logic gates. If we look again at the truth table for NAND, Figure 6.3(b), we see that an alternative interpretation of the NAND function is possible, if we assume that the active state on the input is a 0 and that the active state on the output is a 1. If either input, A *or* B is active at a 0, then the output is active at a 1. This functional description directly corresponds with the alternative symbol for a NAND gate, shown in Figure 6.4(b). It is made up of an OR symbol with 'bubbles' at the *inputs*. Either symbol for a NAND gate (Figure 6.3(b) or 6.4(b)) can be correctly used in the same circuit diagram, depending on the active logic levels at particular points in the circuit.

The symbols for the other gates which we have met above, can also be represented in their negative logic equivalents, as shown in Figure 6.4. We shall see examples later where either formulation (negative or positive logic) will be useful.

6.3 Functional description of the operation of gates

It was indicated above that a useful way to describe the operation of gates is to understand the way in which they propagate logic signals (1s and 0s) from their inputs to their outputs. The symbols shown in Figures 6.3 and 6.4 help us to do this. In all cases, it might be useful to imagine two kinds of input: control and data. The function of the control inputs is to *enable*, or 'open' the gate so that the data signal can flow through it. If there are three (or more inputs), there will be two (or more) control inputs and one data input. Consider the case of the NOR gate, in the form shown in Figure 6.4(a). For the gate to be enabled, *both* control inputs (say inputs A and B) need to be at a logic 0 before the data (say on input C) will be transmitted through the gate in

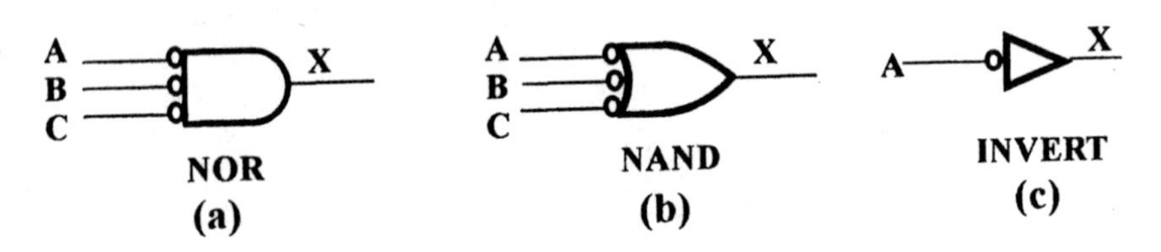

Figure 6.4 Alternative symbols for common gates (three inputs shown).

inverted form (because of the bubble on input C). In the case of a NAND gate all control inputs need to be at a 1 (refer to Figure 6.3(b)) before the data will be transmitted through the gate, again with inversion, this time because of the bubble on the output of the gate. It does not matter which inputs we label as 'control' and which as 'data', the operation is the same, because the gates are symmetrical with respect to their inputs.

If any of the control inputs are not set to enable the gate, it is 'closed' (or *inhibited*), and the output of the gate moves to its inhibited state: 0 for a NOR gate, 1 for a NAND gate. Data is not passed through.

Similar functional descriptions of the operations of AND and OR gates can be provided. You might like to work these out. The functional description of the XOR gate (Figure 6.3(c)) is a little more unusual. There are at least two such descriptions. One is a programmable inverter: if the control input is input A, and the data input is B, (again it does not matter which is which), then if the control input is a 0, data flows through to the output *without* inversion (i.e. the binary information on input B is copied to the output). On the other hand, if the control input is a 1, then data flows through to the output *with* inversion. Sometimes it is useful to 'flip' the binary state of a signal, depending upon the state of a control signal, and the XOR gate can be used to do this. The other functional description of the XOR gate is as a kind of binary comparator: if the two inputs are the same (both 1s or both 0s), then the XOR gate outputs a 0. If they are different then it outputs a 1.

6.4 Some simple examples of combinatorial logic circuits

We complete this section on combinatorial logic by describing two combinatorial circuits which have particular relevance within computers, as will be described in Chapter 7. The first of these is an *n-bit binary comparator.* A 'bit' in the jargon is an individual binary logic signal, 1 or 0, such as the output of a gate. In the circuit shown in Figure 6.5(a), n = 4. It compares two '4-bit' binary numbers, A_{0-4} and B_{0-4}. The output of the NOR gate (gate 5) will be a 1 only in the specific case that inputs A_0–A_3 have *exactly* the same bit pattern on them as that on inputs B_0–B_3 in which case the inputs to gate 5 will all be 0s. A circuit such as that in Figure 6.5(a) (probably with more inputs than four) therefore allows the computer to compare two pieces of information, to judge whether they are the same, and to take appropriate subsequent action if they are. This kind of logical operation is useful,

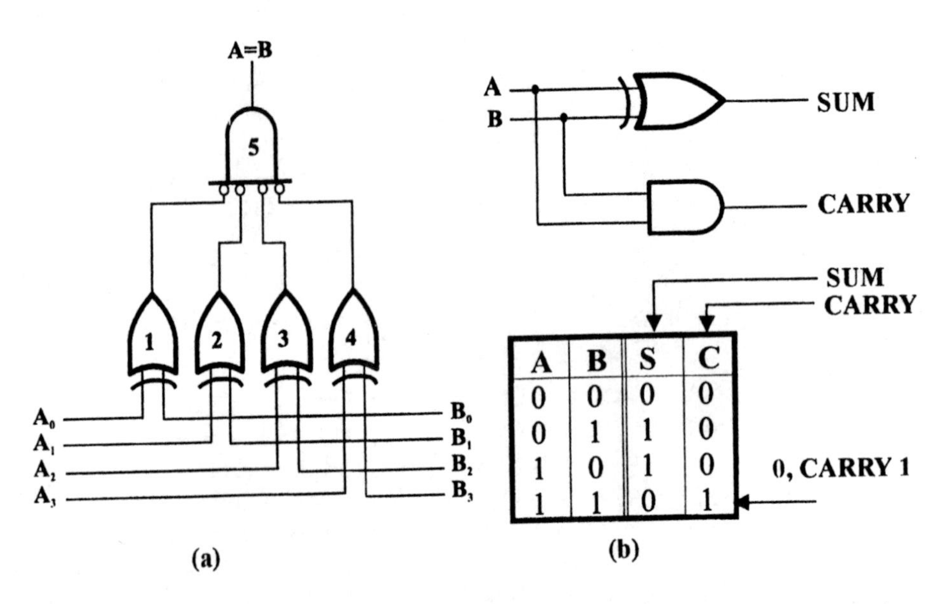

A	B	S	C
0	0	0	0
0	1	1	0
1	0	1	0
1	1	0	1

Figure 6.5 (a) and (b) Some important combinatorial circuits.

since information in a computer, such as a person's name or other information to be processed in a program, is stored as collections of bits, as we shall see in Chapter 7. This is a fundamentally important operation in a computer, since it allows the computer to make a *decision* within its processing of the information. We shall return to this point in Chapter 7.

Figure 6.5(b) is a simple form of adder circuit, known as a *half adder*, which might form part of the arithmetic processing parts of a computer. It carries out binary addition of two binary digits applied to inputs A and B, as shown in the accompanying truth table. The final entry in the table may be interpreted as '1 add 1 gives 0, carry 1', which is consistent with the operation of binary arithmetic. Developments of this circuit form the basis of the arithmetic processing capabilities of computers, as will be described in Chapter 7.

6.5 Sequential logic

As stated above, sequential logic is a branch of binary logic where the output of the circuit or system is a function of the *sequence*, or order of application of the binary inputs to the circuit. Like any other logic circuit, sequential circuits are made from the fundamental logic gates AND, OR and INVERT. They are important within the context of computer systems because they provide the vital components of *memory* (used for the storage of programs and other information), and also the general

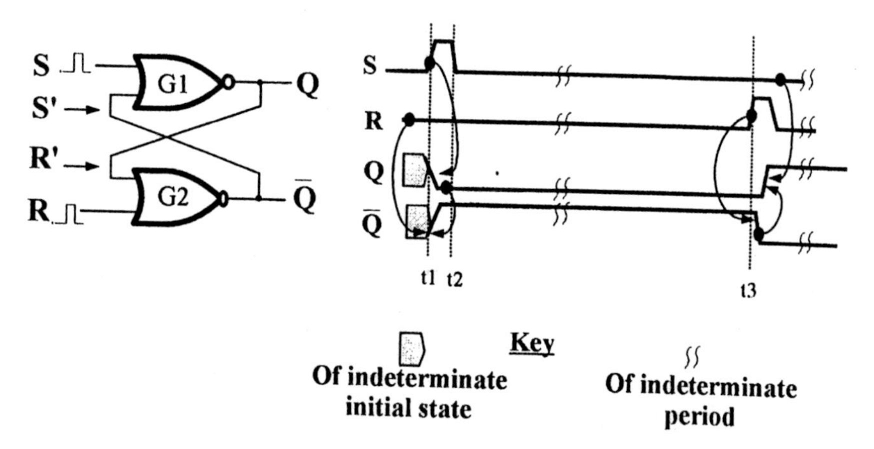

Figure 6.6 The SR bistable.

controllers which lie at the heart of the internal operation of a computer, as will be described in Chapter 7.

We start our brief introduction to sequential circuits by taking a look at a configuration of gates which forms a fundamental building block in sequential systems. It has several names: SR (for 'Set Reset') bistable, SR latch, or SR flip-flop.

The circuit diagram of the archetypal SR bistable is shown in Figure 6.6. As can be seen, it is formed by a cross-connected pair of NOR gates. It is a basic form of memory circuit, as will be shown below. First of all however, its basic operation will be described.

It is important to note that the normal mode of operation of this circuit is to pulse only *one* input, S or R high (i.e. place a momentary logic 1 on one of these inputs, as shown in Figure 6.6), whilst the other input is held low. Let us consider the case that S is pulsed high, as shown at instant t_1 in Figure 6.6. The shape of the NOR gate symbol tells us that if *either* input is high, then the output of the gate will be forced low by the operation of the gate. The consequence of pulsing S high therefore is to set output Q low. However, by the cross-coupled connection, this results in input R' to gate G_2 also being set low. Now, since R at this point in time is also low, *neither* input to NOR gate G_2 is active (high), and so the output of this gate $\bar{Q}$ is forced to its *inactive* state – a logic 1. By the cross-coupling, input S' to gate G_1 is therefore set high. The next point is important, since it provides the memory functionality of the circuit. *The original input pulse on S may now be removed without affecting the output states of the circuit.* This is because the logic 1 fed to S' by $\bar{Q}$ takes over' from the pulse on S. Gate G_1 does not mind which input is a 1; if either (or both) inputs are at a 1, then the output will remain a 0. The circuit will remain indefinitely with the outputs **Set** in the state Q, $\bar{Q}$ = 0,1 until either the power is turned off to circuit, thus destroying its functionality, or until a pulse is applied to R, in which case the sequence of operations described in the next paragraph takes place. The sequence of operations described thus far takes us to instant t_2 in Figure 6.6.

If input R is subsequently pulsed high (with S held low), as shown at instant t_3, $\bar{Q}$ is set low and Q is set high by a sequence of operations analogous to that described above, where the roles of gates G_1 and G_2 are now interchanged. The circuit will now remain **Reset** indefinitely in the state Q, $\bar{Q}$ = 1,0 until either the power is turned off, or until another pulse is applied to S, in which case Q, $\bar{Q}$ will revert to 0,1 again.

The typical speed of operation of this circuit is high, being determined primarily by the time it takes for the effect of a pulse at S (or R) to result in a subsequent 1 being established at S′ (or R′). This is determined by the time it takes for the logic signals to propagate through the two gates G_1 and G_2 – about 5 ns for typical gates (1 ns 'nano-second' is 10^{-9} seconds: one thousandth of a millionth of a second!).

An SR bistable can also be formed from two cross-coupled NAND gates, and is often seen in this form. The operation is directly analogous to that described above, except that the inputs S and R are normally held high, and are pulsed low to operate the circuit. The reader might like to analyse this form of bistable as an exercise.

The SR bistable is a very elementary form of memory. A way of summarising the operation of the circuit in Figure 6.6 is as follows: the position of the 0 on the output (Q or $\bar{Q}$) indicates which input (see Figure 6.6) was last pulsed high (i.e. set momentarily to a '1'). If Q is 0, then input S was last pulsed high, whereas if $\bar{Q}$ is 0, then input R was last pulsed high. In this way, the circuit 'remembers' which input was last pulsed.

This, unfortunately, is not a very useful form of memory for use in computer systems. Computers tend to operate in an iterative cycle: calculate a result, store it in memory, calculate a result, store it in memory, etc. A 'result' in this context is a binary number (i.e. a collection of 1s and 0s) produced as a consequence of the operation of the computer's program – adding two numbers together for instance. A more useful form of memory therefore would be one where the 'result' (a binary 1 or 0) could be presented by the computer at a 'data' input to a memory, and where the computer could subsequently pulse a 'store' input to

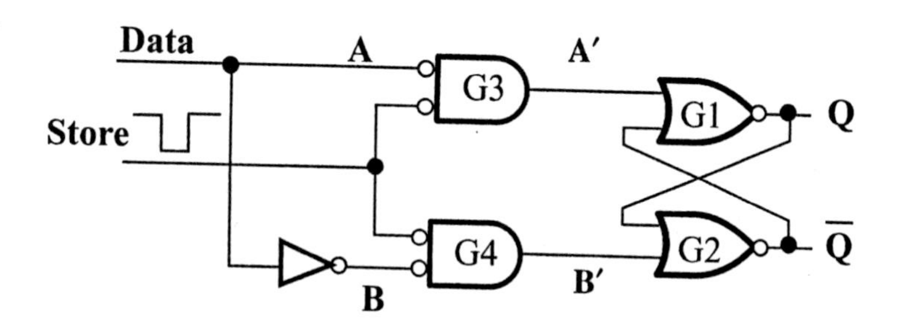

Figure 6.7 The D-type latch.

the memory, which would instruct the memory to hold (or store) that datum value. It would be stored by the memory until the computer wishes to over-write this value at some subsequent time, by means of another 'present/store' cycle, when the original datum value is no longer needed.

The discussion below describes the so-called 'D-type' latch which develops the SR bistable into a form which is ideally suited to this 'present/store' operation of computer memory.

The D-type latch extends the configuration of the RS bistable (see Figure 6.7) by adding the NOR gates G_3 and G_4. Notice the alternative symbols used for these gates. They imply that they will be looking for a logic 0 AND another logic 0 at the inputs, generating a logic 1 at the output when this combination of inputs is found. This is a valid interpretation of the operation of an NOR gate (see the truth table in Figure 6.3(a)). The data is present in its true (direct) form at A, and in its inverted form at B, so one or the other of these inputs will be a logic 0 depending on the polarity of the data input.

When the 'store' pulse is present (notice it is a logic 0 pulse), the logic 0 at A or B will be transmitted through its associated gate, G_3 or G_4, to produce a logic 1 at A′ (for 'data' = logic 0) or at B′ (for 'data' = logic 1). This logic 1 then activates the SR bistable, G_1, G_2 in the way described above, so that it 'remembers' which input was last pulsed high. In particular, bistable output Q will store a logic 1 if the binary information on the 'data' input is a 1, and it will store a 0 if the information is a 0. To summarise: the output Q *stores* the information which was present on the 'data' input when the store pulse was present. The information thus stored at the output will be retained indefinitely, either until the power is turned off, or until another 'store' pulse with different information present on the 'data' input. This exactly conforms with the kind of operation required of the computer memory described above; the computer presents one bit of the information it requires to be stored to the 'data' input, and then it generates a 'store' pulse, moving and retaining the data to the memory output.

6.6 Summary

In this chapter we have considered the basic building blocks from which digital systems are built. We have identified the two main branches of digital system: combinatorial and sequential, and have considered the operation of the logic gates which are used to construct these systems. We understand how the operation of these gates relates to the processing of binary signals. We have seen a few key examples of circuits which will be useful in later chapters. We now have a sufficient understanding of these logical fundamentals so that we can follow the operation of basic computer systems. This forms the subject of Chapter 7.

Suggested further reading

Gajski, Daniel D. (1997) *Principles of Digital Design*, Prentice-Hall, ISBN 0-13-242397-9.

Johnson, E.L. and Karim, M.A. (1987) *Digital Design, A Pragmatic Approach*, PWS, ISBN 0-534-06972-X.

7 Computers and programs

Overview

In this chapter, we build on the logic circuit fundamentals established in Chapter 6, to gain an understanding of the basic structure and operation of computer systems. The manipulation of digital audio signals in music technology and audio systems is almost universally carried out by computers, and so some understanding of these important audio processing components is valuable.

In this chapter, we therefore learn what it is that constitutes a computer program, from the point of view of the machine (the computer), and also from the point of view of the programmer, and how the data manipulated by such programs is represented. We will examine the essential building-blocks from which all programs are made, and consider how these program constructs are supported by the fundamental operations of the machine. The approach is generic: no specific computer is used as the basis for this discussion – rather the fundamental features which are common to *all* computers are described.

The material presented in this chapter will set the basis for Chapter 8, where we will study the use of a computer as a *component* in digital audio and music technology systems. In Chapter 8 we will be looking at PC audio cards and digital signal processors (DSPs) as examples of the application of material covered in Chapter 7.

7.1 The essential architecture of a computer system

The layout of the simplest form of conventional computer is shown in Figure 7.1.

7.1.1 The CPU

The central processing unit (CPU) carries out all the manipulation of information required by the computer. The computer follows a sequence of instructions, known as a *program* to carry out this manipulation. For instance, if a collection of numbers is to be added together, the program will contain a sequence of 'addition' instructions. The actual addition will be done in the CPU, although the numbers themselves would be fetched from the memory unit, where they are stored.

This process of program execution acting on memory data is typical of computer applications. This is true whether the system is intended to be a word-processor, Internet browser, digital audio editor or any other computer based system.

The instructions forming the program are stored as binary information (1s and 0s) in the memory unit. The memory also normally stores in binary form all of the information ('data') which will be manipulated by the program, whether this is textual information in a word-processor, or audio samples for a sound processing system, or the collection of numbers in the addition program referred to above.

7.1.2 The memory unit

The memory unit typically is a vast array of D-type latches. The D-type latch was described in Chapter 6. Each latch stores one

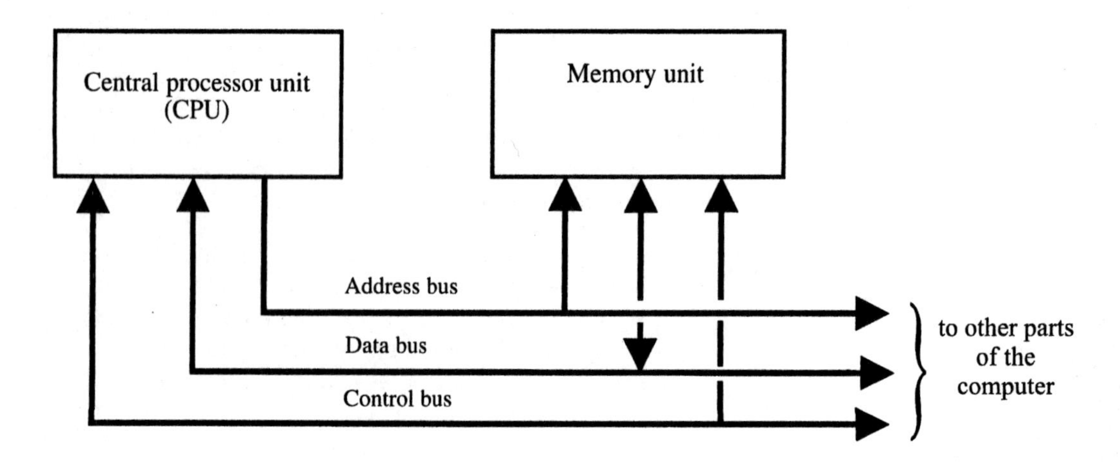

Figure 7.1 The architecture of a simple computer system.

piece of binary information (a '1' or a '0') known as a 'bit'. Often a number of 'bits' of information are collected together into larger groups; 8 bits of information grouped in this way is known as a 'byte', 16 bits is often described as a 'word', and 32 bits as a 'long word', or sometimes as a '32-bit word'. Thus, in a computer, the memory unit might be described as a '32 Mbyte memory'. The 'M' stands for 'Mega', meaning one million. In this example (quite modest by contemporary standards), the memory would contain the equivalent of $32 \times 8 \times 1$ million D-type latches!

In many computers there are in fact two main forms of memory. One, known as 'RAM' (random access memory), is the D-type latch form described above. This kind of memory is often known as 'volatile storage', since the content is volatile (lost) when the power is switched off, as described in Chapter 6. This happens unless a small battery is used to keep the data content alive, in which case this would be described as 'battery-backed RAM'. Actually, the description of this memory as 'random access' (meaning that information can be accessed at will, from any part of the memory) is something of a misnomer, since most forms of memory have 'random access' in this sense. A better name for this kind of memory might be 'read/write memory', since the computer can change (write) the information stored in a latch, as well as inspect (read) the content.

The other common form of memory is known as ROM (read-only memory), implying that information cannot be written into ROM by the computer. It can only be read. The information stored (binary 1s and 0s) is 'burnt' into tiny fuses within the ROM. A fuse which is burnt in this way might be deemed to contain a binary 1, whilst a fuse not burnt might be deemed to contain a binary 0. Once 'burnt' in this way, the content of the memory cannot be changed by turning the power off, or by any other means for that matter. Hence this kind of memory is described as 'non-volatile'.

ROM is used to store programs which must not be lost once the computer is switched off. You might be aware that programs and data can be stored on magnetic media such as magnetic disks and tape, which also retain their information once the power is turned off. However, a computer fitted with disks will still need ROMs, *because the computer will need to run a program just to extract information from disk.* This program would tell the computer how to control the disk, find the relevant information on it, and how to place the information found there into main (RAM) memory, where it can be used to further direct the computer. This information stored on the disk may itself form the program for some digital audio processing task.

The disk controlling program must not be lost between sessions on the computer, because it is responsible for bringing the computer to life, as described above. This process of bringing a computer to life under control of a ROM-based program is known in the jargon as 'booting' the computer(!), and the ROM is often known as the 'boot ROM', or sometimes as the 'basic input output system' (BIOS) ROM.

7.1.3 Buses

Referring to Figure 7.1 we see that the memory (ROM or RAM) is connected to the CPU by interconnection paths known as 'buses'. A bus is simply a collection of wires grouped together according to their function. In most computers there are three of them – a 'data bus', an 'address bus' and a 'control bus' – whose functions will be described shortly. We recall that instructions telling a computer what to do (the 'program') and the information to be manipulated by the program (the 'data') are all stored as binary information in the memory. The data bus in some early computers had eight wires in them, allowing the CPU to collect information from the memory 8 bits at a time. These were known as '8-bit processors'. These days the data bus is more often 32 or 64 bits.

The information (program or data) is grouped into 8, 16, 32 or 64 bit 'chunks' in memory known as 'memory locations', or 'memory addresses', each of which is identified by a unique number known as the 'address' of the location.

7.1.4 The fetch-execute cycle

When the processor (i.e. the CPU) goes through the process of executing a program, it carries out the following cyclical sequence of operations, one cycle for each instruction within the program:

1 The CPU puts a binary number (typically 24 or 32 bits in extent) onto the *address bus*. This identifies the address location in memory where the instruction is stored.
2 The CPU then activates one of the lines on the control bus (by writing a binary signal to it) to tell the memory exactly *when* it should respond. Let us call this line the 'memory read enable' (MRE) line. Since the whole transaction will typically take place within about one ten-millionth of a second (100 ns), it is impor-

tant to carefully synchronise the operation so that it happens within this time. This is the job of the control bus signals.

3 When instructed by this control bus signal, the memory places the *contents* of the memory location specified by the binary number on the address bus (step 1) onto the data bus. This will be 32 or 64 bits, depending on the width of the data bus. At this stage in the cycle, this binary information fetched from memory via the data bus will represent an instruction, telling the CPU what to do at this point in the program.
4 The CPU will collect this *instruction word* (known as the 'machine code instruction') from the data bus, and will decode the individual 1s and 0s to determine what this particular instruction is telling the computer to do. It then carries out the required operation.
5 This particular instruction may require the CPU to fetch information (data) from memory. Suppose the instruction requires two numbers located in different memory locations to be added together for instance. In this case, the CPU would go through two cycles of placing the address of a number in memory onto the address bus, activating the memory read enable line, and reading the binary number fetched from the specified memory location from the address bus into the CPU.
6 The two numbers would then be added together in the CPU. The instruction might require that the result should be sent to memory, in which case the CPU would place the address of the destination location in memory onto the address bus. It would then place the result of the addition process onto the data bus, and activate a 'memory write enable' (MWE) line on the control bus. This control bus signal would instruct the memory to collect the information now present on the data bus (the result) and place it into the memory location specified by the number now present on the address bus.

This whole process 1–6 is known as the *fetch-execute cycle* for this particular instruction (add). The process of running a program in a computer is a process of cycling through the stages of the fetch-execute cycle, 1–6, until all of the operations required by the program are complete.

The description of the operation of a computer system given above is a sufficient model of what a computer does for many purposes. The remainder of this chapter amplifies this model. It describes in more detail how information is represented within the computer system and explains the different kinds of instructions available within a typical computer. You only need to read on if you wish to understand this extra detail!

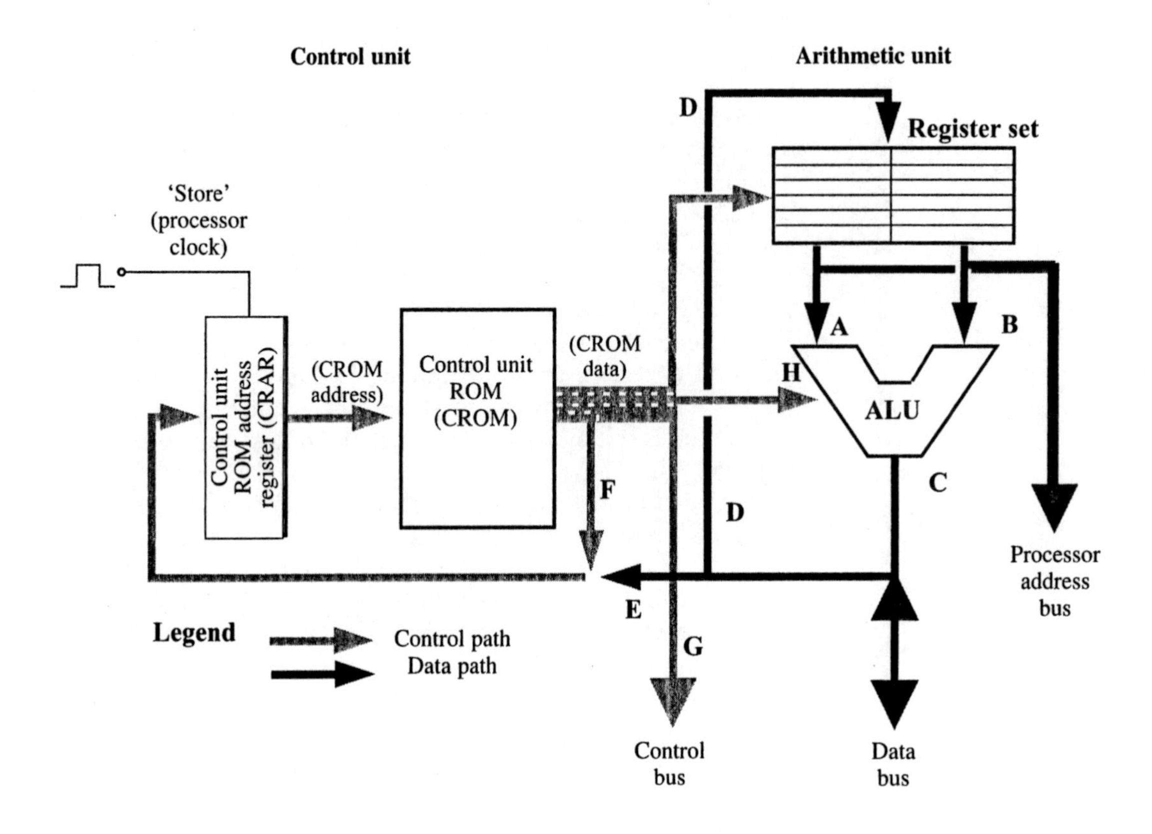

Figure 7.2 A typical central processor unit (CPU).

7.2 A look inside the central processing unit

We now take an overview of the operation of the central processing unit. This will help us to:

- understand how a computer works, at least in outline, and
- understand how the main types of instruction interact with the CPU, and hence what the outcome of these instructions is.

Figure 7.2 shows a block diagram of the main components within a typical CPU. The circuitry represented in the diagram would typically be contained within a single microprocessor integrated circuit (or 'chip'), or within a digital signal processor (DSP).

There are two main sections in the CPU: the arithmetic unit and the control unit. The points where the buses (address, data and control) emerge from the CPU are shown. We will consider the arithmetic unit first.

7.2.1 The arithmetic unit

Within this unit, there are again, two main components: the register set and the arithmetic and logic unit (ALU).

The ALU is the part of the processor which manipulates data as required by the program. Supposing two numbers are to be added together. The first number would be input into the ALU via input path A, the second on input path B. The result of the operation appears on output path C. In many processors, the ALU is a relatively simple combinatorial logic circuit, being an extended development of the half adder circuit presented in Chapter 6. A typical ALU will also carry out subtraction, various logical operations such as AND, OR, INVERT on a bit-by-bit basis (see Chapter 8), negation and any other manipulation of data required by the program.

Note that extended arithmetic operations such as multiply and divide are often not directly implemented in the ALU. For instance when numbers are to be multiplied, this is achieved in many cases by a process of repeated addition, involving many cycles through the ALU. This means that these kinds of arithmetic operation are relatively slow. In fact, one of the distinguishing features between a DSP and a conventional microprocessor is that a DSP's ALU usually *does* contain a multiplication unit (although often no division unit) and this achieves the desired result in one pass through the ALU. This makes a DSP particularly suitable (i.e. fast) for signal processing operations such as digital filtering.

The particular arithmetic operation to be performed at any point in time by the ALU is specified by a binary bit pattern applied to the ALU control line inputs by the control unit. This is shown as point H in Figure 7.2.

7.2.2 The register set

The register set is essentially a small block of RAM contained within the CPU. A 'register' is one particular location within this RAM, typically 32 or 64 bits wide. There may be 20–30 registers in a typical processor, and they are directly involved in the execution of the instructions, as we shall see. There are usually two main types of register: data registers and address registers.

Data registers hold numbers which are about to be processed by the ALU, or hold the result of a processing operation, loaded via the output path D from the ALU. Numbers can also be loaded into registers from the data bus (e.g. from the main memory unit) ready for processing in the ALU. Suppose our register set contains eight data registers; each would typically be labelled with a number in the range 0–7, e.g. DR7 to DR0.

The primary function of the address registers is to drive the *address bus* during the operation of programs, hence the association of the address bus with the output of the register set in Figure 7.2. When the processor needs to fetch information (instruction or data) from memory, one of the address registers is loaded with a binary number (32 or 64 bits) which is the number of the location in memory (the address) which contains the information. Similarly, if the processor needs to send information to memory, an address register is loaded with the address to be used. The control unit then arranges for the address register to be connected to the address bus at the appropriate time. This will be when the processor is accessing memory during the fetch or execute of an instruction. The address registers might typically be labelled AR0–AR7 for an eight-address register machine.

7.2.3 The program counter

One of the address registers, known as the program counter (PC), deserves special mention here. Its purpose is to contain the address in main memory of the *next instruction* to be fetched by the processor. It is automatically updated by the control unit after the fetch of every instruction so that it always 'points' to the address of the next instruction in memory.

Sometimes the address of an instruction or item of data in memory needs to be *calculated* as part of the execution of a program. For instance suppose the contents of every *tenth* location need to be added into an accumulated sum. In this particular example, the program would use one of the address registers to hold the memory address to be used. It would place the contents of this address register onto the address bus, thereby fetching a datum from memory. It would then add this into the accumulating sum, which would probably be held in a data register. The processor would finally *add 10* to the contents of the address register (so that it 'points' to the next item of data), *using the ALU* to do the addition. It is therefore convenient that the address registers also have ready access to the ALU via paths A and B, as shown in Figure 7.2.

7.2.4 An example of the fetch-execute cycle

The example includes a description of the way in which the buses are used during the fetch-execute cycle, and also how the

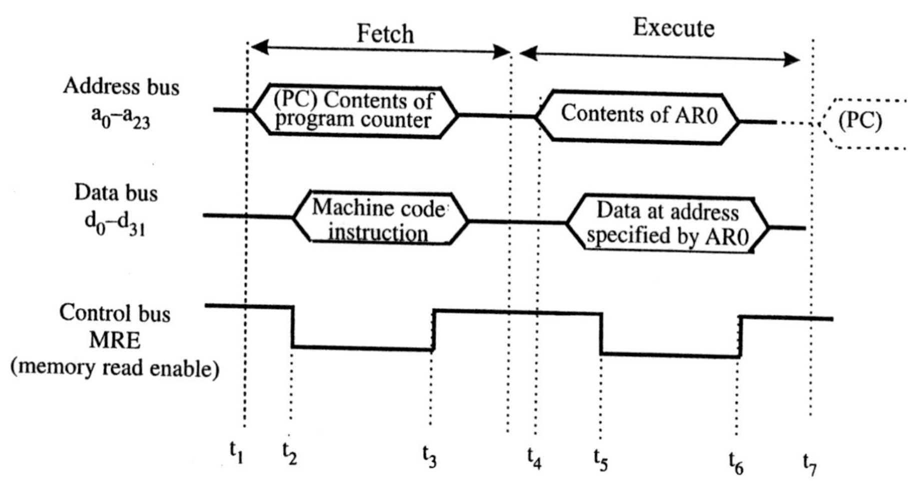

Figure 7.3 Logic signal activity on the processor bus (for instruction add *AR0, DR0).

arithmetic and control units are used. The instruction which is used in this example is given the mnemonic:

add *AR0, DR0

The meaning of this hieroglyph is as follows: add the *contents of the memory location* (hence the '*') whose address is currently held in address register AR0 to the datum already held in data register DR0. The result will be left in DR0. (Do not worry if the meaning of the instruction is not immediately obvious; the written form of instructions will be explained later – just accept the explanation for now.)

As explained above, the processing of the instruction takes place in two phases: fetch and execute. Figure 7.3 shows the effect of the execution of this instruction in terms of the signals appearing on the three buses of the processor.

The operation starts when the processor initiates the *instruction fetch*. The control unit places the contents of the program counter onto the address bus at time t_1, so that the memory is supplied with the address in memory of the next instruction to be executed. Figure 7.3 indicates that there are 24 individual lines in this particular processor's address bus, labelled a_0–a_{23} The memory is given a short period of time, between t_1 and t_2, to find the required information. The control unit then asserts one of the lines of the control bus, 'memory read enable' (MRE), which instructs the memory to place the information (i.e. the instruction) onto the 32 lines of the data bus at time t_2. The control unit allows this information to settle on the data bus for a short time t_2–t_3, before it collects it at time t_3. Having done this, it also switches off the MRE signal and removes the program counter from the address bus. This marks the end of the fetch phase.

During the interval t_3–t_4, the control unit decodes the instruction, to determine the appropriate operations which will take place during the execute phase. For this particular instruction, the execution entails placing the contents of address register AR0 onto the address bus at time t_4, to specify the location in memory of one of the numbers which is to be processed (added in this case) by the execution of the instruction. Such a number which is to be processed by an instruction is called an 'operand' in the jargon. In the case of this particular instruction, the other operand is located in data register D0. We are now into the instruction execute phase. Again, the memory is given a short time, t_4–t_5 to find the datum, before the control unit again asserts MRE at t_5, instructing the memory to place the datum onto the data bus. The processor reads the information off the data bus at time t_6. The address and data buses are then switched off by the control unit, their services not being required for the remainder of this instruction. The processor completes the addition of the two operands during the time interval t_6–t_7, leaving the result in DR0. It also updates the contents of the program counter, so that it points to the *next* instruction in memory. The processor would then initiate the fetch of the next instruction at t_7. The whole process of the fetch and execute of one instruction would typically take only a few hundred nanoseconds to carry out.

7.2.4 The control unit

We complete this overview of the operation of the central processing unit by examining the function of the control unit (CU) during the fetch-execute cycle. Although there are many ways of designing the CU, perhaps the simplest to describe is one oriented around a ROM based sequential controller, as shown in Figure 7.2. The output of the control unit ROM (CROM) orchestrates the operation of the elements within the arithmetic unit such as the ALU, a register or data path, which will participate in the execution of a particular phase of the fetch-execute cycle. The address inputs to the ROM in the example shown in Figure 7.2 are provided by a group of D-type latches (one latch per address bit) known as the control unit ROM address register (CRAR). The contents of the CROM constitute a set of small programs (known as *microprograms*) which steer the various parts of the arithmetic and control units through the different operations *within* the phases of the fetch-execute cycle. The operation of this microprogrammed control unit is outlined below.

7.2.6 Control unit microprogram

As has already been explained, the processor operates in a continuous cycle of fetch and execute. There are sections of the content of the CROM (the microprogram) which are concerned with instruction fetch, whilst other parts are concerned with the execution phase of specific instructions. Let us follow the operation of the microprogram for the fetch-execute of one instruction. The processor is initialised (or 'reset') by forcing the address in CROM of the start of the 'fetch' microprogram into the CRAR. This is done by an agent not shown in Figure 7.2, although it is a simple matter of pre-loading the CRAR with a fixed bit-pattern corresponding to this start address.

In this first step of the fetch microprogram, the CROM output arranges for the program counter (PC) to be displayed on the processor address bus and for path F to be switched to the input of the CRAR. The outputs of the CROM driving F provide the *next address* in the fetch microprogram. When the CRAR 'store' input is pulsed by an external logic signal known as the 'processor clock', this new address is loaded into the CRAR, thus stepping the CROM onto the next phase of the fetch-execute cycle.

This next phase activates the memory read enable (MRE, see above). MRE is one of the outputs of the CROM (via path G), as are all other members of the control bus. MRE is 'activated' simply by having the MRE output of the CROM at this microprogram address set to a logic 0 (assuming 0 is the active state). The CROM at this point also arranges for path E to be switched to the input to the CRAR. Recall that we now have PC on the address bus and that MRE is active. The external memory is therefore at this instant in time placing the contents of the address location onto the *data* bus.

This location contains the binary code representing an *instruction*. This instruction code is therefore loaded into the CRAR on the next processor clock via the data bus and path E. *The instruction's binary code provides the start address in the CROM of the microprogram associated with the execution phase of this particular instruction.* On the next, and subsequent processor clocks, the outputs of the CROM step through the sequence of the instruction's *execute microprogram*, to manipulate the arithmetic unit to implement the various operations required by this instruction. Perhaps an address register is switched onto the address bus, or perhaps two data registers in the register set are sent to the ALU for addition via paths A and B. In this particular example, the ALU control would be set to 'add' and the ALU output sent back to the register set via paths C and D.

In each microprogram step, path F is used to load the CROM address of the next stage in the microprogram into the CRAR. This ensures that the control unit as a whole steps through a coherent sequence of operations to carry out the instruction.

At the end of the execute microprogram for each instruction, the program counter is sent through the ALU so that an appropriate number, corresponding to the size of the instruction (in memory words) can be added to it. This ensures that the PC is left at the end of each instruction 'pointing' to the location in main memory containing the next instruction. Finally, path F is used to force the start address of the 'fetch' microprogram back into the CRAR, so that the fetch cycle of the next instruction is initiated.

In this way, the processor cycles through the fetch-execute process, stepping through a sequential list of instructions in main memory.

In the design of any specific processor, there will be elaboration and various optimisations of the concepts described above. However, the information presented above is a sufficient, if simplified model for understanding the internal operation of a central processing unit. We are now ready to consider the way in which programs, instructions and data, are represented in memory.

7.3 The representation of instructions in a computer

It will be clear from the previous section that *both* instructions and data are made up of binary words consisting of 1s and 0s, stored at appropriate locations in memory. It is entirely possible to program a computer by putting these 1s and 0s into memory, and then launching the program by forcing the start address (also in 1s and 0s) into the program counter. Indeed, some early computers were programmed in just this way. They had a row of switches on the front panel, one for each bit in an instruction/data word which could be used to load the binary information into memory.

The problems with this particular 'switch' approach to programming computers were that (i) it was impossibly tedious, and (ii) more importantly, it was prone to error. A mistake in even one bit in any instruction could lead to radically different and confusing operation of the computer program. The program could even destroy itself by over-writing the memory locations containing the program! It was therefore essential to find ways of generating programs and data more *accurately* and conveniently. Various ways of doing this have been devised. They include machine code programming, assembler programming, interpreted code and compiled code, all of which will be described briefly below.

7.3.1 Hexadecimal machine code programming

In this form of programming, the binary 1s and 0s representing an instruction or data word (or indeed any binary information such as a MIDI message), are split into groups of four. Then the bit pattern of each group is represented by writing down its decimal equivalent:

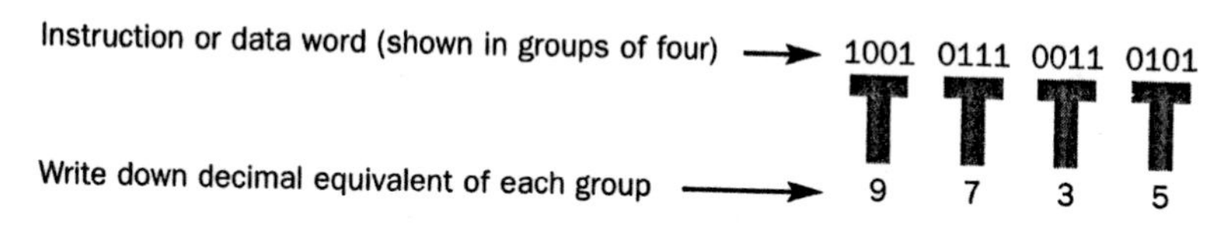

Thus, the collection of bits 1001011100110101, can be represented *completely accurately and unambiguously* by the hexadecimal number 9735.

A small problem however arises with the simple example given above. Four binary digits (bits) can represent all of the decimal numbers in the range 0–15. How would we represent the binary number 1100 (equivalent to decimal 12)? If we simply wrote '12' as our hexadecimal representation, carrying on our idea of writing down the decimal equivalent of each group, this would be ambiguous, since '12' could also represent the group of binary digits '0001 0010'. To avoid this ambiguity, binary digit groups corresponding with 10 through to 15 are labelled A through to F respectively. To take another example:

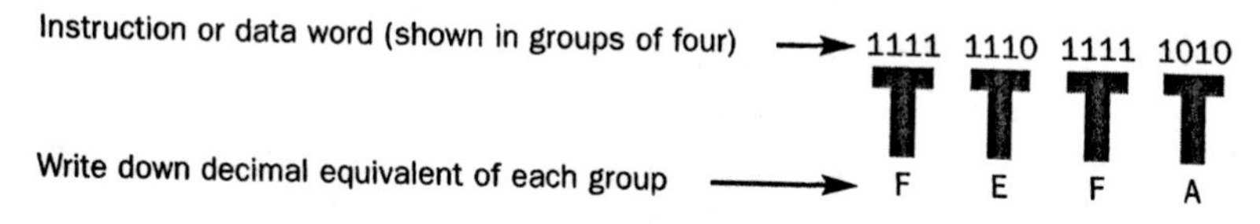

Thus, the collection of bits 1111111011111010 can be represented unambiguously by the hexadecimal number FEFA. Very often a hex number is prefixed by '0x', so that it is clear that this is a hexadecimal number, and not some acronym for a football club! Thus the number above would be shown as 0xFEFA.

We now have a way of representing binary information (for example computer programs) in a way which is less prone to human error, and indeed, some relatively large programs have been written in hexadecimal machine code. However, whilst some people might be able to tell you that 0x9356, 0xFACA,

0xDEC3, 0xFFF3 divides one number by another on a particular computer, and could write programs rapidly (and accurately) in this form, most of us need a little assistance to help us with understanding and writing such programs! Assembler, and interpreted and compiled code, as described below, are different ways of doing this.

7.3.2 Assembler programming

Because of the difficulty in understanding programs written in binary or hexadecimal machine code, manufacturers of processors produce a system of *mnemonics* to act as *aide memoire* for the meaning of a set of machine codes. Here is an example, where the left-hand column contains the (hexadecimal) machine code, and the middle column contains the corresponding mnemonics for a certain processor. The right-hand column is simply a *comment* explaining the effect of each instruction.

```
CD00   load @0x1000, DR0    ;fetch a number from memory loca-
                            tion address 0x1000 and put it in
                            data register DR0:

CE01   load @0x1001,DR1     ;fetch a number from address
                            0x1001,and put it in DR1:

E40C   add DR0,DR1          ;add the contents of the two reg-
                            isters, leave the result in DR1:

CC02   store DR1, @0x1002   ;write the result into memory
                            location address 0x1002:
```

This program simply adds two numbers, initially stored in two successive memory locations, and writes the result to a third successive memory location. The '@' sign indicates that the associated number (1000, 1001, 1002 in this example) is to be used as a *memory address* and hence this number will be placed onto the address bus by the processor during the program's execution. The use of memory to store or provide operands is known as 'indirection' in the jargon.

We have already seen an example of the use of mnemonics to represent an instruction in Section 7.2:

add   *AR0, DR0

Here the '*' indicates that the contents of address register AR0 is to be used as an address – in this case to tell the processor where one of the operands is stored in memory. The way in which the processor identifies the location of an operand (e.g. by the use of '*' and '@' in the examples above) is known as an *addressing mode*. It is important to remember that in this form of program representation, there is *one mnemonic* for each elemental machine instruction. We shall see examples later of programming languages where this is not the case.

Manufacturers produce a utility program known as an *assembler* which converts the mnemonic representation produced by the programmer into the binary machine code form. This can then be loaded into the computer's memory and run as a program. For this reason, programs written in this mnemonic form are known as *assembler programs*. The programmer writes the mnemonics into a text file using an editor or perhaps a word-processor, and then calls up the assembler to convert the mnemonics into the binary form. The binary is often stored in another file by the assembler, and a program known as a *loader* is used to put the binary into memory, and run the program. Sometimes the loader is provided as an integral part of the assembler, so that the user may not be explicitly aware of its presence.

7.3.3 Limitations of assembler programming

Assembler code is useful in the sense that it allows the programmer to manipulate the computer directly within his or her programs, and thus optimise them for a particular processor in terms of speed of execution, memory usage, etc. This is possible because the instructions explicitly refer to processor components such as registers and memory locations, as the examples above illustrate. However, since the instructions do refer to processor specifics, and since these specifics vary from processor to processor (they may have different numbers of registers for instance), it immediately follows that a program written for one processor may not be compatible with another. The programs are not 'portable' between processors. Also, there is no standardisation of the format of assembler instructions; they may use different mnemonics. Programs carrying out the same operation on different processors may look completely different, and this further compromises the portability of programs.

This portability issue is important. If a program is written in assembler for one processor, which is then superseded by another newer design, then the program will have to be completely re-written. Often this is not feasible for commercial and logistical reasons, and as a result, many useful musical programs (for instance) have died with the processors which ran them.

Another disadvantage of assembler programs is that the power of each instruction is normally considerably limited, being constrained to very elemental operations. A simple calculation such as y = 1/x, or y = sin(x) may translate into many assembler instructions, as will be shown below.

7.3.4 High level languages, compilers and interpreters

High level languages (HLLs) deal with these problems. Examples of HLLs include C, C++, Java, Pascal, Prolog, Visual BASIC and so on. These languages provide a standardised set of programming constructs (or 'syntax') which is common across a range of processors. The constructs do not refer directly to machine components such as registers, but operate at a higher, more generalised level, where the details of the processor are not visible and are therefore irrelevant. This 'abstraction' from the details of the processor also allows an apparently simple HLL instruction to be translated into a large number of elemental assembler instructions. This is helpful, since the details of the implementation of a program statement such as y = sin(x) on a particular processor are of no particular concern to the programmer. He or she thus writes in one, standardised language which is available for a number of processors, thereby dealing with the portability problem. There is a separate 'translator' for each different processor which produces the appropriate machine code. A compiler and an interpreter are examples of such translators. The example below shows how a simple instruction written in a typical HLL may be translated into assembler code by a compiler.

Suppose the HLL instuction is: Z = X – Y; here, 'variable' Z has written into it the result of the subtraction of 'variables' X and Y. A *variable* in a computer program is a memory location which holds the value of a number used in (or resulting from) a calculation. It usually corresponds to a term in an algebraic equation, as the example above shows. This particular instruction might be translated into the following assembler code by a typical compiler:

```
load   *AR0(1), DR0    ;put the contents of number X into
                       DR0
load   *AR0(2), DR1    ;and the contents of Y into DR1
sub    DR1, DR0        ;do the subtraction, leave the
                       result in DR0
store  DR0, *AR0(3)    ;copy the result into the memory
                       location
                       ;allocated to variable Z.
```

In this example, the variables are stored in a block of consecutive memory locations and the address of the beginning of this block would have been loaded into address register AR0 at an earlier point in the program. The construct *AR0(2) means: 'use the contents of address register AR0 as the base address to be used to locate the block of memory where the variables are stored. Add two to the contents of AR0, to give the address *within* the block for this particular variable (Y in this case).'

It is clear that one simple instruction in the HLL translates into several assembler statements. A statement such as y = sin(x) might translate into many more lines of assembler code. This illustrates the labour saving gains provided by the HLL. It takes about the same amount of time to write and find errors (or 'debug') a line of program code, no matter what programming language is used, whether it be assembler or HLL. Therefore the use of HLLs saves a lot of time for programmers. Note however that this may not translate into programs which run faster on the processor. One of the advantages of assembler programming is that the programmer can fine tune the operation of the program as it will run on the processor, because he or she can manipulate the processor architecture directly. The tuning may be to do with increasing execution speed or to reducing the amount of memory used by a program, but as indicated above, this will incur a cost in reduced portability.

To prepare a program for a compiler, the programmer writes HLL program statements into a 'source file', using an editor, and then runs the compiler. This checks the contents of the source file to see whether the HLL program obeys the grammatical rules (or 'syntax') of the language. Any errors are reported to the programmer, and the process is suspended until these errors are rectified by using the editor. When the compiler is satisfied that the program obeys the syntax rules, it translates the HLL statements into equivalent assembler mnemonic code. It then uses an assembler to generate the binary machine code which will run in the computer.

Sometimes the program will use pieces of machine code which have been compiled at an earlier time, and perhaps provided with the compiler. For instance, many of the utility functions provided with the compiler for doing things like printing the results generated by programs on the computer's screen, or drawing graphics images fall into this category. So are mathematical functions such as 'log' and 'cos'. There is no point in compiling each of these functions every time they are used (some of them are very large), so they are often provided in a *pre-compiled* form, held in function 'libraries'. It is the job of a program provided with the compiler known as the *linker* to search these libraries to find the machine code for the functions which the compiler discovers are used within the program. The linker then appends this machine code to that generated by the compiler from the programmer's HLL code, ready for loading into the computer's memory.

An *interpreter* is also a kind of translator. In this case, the HLL language statements are typed directly into the computer's memory (although they can also be stored in a disk file as well, in most cases). When the program is ready to be run, the interpreter scans each line of the HLL program, translates it into machine code and immediately runs the code on the computer. This is done line by line – in other words, the translation is done *as the program is run*, and each time it is run. This gives a very satisfactory feeling of immediacy to the programmer – any changes made to the program take immediate effect. However, the scanning and translation are lengthy processes, so the program runs dramatically slower than an equivalent compiled program. This is because of the previous translation to machine code in the compiled case. The execution speed of the compiled program can be comparable to that of assembler code. Many dialects of the BASIC HLL are interpreted systems.

7.4 The representation of data in a computer

It is clear from earlier explanations that numbers to be manipulated by programs are represented as *binary* numbers in the computer, as indeed are the instructions. However, numbers come in a variety of shapes and sizes – negative numbers, integer (or whole) numbers, fractional numbers, exponential forms such as 3.664×10^{-9} – so how can these number formats be represented in binary form?

7.4.1 Representing integers

The integer representation has already been used in this chapter. Table 7.1 shows how some examples of integer binary numbers are constructed.

As Table 7.1 shows, to construct the binary representation of 97 (for instance), we need to add 64 + 32 + 1. This is specified in the binary representation by marking the relevant columns with a '1', all others being set to '0'. Thus 97 as an 8-bit binary number is 01100001 and as a hexadecimal number, this is 0x61.

Table 7.1 Binary representation of positive integers

128	64	32	16	8	4	2	1	Decimal equivalent
0	0	0	0	1	0	1	0	10
0	0	0	0	1	0	1	1	11
0	0	0	0	1	1	0	0	12
	...	...	...	...	...	...	...	...
0	0	0	1	0	1	1	1	23
0	0	0	1	1	0	0	0	24
	...	...	...	...	...	...	...	...
0	1	1	0	0	0	0	1	97
0	0	0	0	0	1	0	0	4
0	0	0	0	0	0	1	1	3
0	0	0	0	0	0	1	0	2
0	0	0	0	0	0	0	1	1
0	0	0	0	0	0	0	0	0

7.4.2 Representing negative numbers

How do we represent negative numbers in this binary scheme? Let us extend the lower part of Table 7.1 in a logical, sequential manner.

Table 7.2 Binary representation of positive and negative integers

128	64	32	16	8	4	2	1	Decimal equivalent
0	0	0	0	0	1	0	0	4
0	0	0	0	0	0	1	1	3
0	0	0	0	0	0	1	0	2
0	0	0	0	0	0	0	1	1
0	0	0	0	0	0	0	0	0
1	1	1	1	1	1	1	1	–1
1	1	1	1	1	1	1	0	–2
1	1	1	1	1	1	0	1	–3
1	1	1	1	1	1	0	0	–4
1	1	1	1	1	0	1	1	–5

Here, we move from one row downwards to the next by *subtracting* 1 from the binary representation of each row. This progression is clearly seen by moving down the rows, following the decimal number sequence 3-2-1-0 in the table. *Applying the same logic* to the row for decimal 0, we see that the next row (–1) would be all 1s. (Subtract binary 1 from binary 0 is 1, borrow 1, across all bit positions in the number.) The sequence continues in this same logical fashion through the other negative numbers.

We see that a negative number is characterised by having the *most significant bit* (msb) set to a logic 1. The msb is the left-most bit in Table 7.2. Each binary representation of a negative number as shown in Table 7.2 could, of course, be described by a hexadecimal number. Thus –4 as an 8-bit number would be represented by 0xFC.

The form of negative number representation shown in Table 7.2 is known as the *2s complement representation* of negative numbers, or sometimes rather more loosely as the *signed integer representation*. We can move up the table (i.e. in the positive direction) simply by adding 1 to the binary numbers, and we can move down (in the negative direction) by subtracting 1. This is a natural mode of operation for digital circuits. For instance it would be quite easy to design a sequential circuit (a counter) to add or subtract 1 in this way. For this reason, 2s complement numbers are widely used to represent numbers in digital systems, including computers.

You will have noticed that we have added a degree of ambiguity in the binary representation of integers. A '1' in the most significant bit position could now indicate a large positive number, or alternatively, a negative number. For instance, the 8-bit number 11111100 (0xFc) could either represent +252, or –4, depending on whether we regard the binary representation as a (signed) 2s complement number (–4), or as an unsigned direct binary number (+252). In fact, either interpretation is valid, and the programmer has to make the appropriate interpretation, depending on the context in which the number is used. To help with this process, many high level languages allow the programmer to specify from the outset whether a number is to be interpreted as 'signed' or 'unsigned'.

We described the process above of generating the 2s complement representation of a particular negative number as a process of extending Table 7.1 downwards by successively subtracting 1 from each entry. Whilst this will always work, as a method, it will be impossibly tedious for large negative numbers. Imagine trying to find the 2s complement representation of the 16-bit number –600 by extending the table downwards! We need a more concise method – a rule of thumb. Fortunately one is available. It works as follows:

Write down the binary representation of the corresponding positive number. For –600 this would be:

0000001001011000 (+600)

Invert all of the bits:
1111110110100111

Add binary 1 to this result:
1111110110101000 (–600)
(0xFDA8)

7.4.3 Floating point numbers

With a little more work, it is also possible to use binary digits to represent a number with a decimal fraction, such as –10.654 or 2.345 – so called 'floating point' numbers. One scheme commonly used in digital signal processors is illustrated below.

The binary word representing the floating point number is partitioned into two main parts: the 'exponent' and the 'mantissa', which is further divided into 'sign bit' (S) and '.fraction' fields.

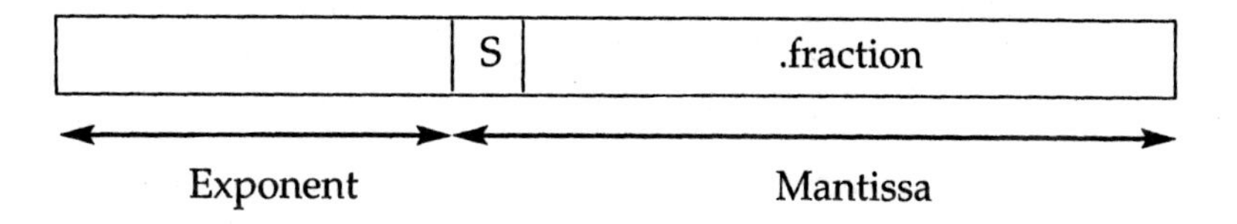

The binary digits in the '.fraction' field are used a little differently compared with the integer representations we have seen so far. If the most significant (left-most) bit of this field is set to a binary 1, this is taken to represent the (positive) fractional number 0.5. The next most significant bit (to the right) represents half of this, 0.25, the next 0.125, and so on. Thus for an 8-bit mantissa, the fractional number 0.75 would be represented as 11000000, whilst 0.625 would be 10100000.

The complete floating point number is represented by the equation:

$$\text{Number} = \{(-2)^S + .\text{fraction}\}2^{\text{exponent}}$$

The values of the 'sign bit', S, can only be 0 or 1 (it is a binary digit). Thus $(-2)^S$ will have only two values: 1 (for S = 0) and –2 (for S = 1). The '.fraction' field always evaluates to a positive fraction, as shown above, and its value will always be less than 1. This positive fraction will therefore be added to 1 (with S = 0), to give a total number which lies between +1 and +2. With S = 1, the positive fraction will be added to –2, to give a total value between –2 and –1. The exponent is a *signed* (2s complement) number, so the result of the above addition is multiplied or divided by powers of 2. This allows a full range of numbers to be represented, including those in the range between –1 and +1.

Thus a number such as 0.75 would be represented by the binary number 1111111101000000 (0xFF40 assuming 8-bit fields for exponent and mantissa), which corresponds with S = 0, .fraction = 0.5 and exponent = –1 (divide by 2).

We now have some idea of the way in which instructions and data are represented in the computer's memory. We are therefore ready to look at the way in which sequences of instructions – programs – are constructed.

7.5 The components of programs

Computer programs can be large, complex structures, sometimes consisting of millions of individual machine instructions. It is therefore perhaps surprising to discover that such programs are built from a very small number of fundamental programming constructs in most normal computer languages. It is the purpose of this section to introduce these constructs, and to describe the facilities within the processor which support them.

7.5.1 Sequence within programs

We have in fact met one of these constructs already: *sequence*. The instructions are placed in a simple linear list in memory, and the processor executes the instructions one after the other, as it finds them there. This is the kind of operation that was described in the fetch-execute cycle described in Section 7.2 above.

It is often useful to have a diagrammatic way of representing the operation of instructions in a program – you may be familiar with the use of flowcharts for instance. We will use a form based on the Design Structure Diagram (DSD) in illustrating our examples. Figure 7.4 shows the different forms for representing *sequence* in DSDs.

The two diagrams in Figure 7.4 are alternative forms for the same program structure: the sequence of program steps A→B→C→D. Notice that, as shown in Figure 7.4(b), DSD diagrams can 'grow' sideways. This makes them more adaptable than conventional flow charts, which are difficult to extend.

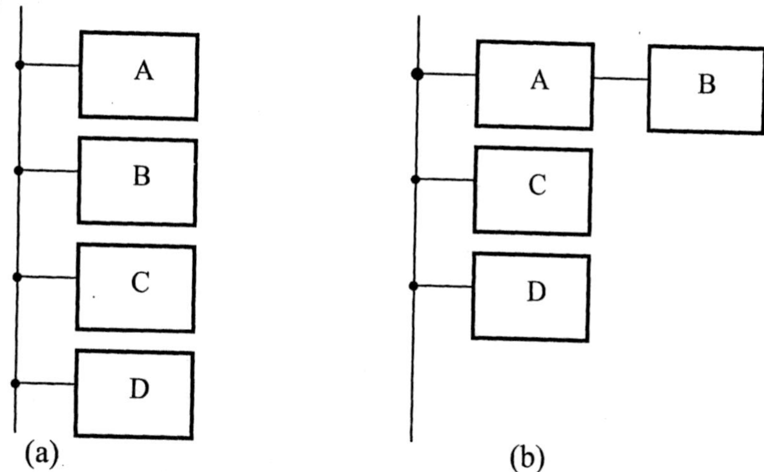

Figure 7.4 DSD representation of sequence in a program.

7.5.2 Choice within programs

The second of the fundamental programming constructs is *choice*. The program evaluates some condition or result arising from its progress thus far (see below for more details of this). Depending upon the result of this evaluation, it chooses between two pieces of code which will define its next course of action. Figure 7.5 shows a DSD illustrating this process of choice in an example drawn from 'everyday life'.

Figure 7.5 illustrates another aspect of the DSD, in addition to the choice construct. The use of the round-edged box provides an explanation of the sequence of constructs attached to it. In this case, we are asking if it is cold *because* we are deciding what to wear today. To look at it another way, a square-edged box describes an action or instruction to be carried out. A round-edged box does not; it merely provides explanation. It is therefore possible to include documentation of how and why a

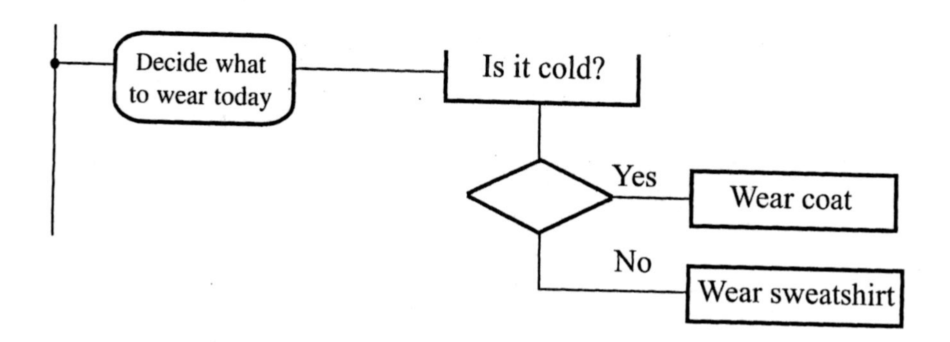

Figure 7.5 DSD for the 'choice construct'.

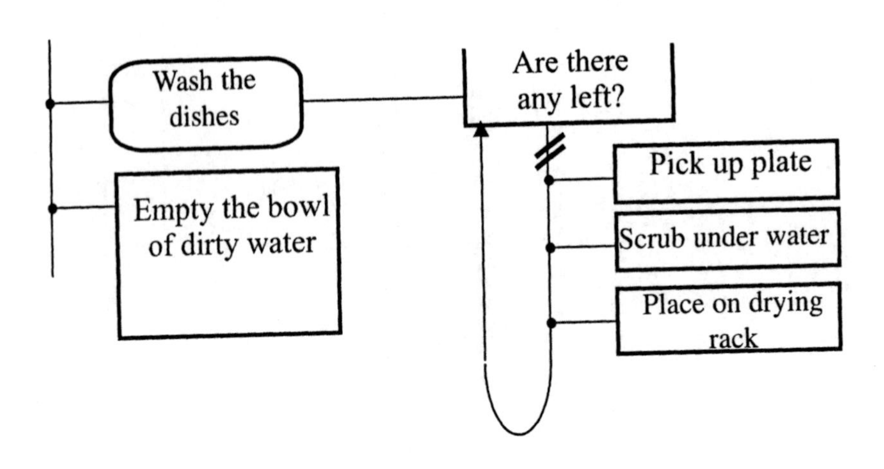

Figure 7.6 DSD representation of program looping.

program operates as it does, within the DSD. This can be very useful if the programmer needs to refer back to a program some time after it was designed and coded. This documentation facility within DSDs is another advantage compared with conventional flow charts.

7.5.3 Iteration within programs

The third important construct is *looping* or *iteration*. In a looping construct, a sequence of operations or instructions is carried out repetitively. Each time around the loop (i.e. on each repetition), a condition is evaluated and if the condition is found to be false, then the loop is terminated and the next operation in the overall sequence (Empty the bowl in the example shown in Figure 7.6) is carried out. In this case, the condition is 'Are there are any left?'

The evaluation can be made right at the beginning of the loop (indicated by the two small parallel lines there), in which case the contents of the loop may not be executed even once. Alternatively the evaluation can be carried out at the end of the loop (the two parallel lines would then be moved to the bottom of the loop), in which case the contents of the loop are guaranteed to be executed at least once before the evaluation takes place. The programmer has to decide which of these two options suits the logic of the program best. Most programming languages provide both options.

7.6 Conditional branching and the program status word

The typical processor contains facilities which directly support the choice and iteration constructs described in the previous section. These facilities comprise principally the *program status word* (PSW) and *conditional branching mechanisms.*

The PSW is a specialised register in the arithmetic unit whose individual bits are dealt with by the processor in a way which helps it in the decision-making process, which is described below.

During the execute cycle of an instruction, the processor sets the *individual bits* of the PSW according to the result or outcome of the instruction. These individual bits are known as *condition flags,* or *status flags.* For instance if an instruction adds two numbers together, and the result is zero, then the processor automatically sets the *zero flag* so that subsequent instructions can be made aware that this result was zero. A typical processor contains as a minimum set, a zero flag, a negative flag (result of previous operation is negative), a carry flag (two numbers added together resulted in a number which was too big to fit into the allocated space) and an overflow flag (indicating that two positive numbers were added together which resulted in the most significant bit being set, leading to the danger that the programmer might interpret the result as being negative).

These flags are used in conjunction with *conditional branch instructions.* Together, the flags and the conditional branch instructions implement the decision-making capability of the processor.

A branch instruction simply loads a binary number into the Program Counter. You will recall that the Program Counter is an address register, whose contents define the location in memory of the next instruction to be carried out. The binary number (the branch address) is built into the machine code of the instruction by the assembler. The branch address is used by the program to leave its present position in a sequential list of instructions, and move to another place in memory to pick up its next instruction. Subsequent program execution will continue in a sequential order from that address.

There are unconditional and conditional branch instructions in the instruction set of a typical processor. The *unconditional* branch *always* goes to the new (branch) address for the next instruction. The mnemonic for the *conditional* branch specifies which particular condition flag should be inspected by the processor during

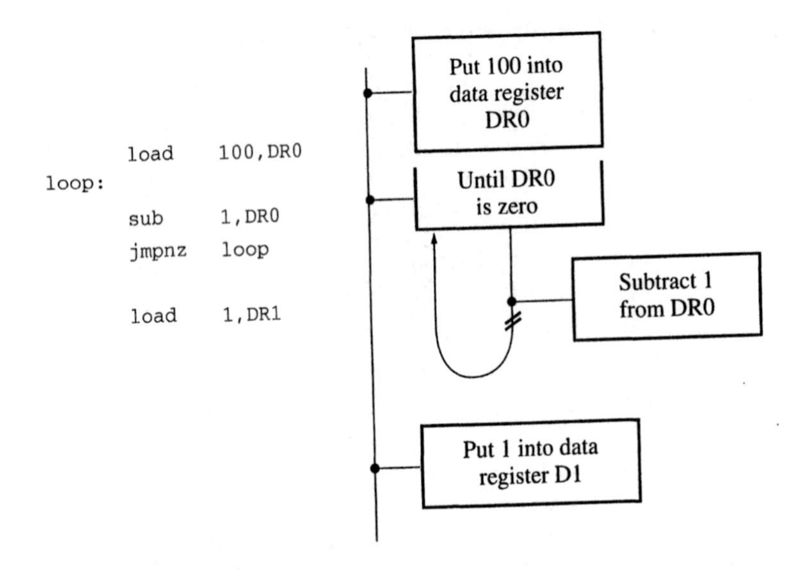

Figure 7.7 An assembler program containing a conditional branch, and its DSD.

the execution of the conditional branch instruction. If the flag is set in the condition specified by the instruction, the branch to the new address is taken. Otherwise program execution continues with the instruction which *immediately follows* the branch instruction in memory. The example shown in Figure 7.7 will help to clarify this. The program shown in the figure sits in a loop, decrementing the contents of a register by 1 each time. It exits the loop when the number in the register reaches zero.

The program ends with DR0 set to zero, reduced from its initial value of 100, and with DR1 set to 1. The instruction mnemonic 'jmpnz loop' is a conditional branch, and it means 'jump to the instruction located at memory location indicated by the label "loop", provided that the result of the previous instruction (sub 1,DR0 in this case) was *not* zero'. The zero flag of the PSW will be inspected by the processor during the execution of this instruction each time the program passes around the loop, in order to decide when to exit the loop.

The assembler will insert a binary number into the machine code of the conditional branch instruction corresponding to the *address* of the instruction at the label 'loop'. The assembler keeps track of these addresses for the programmer, so that s/he does not need to work out these absolute addresses for the branch instructions. Instead, s/he can conveniently refer to the addresses symbolically (e.g. 'loop'), as the example in Figure 7.7 shows.

These conditional branch instructions have a particular importance in our study of computers. Although a computer has no real 'intelligence' in the conventional cognitive sense, the use of conditional branch instructions allows the computer to make a *decision* regarding its next course of action, given the data it currently has. This is the fundamental origin of the flexibility which the computer uses to analyse and navigate complex problems.

7.7 Subroutines and stacks

It is necessary sometimes in the design of a program to include an identical piece of code at several different places in the body of the program. This might occur for instance when identical processing operations need to be applied to data which becomes available at several different points in the program. It would be quite feasible to write a copy of the appropriate code at each of the several points where it is required. However, it might be more convenient to have a single copy of the code, and call it up wherever it is needed. Only one copy of the shared code would need to be held in memory, resulting in efficient use of this valuable resource. This single copy of the code is called a *subroutine*. In some languages it is known as a *procedure*, and in the language C, it is known as a *function*. The purpose of this section is to describe the mechanisms in a typical processor which support the use of subroutines.

7.7.1 The subroutine call

A special instruction is usually provided in the instruction set, often given the mnemonic 'call' which causes the processor to branch to the subroutine code, whenever it is needed. The address in memory of the subroutine's first instruction is included by the assembler in the machine code of the call instruction. This causes the processor to branch to the subroutine whenever it encounters the call instruction carrying this address. When the execution of the subroutine is complete, the processor returns to the instruction in memory *immediately following* the initial call instruction. This is illustrated in Figure 7.8 below.

Each of the boxes in Figure 7.8 represents a block of instructions in a program. The same piece of program (the subroutine) is needed after blocks 1 and 3. Therefore the last instruction in each of these blocks is a call instruction, causing a branch in program flow to the first instruction in the subroutine. When the subroutine completes its execution for the first time, program flow returns to the first instruction in block 2. The second time it is called, after the end of block 3, it returns to the first instruction in block 4.

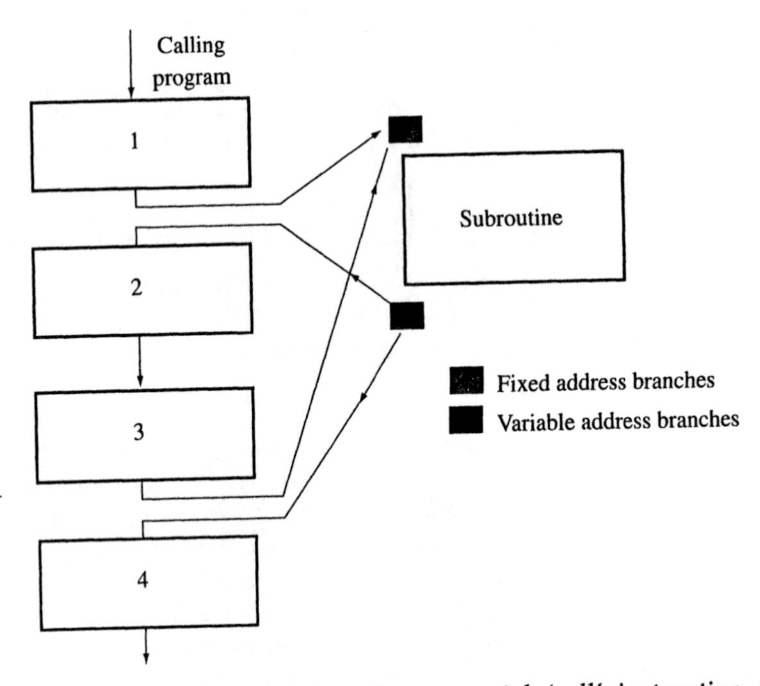

Figure 7.8 The use of a subroutine in a program.

You might be wondering why a special 'call' instruction is needed at all. In particular, why is it not possible to use ordinary branch (jump) instructions to go to the subroutine? The answer lies in the way in which the program flow must return to the main 'calling' program at the end of each call of the subroutine. By looking at Figure 7.8, you will see that the address used to cause the branch at the *end* of the subroutine is different each time it is used. In one case it returns to instructions at the beginning of 2, in the other case to those at the beginning of 4. *The branch address used at the end of the subroutine therefore potentially changes each time it is used.* It would be difficult, in most cases impossible, to change the address in the machine code instruction of a normal branch instruction (after all, it might be burned into ROM). It is therefore impossible to use these conventional branch instructions to terminate a subroutine. The purpose of the call instruction is to set up a *variable address branch process* at the end of the subroutine, using an alternative mechanism. This mechanism is known as a *stack*.

7.7.2 The stack mechanism

Figure 7.9 shows how the stack is used to control this variable address branching in a program which contains two subroutines. The first subroutine is called from the main program and the second is called from within the first. These are called *nested sub-*

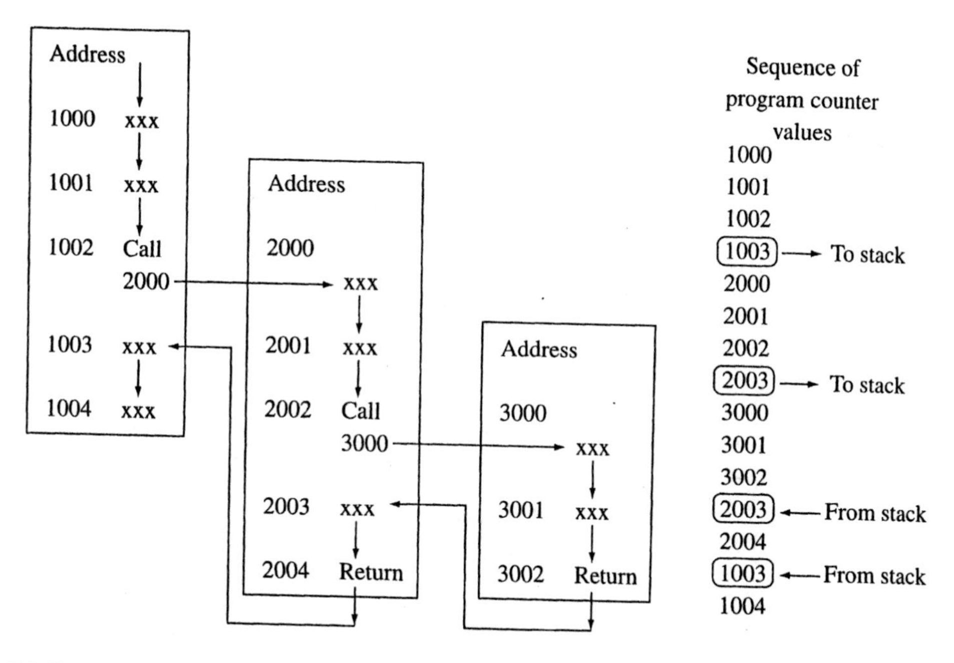

Figure 7.9 Program counter values for nested subroutines.

routines. The diagram shows the outline of the program mnemonics, the sequence of program counter values, and the content of the stack. In order to simplify the discussion, let us assume that the main program starts at address 1000, subroutine 1 at 2000 and subroutine 2 at 3000.

On the 'outward' journey through the subroutines, whenever the processor encounters a 'call zzzz' instruction, it stores the *address* of the *next* instruction it would have carried out following the call instruction in a safe area in memory – the stack. The way in which the stack is placed and managed in memory is described below. Next the processor forces the value zzzz into the program counter, so that program flow branches to the instruction located at address zzzz (the beginning of the subroutine). In the example shown in Figure 7.9, this happens twice, so that we have addresses 2000 and 3000 loaded into the program counter by call instructions. We therefore have address values 1003 and 2003 located in the stack.

You will notice that each subroutine is terminated by a 'return' instruction. This instruction implements the variable address branch mechanism described above. The execution of the *return* instruction causes the processor to extract the *last* address stored in the stack and to force this into the program counter. Therefore program flow branches to this address, which corresponds to the instruction which immediately follows the initiating subroutine call, as shown in Figure 7.9. In the example shown in the figure, this happens twice on the 'return' journey back through the two subroutines. The information (addresses) stored on the stack thus controls the variable address branch mechanism required for the subroutine.

7.7.3 The push and pop instructions

It will now be clear that the stack acts as a 'safe haven' for the storage of the return addresses. It can also be used to store other useful information used by the subroutines and by the main program. For instance, if the main (calling) program and subroutine 1 both use data register DR0, then information placed in this register by the main program will be *overwritten* by the subroutine. This could cause a malfunction in the main program when it resumed following the execution of the subroutine, if the original information in DR0 was critical to the subsequent operation of the main program. Most processors therefore provide instructions which allow the contents of registers to be stored in the safe haven of the stack, alongside the return addresses. The mnemonic 'push' is often given to these instructions and one therefore speaks of 'pushing' information onto the stack. For instance, if we wished to preserve the contents of DR0 on the stack, so that subroutine 1 could use this register safely, then the first instruction in subroutine 1 would be 'push DR0'. The reverse process, collecting information from the stack and reinstating it in a register, is called 'popping' the information from the stack. This is because the mnemonic for the instruction carrying out this operation is typically something like 'pop DR0'.

An extension of this idea of using the stack for the storage of data is often used to pass data into and out of subroutines, between the calling program and the subroutine. Typically the subroutine is asked to carry out some processing on an item of data, on behalf of the calling program (which may possibly itself be a subroutine). In these circumstances, the data (called an 'argument', or 'parameter' in the jargon) is pushed onto the stack by the calling program. It is popped off by the subroutine, when it carries out the required processing.

Although the stack is clearly a valuable mechanism for use in connection with subroutines, it needs to be used with care, since programming errors can occur if programmers do not fully understand how the stack is being used.

7.7.4 An example of the use of subroutines

The example in Figure 7.10 shows how the stack is used and controlled in a program where a number is passed into a subroutine on the stack for processing. See below for further explanation.

Calling program: 'main'

```
load    SP,10000     ;start the stack at address 10000
...
push    D0           ;put the data on the stack for pro-
                     cessing in sub1
call    sub1         ;call the subroutine to do the pro-
                     cessing
pop     D0

(used the data now - pop it off the stack and throw it
away)

...                  ;continue in 'main'
```

Subroutine 'sub1'

```
push    D1           ;save contents of D1
push    D2           ;and D2

(It is assumed that D1 and D2 are used by both 'main'
and 'sub1')

load SP(+3),D1

(step around everything on stack to get input data orig-
inally contained in D0 into D1)
...                  ;process the input data
call    sub2         ;for further processing
pop     D2           ;re-instate D2
pop     D1           ;re-instate D1
return               ;back to 'main'
```

Figure 7.10 illustrates how the stack operates with this program. The calling program, 'main', starts by loading an address register known as the stack pointer (SP) with the address 10000. The function of the stack pointer is to show the processor whereabouts in memory the stack is located, in this case, starting at address 10000. The instruction 'push D0' in 'main' places the

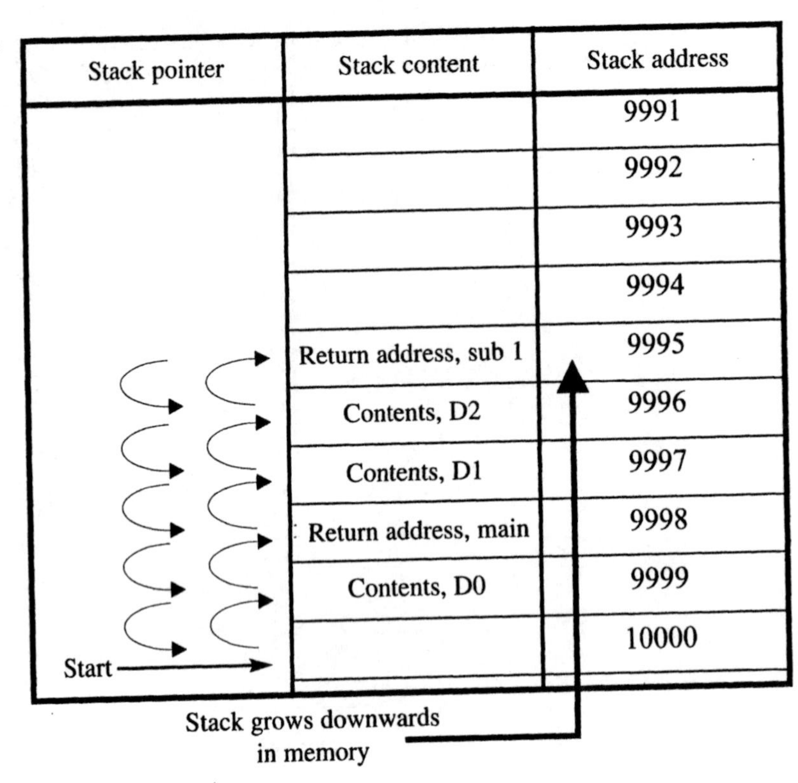

Figure 7.10 Operation of the stack for the programming example on page 213.

contents of D0 onto the next free location on the stack (9999), and updates the stackpointer (by subtracting 1 from its current contents) so that it points to the newly occupied location on the stack.

The 'call sub1' instruction uses the stack in a similar way. It stores the contents of the program counter onto the stack (this will be the return address to 'main', where processing will resume when the subroutine is complete) and again updates the stack pointer (subtracts 1) so that it continues to point to the last occupied location on stack. All of the 'push' and 'call' instructions operate in this way, so that the contents of the stack typically grow *downwards* in memory, starting (in this example) at address 100000.

By the time that 'sub1' wishes to collect the information transmitted by 'main', the information originally contained in D0 is overlaid by a return address (to 'main') and the contents of two data registers (D1 and D2) placed further down the stack. The instruction 'load SP(+3), D1' therefore temporarily adds 3 to the current contents of the stackpointer (SP) and uses the result to point to the address in memory which it should access to fetch the data. Referring to Figure 7.10, you will see that this is the location used by 'push D0' in 'main'.

When all of the processing in the subroutines is complete, 'pop' instructions are used to re-instate the contents of registers D2 and D1. A 'pop' instruction uses the current contents of the stack pointer to indicate the last occupied location on stack. It then transfers the information from the stack into the data register specified in the 'pop' instruction. Finally it updates the contents of the stack pointer so that it continues to point to the last occupied location on stack. Since we have now consumed information from the stack, the last occupied location now lies one location *higher* in memory. The 'pop' instruction therefore adds 1 to the current stack pointer content so that it continues to point to the last occupied location on the stack. The stack consequently recedes to higher memory locations as information is consumed from it.

A 'return' instruction uses the stack in a similar way to the 'pop' instruction. The 'return' instruction takes whatever it finds on top of the stack (as indicated by the stack pointer), and puts this information into the program counter (thus defining the address of the next instruction to be carried out). It then updates the stack pointer by adding 1 to its contents, so that it continues to point to the last occupied location.

Eventually, processing will return to 'main' by the operation of the 'return' instruction in 'sub1'. At this point, the only thing left on the stack is the parameter passed into 'sub1' by 'main'. This was originally contained in register D0. This item is cleared off the stack and discarded, since it has been used and is now no longer needed, simply by 'popping' it into D0 at the end of 'main'.

Since the machine code of a *program* is usually loaded into lower addresses in memory, the *stack* is correspondingly loaded into higher addresses in memory, so that a clear address space exists between them. There is thus no danger that they will collide. If a poorly controlled stack allows the stack and program to collide, the use of the stack by subroutines will overwrite program machine code as data is written into stack memory areas, with disastrous consequences.

You will observe from Figure 7.10 that when all of the processing is complete, the stack pointer returns to its initial value, indicating an empty stack. In order to avoid the problems described in the previous paragraph, the use of a stack should always be *balanced*. There should always be as many 'pops' as 'pushes', as many 'returns' as calls, and any extra information placed onto it should be cleared off. If, for instance, we had omitted the final 'pop D0' instruction in 'main', the program would have worked as intended, except that each time 'main' is used, one *extra* location on the stack would remain as occupied. If 'main' is used repetitively under these conditions, the stack will gradually encroach downwards in memory, until it eventually collides with useful code or data, causing a program malfunction. This kind of error can be pernicious. There may be millions of memory locations between the initial location of the stack, and other useful information lower in memory. So, for millions of operations of 'main', the program apparently functions correctly. It is only when the stack finally reaches the useful information that the program crashes, to the confusion and consternation of the user (and often the programmer!).

A common way of abusing stacks in this way is to branch or 'jump' directly out of subroutines, bypassing the return and pop instructions. This would leave return addresses and push(ed) data on the stack. For this reason, amongst others, many modern programming languages do not provide a direct branch instruction, making it impossible to jump out of subroutines.

7.8 Summary

We have now completed our preliminary introduction to the operation of computers. We have looked at the internal operation of a simple computer, and have some understanding of the data structures which it might process. We understand the fundamental programming constructs from which programs are built, and the features built into the computer to support these constructs.

Virtually all processors have the kind of features described in this chapter. We therefore have an adequate understanding of the nature of the 'beast', and we are now ready, in Chapter 8, to see how computers are used as components in music technology and digital audio systems. We are also in a position to study the layout, or 'architecture' of specialised digital signal processors (DSPs) and other computer based systems used for this work.

Suggested further reading

Chamberlin, Hal (1980) *Musical Applications of Microprocessors*, Hayden, ISBN 0-8104-5753-9.

Watkinson, John (1998) *The Art of Digital Audio*, Butterworth–Heinemann, Oxford, ISBN 0-240-51270-7.

8 Interfacing: the use of the computer as a component

Overview

Chapter 7 described how the computer worked as a stand-alone component. However, few objects of any kind are useful if they operate entirely in isolation. It is their *interaction* with the environment in which they are immersed which makes them valuable. This is certainly true of computers. A machine which runs programs in splendid isolation without communicating with the outside world is in reality of no use at all.

In the case of music technology systems, interaction might mean collecting performance gestures from a musician's keyboard or other performance device, and translating this gestural information into corresponding sound through further interaction with a sound source. For digital audio systems, interaction might mean sampling a sound signal, transforming the sampled signal in some way within a program, and reconstituting the transformed signal to produce a different sound to that originally sampled.

This chapter is concerned with the business of *interfacing*, meaning the ways in which a computer can be made to interact with its surroundings. The computer thus becomes a *component* in a wider system.

We know that the computer processes information in binary form, where the individual 1s and 0s constituting the binary information are represented by voltages respectively of the order of +5 V and 0 V. If information is to be exchanged with the outside world, it will have to be converted to and from this binary form, where given the speed of most processors, the transaction will typically take place in less than a millionth of a second.

By contrast, the signals in the world of sound with which the computer might need to interact are of a very different order. Assuming that the sound has been converted into voltages using a microphone as described in Chapter 2, the strength of the signal will be continually varying, ranging from a few thousandths of a volt to perhaps a few volts. It may vary over a frequency scale ranging from perhaps 10 Hz to 20 kHz.

There is clearly a significant difference between the way in which a microphone presents audio information and the way in which the computer expects and requires the information to be presented. It is the job of the *interface* to carry out the conversion between these two perspectives. An interface is the electronic circuitry connecting the computer to the outside world, together with any software running on the computer necessary to enable this circuitry to interact properly with the computer.

Topics covered

This chapter looks at the process of interfacing from several points of view: from that of the programmer; from that of the computer and from that of the interface. It then looks at several different ways of organising interfaces, before describing some case studies in the form of the digital signal processor (DSP), a MIDI interface and a typical audio sound card for a personal computer.

8.1 Interfacing from the point of view of the programmer

The process of interfacing from the perspective of the programmer is particularly simple. In many cases, interfaces are designed so that their operation *emulates* that of standard memory locations. They are called *memory mapped interfaces*. This means that programmers can treat the interface as if it were a location in computer memory. A specific memory address is assigned to the interface, and the electronic design of the interface is carried out so that it responds to the processor as though it were that memory location. To output data *from* the computer, the programmer uses an instruction which *writes* data into the specified memory location. To input data *into* the computer, the programmer uses an instruction which *reads* data from the specific memory location. Examples of such instructions would be:

output data to an interface assigned the address 0x5000:
store DR1, @0x5000

input data from an interface assigned the address 0x5000:
load @0x5000, DR1

In both cases, the processor communicates the data via the data bus and data register DR1. The advantage of the memory mapped scheme is that the full range of facilities available in the processor for accessing memory locations (i.e. the addressing modes) is, by definition, available for communication with interfaces. The disadvantage is that the interface has to respond as quickly as a memory location would.

Since the speed of execution of programs is critically dependent on the speed of the memory (the processor spends most of its time talking to memory), the memory devices themselves are often designed to be particularly fast. Memory mapped interfaces therefore have to emulate this speed, and sometimes this is difficult to achieve. For this reason, some processors provide special machine code instructions (often given the mnemonic 'in' and 'out'), so that interactions with interfaces can take place under different (typically slower) timing regimes compared with that of memory. These kind of interfaces are called 'port mapped' interfaces.

Many processors use memory mapped interface schemes, and the principles of operation of port mapped interfaces are not very different to those of memory mapped interfaces. Consequently, the description of the operation of interfaces given in the next two sections, based on the memory mapped scheme is adequate for our purposes of understanding how interfaces work in general.

8.2 Interfacing from the point of view of the processor

We saw in the last section that a programmer can establish communication between a program and the outside world, simply by writing instructions in the program which exchange data between a memory location (the interface) and the processor, typically via a data register.

As the processor executes these instructions, the only outward manifestation of this activity is the logic signals which appear on the data, address and control buses, since these are the only point of connection of the processor to the outside world. Therefore interfacing from the point of view of the processor consists entirely of executing the program's instructions, and converting the programmer's intentions into a sequence of signals on the buses, such as those shown in Figure 8.1. It is the job of the electronic circuits in the interface to interpret these signals and to effect the transfer of data between the processor and the outside world. This is explained in Section 8.3.

The timing signals shown in Figure 8.1 are those for an input-output cycle carried out by the two instructions:

```
load     @0x5000, DR1
store    DR1, @0x5000
```

8.3 Interfacing from the point of view of the interface

The task of the interface is now clear. It is connected to the

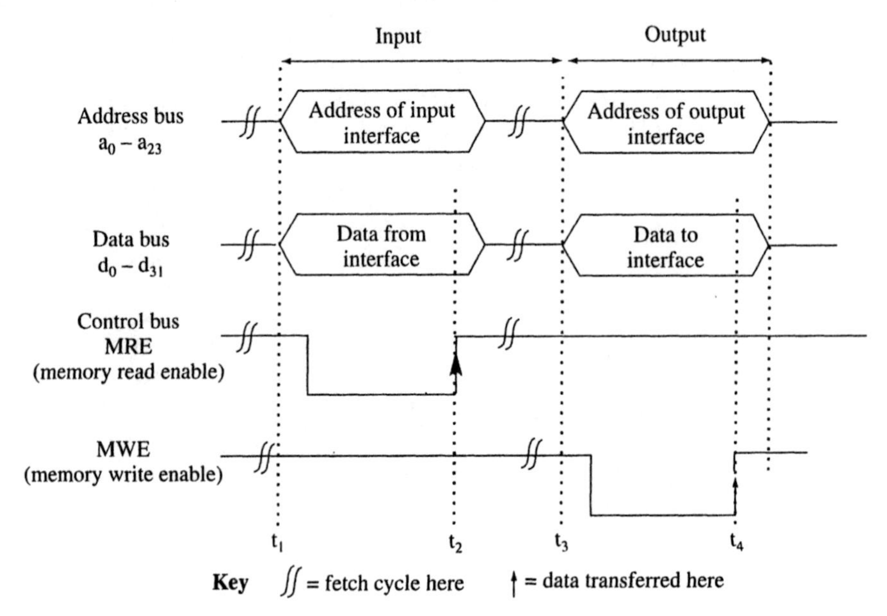

Figure 8.1 Timing signals on the processor buses, to be interpreted by a memory-mapped interface.

processor via the three buses, just like a memory device, and it must respond to the signals described in Figure 8.1. Any given interface will be assigned a specific address, and so when it sees that address on the address bus (e.g. at t_1 and t_3), it must prepare to transfer data. If the addressed interface sees the signal MRE (memory read enable) active on the control bus, indicating that the processor is expecting to read information in from the interface, it must switch the data it has onto the data bus, and hold it there while MRE is active. If, on the other hand, the addressed interface sees MWE (memory write enable) active on the control bus, it knows that the processor is writing data out to the interface, and that it must collect the information being transferred from the data bus, and store it in the interface. In both cases, the processor expects the transfer to take place on trailing edge (marked with an arrow in Figure 8.1) of MRE or MWE.

Not all processors have signals such as MRE and MRW, and even if they had, these control signals would probably be given different names, specific to the particular processor. Nevertheless, control signals equivalent in function to those described will be provided in some form. For example, many processors provide a 'data strobe' on the control bus, which indicates that a data transfer (in or out) is taking place, together with a read/write line which indicates the direction of the transfer. Typically a '0' placed on this read/write line by the processor whilst data strobe is active indicates that data is being written into the interface. A

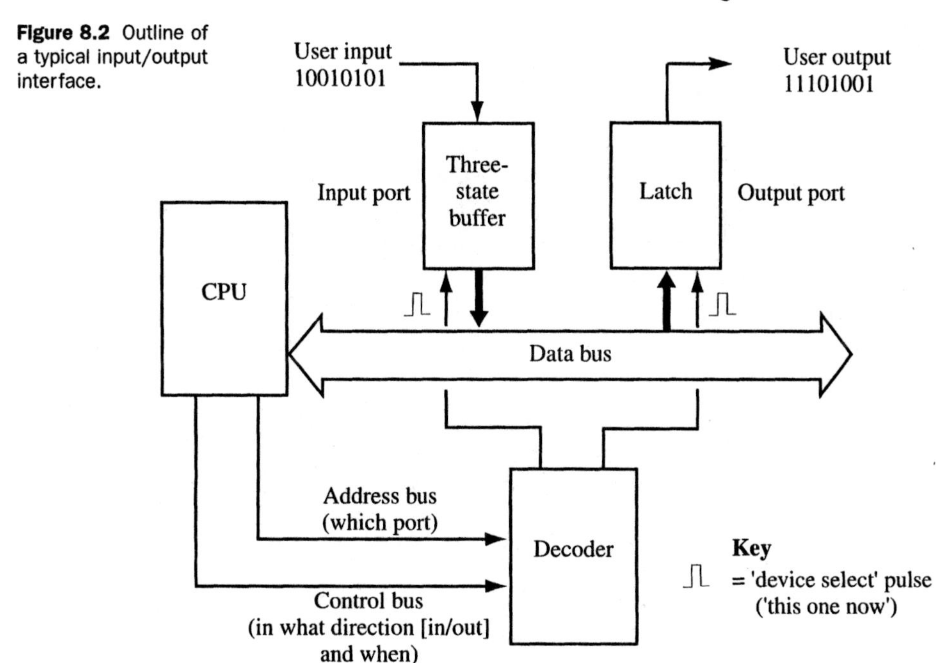

Figure 8.2 Outline of a typical input/output interface.

'1' on the read/write line indicates that data is being read from the interface into the processor. In all cases, it is the job of digital circuitry in the interface to decode these signals, and to respond appropriately. The transaction is quite fast; the width of the MRE and MRW pulses (or their equivalents) is typically about one tenth of a millionth of a second (100 ns), or even faster for modern processors. Figure 8.2 shows the outline of the circuitry used in the interface to respond to these signals.

It is the job of the decoder shown in Figure 8.2 to detect the memory mapped interface's address on the address bus, together with the memory read and write control bus signals (MRE and MWE) and to generate signals known as 'device select pulses' when these address and control bus signals are all active. The device select pulses instruct the actual interface devices (shown as 'three-state buffer' and 'latch' in Figure 8.2) to transfer data to or from the data bus, and hence to communicate data with the processor.

The input port transfers data into the processor via the data bus when a 'read' instruction such as 'load @0x5000,DR1' is executed. The three-state buffer used for this purpose is simply an electronic switch (one switch for each line in the data bus) which connects all of the individual bits in the user's input word (shown as 10010101 in the example in Figure 8.2) onto the data bus when the device select pulse is present. When the device select pulse is not present, the three-state buffer is switched off. This is the origin of the name for this device. Its outputs can have the two normal binary states (transmit a '1' or '0'), as well as a third state, 'off'.

The output port needs to *sample* the data placed on the data bus when the device select pulse is present, while the processor is carrying out an instruction such as 'store DR1,@0x5000'. The port needs to store or hold this data at its output after the store instruction is complete – that is, after the device select pulse terminates. The data would then be permanently available at the output for the user. A D-type latch (described in Chapter 6) provides the exact function which we need here. A D-type latch stores, or 'remembers' the data present on its data input when a pulse is applied to its 'store' input. The device select pulse is connected to each 'store' input of a bank of individual D-types, one for each line in the data bus. The data input of each latch is connected to the appropriate data bus line. This enables the processor to transmit a bit pattern such as 11101001 in one 'chunk', in parallel to the user output.

Figure 8.3 shows more detail of the operation of a typical

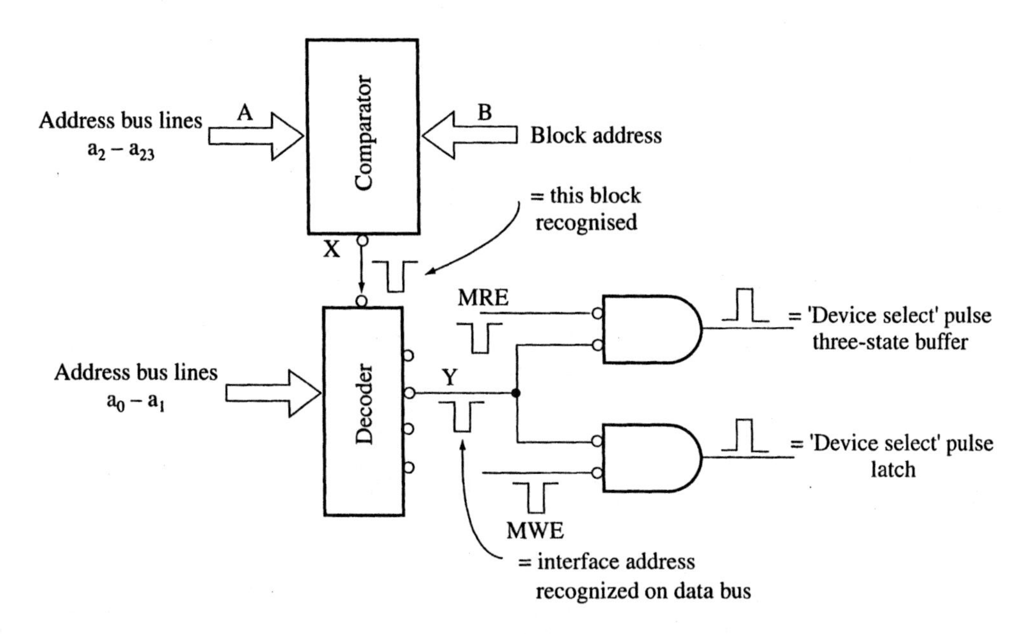

Figure 8.3 The interface decoder circuit.

decoder such as that shown in Figure 8.2. Interfaces are typically arranged into blocks of devices (latches or buffers), containing perhaps between 16 and 64 such devices. For simplicity, Figure 8.3 shows a block as containing just four of these. This corresponds to the number of outputs on the decoder. The first part of the figure which we should understand is the comparator circuit. This is the digital device described in Figure 6.5(a), except that when the binary number on input 'A' matches the binary number on input 'B' its output, X, is activated (switched to a '0' in this case).

Binary number 'B' is a fixed number, often deriving at least in part from a bank of switches located on the interface, which can be physically adjusted by the user. In this circuit, the comparator looks at the upper bits of the address bus, and when these bits match the fixed number 'B', the comparator output is activated, and this enables the decoder circuit (the lower part of Figure 8.3) to operate. When this happens, the decoder scans the lower bits of the address bus (a_0 and a_1 in this case) and switches on the output at Y which corresponds to the binary number present on these lower address bus lines. Thus if a_0,a_1 are 1,1 then output number 3 of the decoder will be switched on. If they were 0,0 then output 0 will be switched on. In the example shown in Figure 8.3, an active output from the decoder is a logic '0'.

Thus a point such as 'Y' in Figure 8.3 is a '0' only if the upper address bits match the fixed number B on the switches, and if the lower bits correspond to the decoder output number of 'Y'. A '0' on 'Y' therefore indicates that a unique address corresponding to that of the memory mapped interface is presently on the address bus. This address is placed on the address bus by the execution of instructions such as:

'store DR1,@0x5000' or 'load @0x5000,DR1'

The number 'B' corresponds to a block number, or block address, and the decoder output corresponds to an address within that block.

It would in principle be possible to have a separate, unique comparator for each interface device, and thus do away with the need for a decoder device. However, the comparators are relatively large circuits. The arrangement shown in Figure 8.3 helps with this problem, and provides useful flexibility by providing a number of outputs which can be used to produce device select pulses for a corresponding number of devices *within* each block. The signal produced at Y uniquely identifies a particular *address*, but it does not include any information about the *direction* in which the information will flow – in or out of the processor. Hence it is not sufficient by itself to activate a buffer or a latch. We need to combine the information contained within the MRE and MWE control signals, which is able to provide this directional information. The two NOR gates connected to the output of the decoders and MRE/MWE do this job. The output of these gates produce a '1' (i.e. a device select pulse) if an address is detected, *and* MRE is detected in the upper gate, to provide a device select pulse for an input device. Similarly when address *and* MWE are detected in the lower gate, it provides a device select pulse for the output device.

In a typical computer system, there will be many such input/output interfaces controlling the many facets of the system's interaction with the outside world. Although the decoding mechanism shown in Figure 8.3 makes provision for the decoding of only four addresses, the general principles outlined in the figure can be extended to accommodate a greater number of ports, as would normally be required in practice.

8.4 Parallel and serial ports

The interfaces described in Section 8.3 would normally be described as *parallel interfaces*. This means that when the processor transfers a 16 or 32 bit word to (or from) the interface, all of the individual bits are available *simultaneously* on the output (or input), each on its separate line. In many cases, this is exactly what is required. However, in other circumstances it may be more convenient to have the individual bits appear one after the other, on one single line. This is known as a *serial interface* or *serial port*.

The use of a serial interface might be convenient where the binary information to (or from) the interface needs to be sent over a long cable run, where extending the individual lines of a parallel interface would result in a cable which was bulky and expensive. The serial interface shown in outline form in Figure 8.4, with its single signal path, would be more suitable under these conditions.

The data is transmitted to and from the processor via the parallel port data registers. The transmitter and receiver registers are essentially collections of D-type latches, arranged so that the output of one D-type feeds the input of the next. Data is thus passed along the chain of (non transparent) D-types, one step for each pulse applied to the transmitter or receiver clock. The data is transferred as a parallel word between the data register and the receiver/transmitter register. The 'load' signal makes this happen in the transmitter's case (this also starts the transmission of serial output data), whilst the 'data received' signal does this in the receiver's case. This signal can also be used to notify the processor that data has been received.

Obviously, the rate of data transfer over a serial interface must be lower than that of an equivalent parallel interface, since all of the individual bits need to be funnelled one at a time down one wire. Nevertheless, many of the data communications interfaces used in digital audio and music technology are serial systems, because of the extensive cable runs often used to interconnect pieces of equipment in the studio or on stage. Examples of these audio and musical interfaces based on serial communications include AES/EBU, IEEE 1494 ('Firewire') and MIDI.

Figure 8.4 illustrates a common feature of computer interfaces. Although the circuit implements a special function (a serial interface in this case), the point of connection to the processor uses a parallel interface connected to the processor bus. This is generally true of even the most complex interfaces. CD ROM and disk drives, local area network (LAN) interfaces, audio input/output devices all connect to the processor via internal parallel interfaces built into these more complex devices. The programmer controls them by writing appropriate bit patterns into these internal parallel interfaces, using the techniques described in Section 8.1. An understanding of the operation of parallel interfaces is therefore fundamental to an understanding of interfaces in the more general sense.

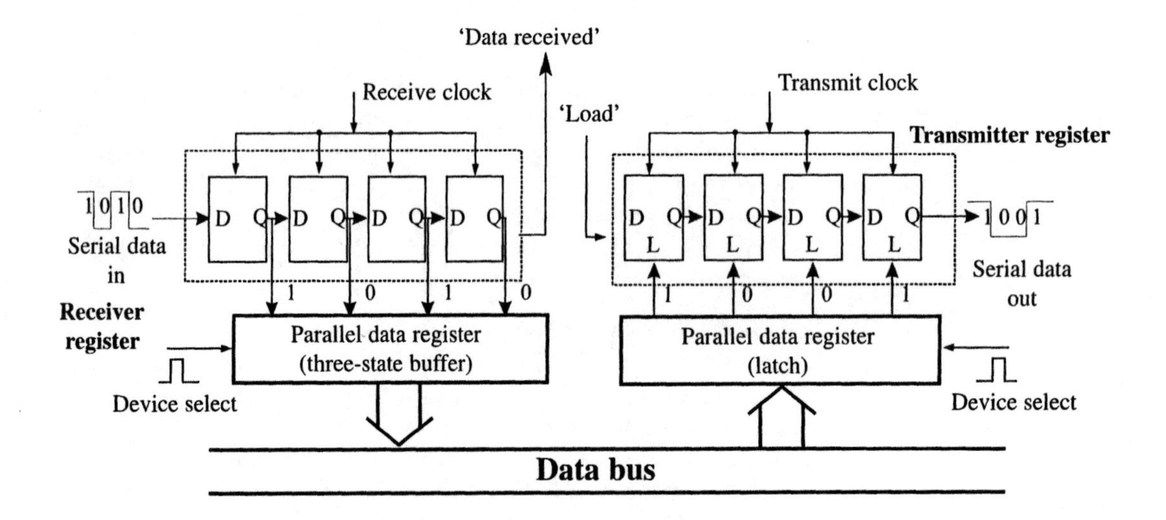

Figure 8.4 Outline description of a serial interface.

8.5 Digital to analogue, and analogue to digital conversion

The interfaces which we have looked at so far have been entirely digital. They have been concerned with inputting and outputting binary signals, which can assume only two states, 'on' and 'off'. These interfaces would be suitable for switching motors on or off for instance, or for dealing with data which is already in binary form, such as MIDI data.

However computers are also required to work with information which does not assume one of these two states, but which continuously varies over a range of values – so called *analogue* data. An important example of an analogue signal in the world of digital audio would be the voltage output by a microphone. As the sound increases gradually in intensity, the voltage increases smoothly through a corresponding range of voltages.

If a computer is to work with such signals, then clearly some sort of conversion device will be required to transform the continuously varying analogue voltage into the two-state binary world of the computer. For instance, if we wish to measure an analogue signal using a computer, we need an *analogue to digital convertor* (ADC). To generate analogue signals from a computer, we need a *digital to analogue convertor* (DAC). The purpose of this section is to understand how these interface devices operate.

Since most analogue signals can usually be converted into voltages using some sort of transducer, we will concentrate on the measurement and generation of voltages.

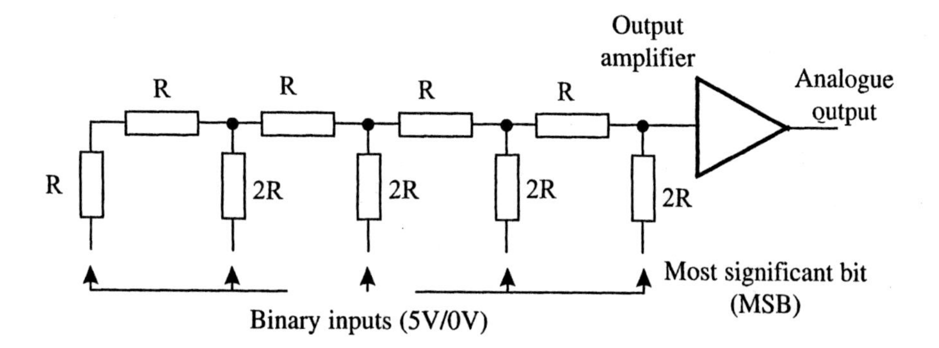

Figure 8.5 An R–2R digital to analogue convertor.

8.5.1 The digital to analogue convertor (DAC)

In many cases it is possible to consider a DAC as a kind of analogue combinatorial device. The binary input signal is connected as a number of two-state (binary) voltages to the inputs of the bottom set of resistors, as shown in Figure 8.5. The closer the position of the input binary voltage is to the output (to the right), the more effect it will have on the output and the greater the output voltage caused by that binary input will be. Conversely, the further down the chain of resistors a binary input voltage is located, away from the output (to the left), the smaller its contribution to the output voltage will be. Its effect will be reduced or attenuated by the chain of resistors lying between it and the DAC output amplifier.

In a typical R–2R DAC, the values of the resistors are chosen so that a binary 1 on the right-most input (nearest to the DAC output) will generate one half of the maximum DAC output voltage. A binary 1 on the next input to the left will generate one quarter of the maximum voltage, the next input to the left again, one eighth of the voltage, and so on down the chain of resistors. If the two most significant input bits (i.e. the two right-most bits) are set to binary '1', then three quarters of the maximum voltage will be generated at the output (adding the two individual contributions together). Different combinations of input bits will therefore generate different output voltages, because the DAC adds together the individual contributions of each bit, and each contribution is scaled to a different voltage, according to its position in the resistor chain. If there were (say) eight binary inputs to the DAC, then there would be $2^8 = 256$ combinations of binary inputs, giving rise to 256 discrete voltage levels which would be available at the output of the DAC.

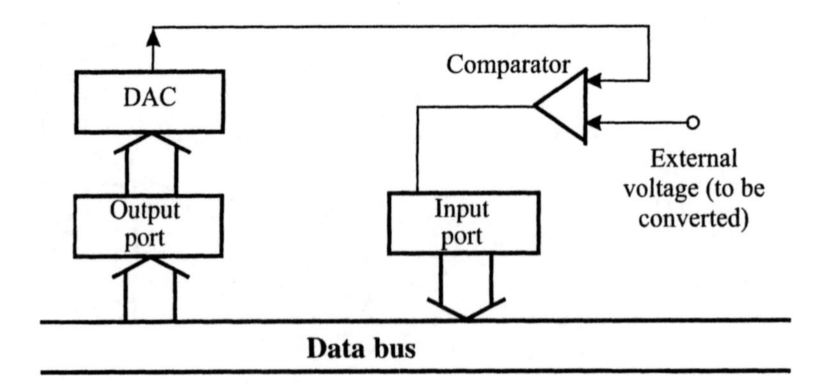

Figure 8.6 An analogue to digital convertor (ADC) interface.

If we wish to use a computer to generate an analogue voltage, for instance so that it can synthesise sound through a loudspeaker, then in principle all we need to do is to connect a DAC's binary inputs to a parallel output port, and to connect the DAC output to a suitable amplifier. The DAC will then produce a voltage which is proportional to the binary number placed in the output port by the computer. Varying output signals can thus be generated by the computer by placing a sequence of binary numbers over a period of time on the output port.

8.5.2 The analogue to digital convertor (ADC)

The converse process of converting analogue voltages to a digital representation is less direct, compared with digital to analogue conversion. It can be carried out within a program, or (more usually) in a hardware device. However, we will present an approach based on a program, since this gives a better impression of the way in which the process works.

Figure 8.6 gives an outline of the interface devices needed to operate in conjunction with the analogue to digital conversion program. You will notice that it involves the use of a digital to analogue convertor as a subcomponent. The program which would be used with this interface to carry out the conversion process works in the following way:

1 The program places an initial 'guess' of the binary conversion value in the output port attached to the DAC. The DAC therefore automatically converts this binary 'guess' into a corresponding analogue voltage.
2 The output of the DAC is routed in the hardware to the input of an analogue voltage comparator device, as shown in the figure. This compares the analogue voltage produced by the DAC with the external voltage which is to be converted into a

binary number. If the DAC output is greater than this external voltage, then the comparator produces a binary 1 at its output, otherwise a 0 is produced.

3 The output state of the comparator is made available to the program by connecting its output to an input port. The program can therefore assess whether its present 'guess' at the binary conversion value is too great or too small. The program then adjusts or refines the guess accordingly, in a subsequent guess–assess–adjust cycle.
4 The program loops, each time outputting its current (refined) guess to the DAC, and reassessing the accuracy of the guess by inspecting the output state of the comparator through the input port. The looping process stops once the DAC voltage is sufficiently close to the externally applied voltage. At this point, the binary number which has been input by the program to the DAC is a binary representation of the external voltage. This is because by definition, the DAC produces an output voltage (now almost equal to the external voltage) proportional to its binary input. Since the program generated this number, it knows the ADC conversion value *ipse facto*.

Note that to all intents and purposes it is impossible for the DAC to come up with a voltage which *exactly equals* the external voltage. The comparator can resolve two voltages to something approaching a few millionths of a volt, whereas the output of the DAC is constrained to take on a set of *discrete* voltage levels, each corresponding to a specific binary input number, as described in Section 8.5.1. The voltage separation of these levels therefore becomes more coarsely defined (widely separated) for smaller numbers of binary input bits. However, even for very large numbers of input bits, any one voltage level is unlikely to *exactly* equal the external voltage magnitude. The controlling program needs to be aware of this limitation and take it into account in the 'guess–refinement' process. For instance, the looping process could stop as soon as the DAC voltage *just exceeds* the external voltage.

There are different kinds of ADC, all of which follow the above outline scheme. The difference between them lies mainly in the way in which they refine the current 'guess' at the binary conversion value.

Perhaps the simplest refinement scheme is that used by the tracking ADC, sometimes described as a form of delta modulator. As shown in Figure 8.7(a), the initial guess is 0. Each time around the refinement cycle, the 'guess' value is *incremented* by 1, until the DAC voltage just exceeds the external voltage. The

DAC voltage therefore ramps up to the external voltage on successive refinement cycles, as shown in the figure. When the DAC voltage exceeds external voltage, the program continues to loop, but whenever the DAC output *exceeds* the external voltage, the 'guess' is *decremented* by 1, and when it is less than voltage, it is incremented by 1. The output of the comparator indicates 'up a bit, down a bit' to the refinement process. Assuming that the external voltage is constant, once on track, the DAC voltage will oscillate either side of the external voltage on subsequent refinement cycles. However, if the external voltage changes, the DAC voltage will follow it around (or 'track' it) because of the dynamic, constantly executing 'guess–refine' nature of the conversion program.

Another refinement scheme gives rise to a different kind of ADC – the successive approximation ADC (SAD), whose operation is described in Figure 8.7(b). In this case, the first 'guess' is to switch on the most significant bit of the input to the DAC. In response, the output of the DAC moves to half of its full-scale voltage in this first guess–refine cycle (see Section 8.5.1). If the assessment of the output of the comparator indicates that the voltage at the output of the DAC is too low in comparison to the external voltage, the SAD algorithm then switches on the next most significant bit. This accounts for one quarter of the DAC's full scale voltage, so that the total voltage now equals three quarters of its full scale. The algorithm cycles down all of the bits of the DAC, deciding whether it needs to be switched on or not. If at any point, the output voltage exceeds the target external

Figure 8.7 ADC operation

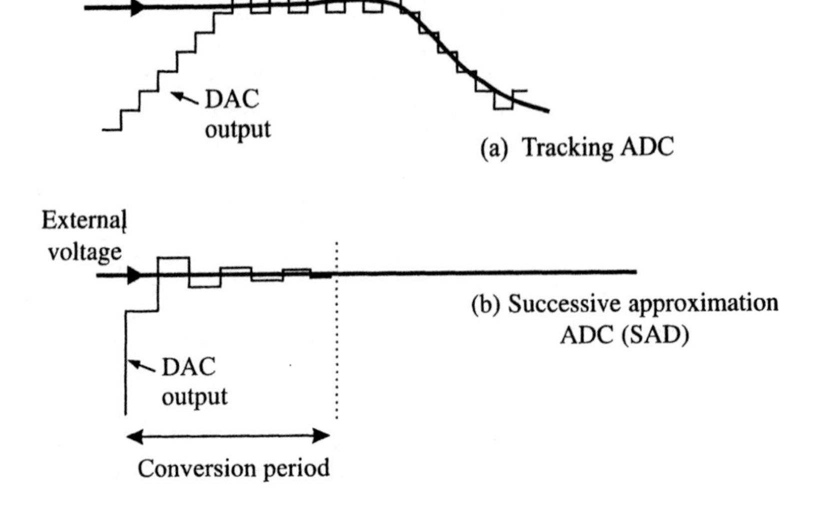

voltage, the last bit switched on is switched off (reducing the DAC output voltage), before considering the required state of the next lowest significant bit.

The SAD gets to its target voltage in as many cycles as there are bits at the input of the DAC – a fixed, relatively small number. By comparison, the tracking ADC may take many cycles to get on track. On the other hand, the SAD will always take the fixed number of cycles to get on target, since it always starts from scratch. Once on target, the tracking ADC will usually require only one cycle to stay on track, since in effect, it retains information about previous conversion values. Also, the SAD requires the external input voltage to be held steady whilst it is converting. If this were not so, decisions it made at the start of the refinement process might be inappropriate by the end of the cycle, leading to an error in the conversion result. The SAD therefore requires a 'sample and hold' circuit on its input to hold the input voltage steady while the conversion takes place.

Both types of ADC have specific advantages and disadvantages. The choice of the ADC type needs to be made in the light of the operational requirements of the system.

We have described above the operation of the refinement process or 'algorithm' for the analogue to digital conversion process. The algorithm could be implemented in software running on the processor. However it is more normal for it to be carried out in hardware, forming part of the ADC device. For example in the SAD, the conversion process is initiated by applying a digital pulse to a 'start conversion' input. The SAD then outputs a 'busy' signal whilst it is converting. The binary output should not be read during this time as it will be only partially complete.

8.6 Polled, interrupt driven and direct memory access (DMA) interfaces

It was mentioned in the introduction to this chapter that one of the jobs of the interface is to match the timing regimes of the processor to those of the external system with which it is interacting. These timing regimes may be very different. In this section we look at ways of setting up and controlling interfaces of different operational speeds. There are various interface control regimes, depending on whether the speed of operation is comparable to the speed of execution of an individual instruction on the processor, or whether it operates faster or slower than this execution rate. Different control regimes are appropriate under these various circumstances. These control regimes are known as *polled, interrupt driven* and *DMA* interfaces.

8.6.1 The polled interface

The polled interface is usually the easiest of the interface schemes to work with, because of its directness. Let us take the example of an interface using an SAD device. As indicated in Section 8.5.2, the interface program must output a 'start conversion' pulse. It must then wait until the conversion is complete, before finally reading the binary conversion value through an input port. The DSD and assembler code representation of this algorithm is shown in Figure 8.8.

The 'AND' instruction in the code in Figure 8.8 carries out a process known as *masking*. The register is pre-loaded in the first instruction with a specific bit pattern known as the 'mask'. The memory-mapped interface accessed by the program inputs the 'busy' bit provided by the SAD, alongside the 'busy' bits from

Figure 8.8 A polled interface algorithm.

```
;load mask word into DR1:
        store  0x02,DR1
;and trigger word into DR0:
        store  0x01,DR0
;start convertor:
        store  DR0,@0x1000
loop:

;get status word. SAD flag is
on second most significant bit:
        load  @0x1001,DR0

;mask status word:
        and  DR1,DR0

;jump if busy flag is set (=1):
        jmpnz loop

;otherwise bring in SAD
conversion value:
        load  @0x1002,DR0
```

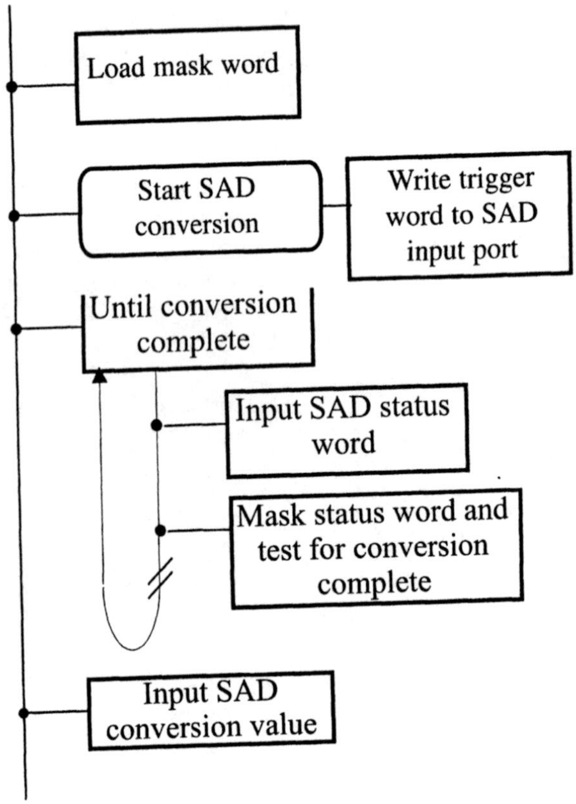

other interfaces. These 'busy bits' are known as 'status flags' in the jargon.

In a typical computer system, there may be many such status flag bits from different interfaces. Each flag is connected to a specific bit position on the status flag input port. When the computer reads this input port, it therefore receives a composite *status word* made up of individual status flags from many interface devices such as the SAD.

The program must wait until this particular bit indicates that the SAD is not 'busy' (i.e. conversion is complete), otherwise it will read an incorrect value. The AND instruction carries out a bit-by-bit AND between individual bits in the mask word and those fetched from the memory-mapped interface by the instruction. Where there is a '0' in the mask word, a '0' is written into the corresponding bit position in the result, as shown in Figure 8.9. Where there is a '1' in the mask word, the logical state of the status flag ('1' or '0') is *copied* into the corresponding bit position of the result. The AND instruction carries out two useful functions:

1 As shown in Figure 8.9, it identifies and isolates the required flag from the others which are not relevant to this interface, by matching the bit position pre-loaded with '1' in the mask word with the status bit of interest, and
2 (most importantly) the AND instruction sets the processor's internal status flags (see Chapter 7), thus forming the basis of the conditional branch in the interface program, as shown in Figure 8.8. It is this step which ultimately lets the computer decide whether the 'busy' bit is set or cleared, and hence whether the interface (the SAD in this case) is ready to transfer the data. It is the point at which the program stops continuously inspecting the status flag, and gets on with inputting the data required from the interface.

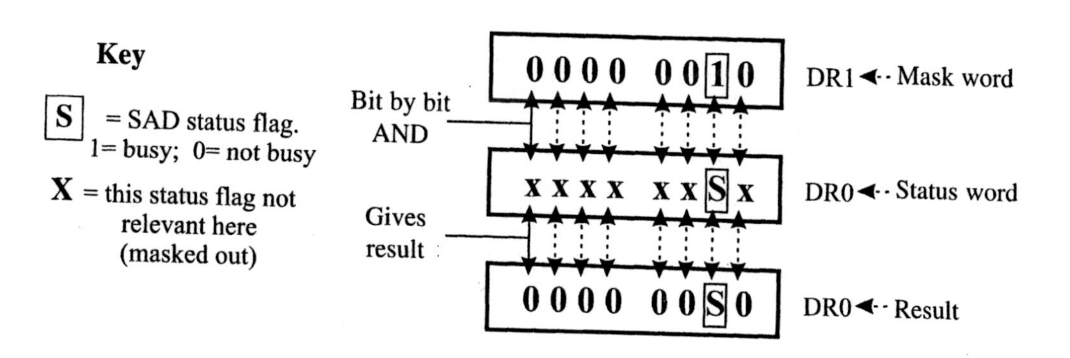

Figure 8.9 The use of the AND instruction to carry out the masking process.

It was stated earlier in this section that one of the advantages of the polled interface is its directness. Unlike other schemes described later, the programmer knows *exactly* whereabouts in his or her program the interface is tested for its readiness, and whereabouts the data is transferred. This makes the operation of a polled interface very *predictable.* This is a considerable advantage in complex programs. The disadvantage of the scheme is that the processor is entirely occupied by continuously looping and inspecting the status bit(s) whilst the interface is 'busy'. If the interface is slow in its operation, this is an inefficient use of the processor, since it is not carrying out any useful computation whilst it is continuously 'polling' the status flags in this way. This polling process is the origin of the name 'polled interface'.

Nevertheless, the predictable operation is such an advantage that the use of a polled interface should not be dismissed lightly. As a guide, use such an interface when its speed of operation (the 'busy' period) is comparable with the execution time of an instruction – a microsecond or so. Then the control program will go around the polling loop zero or one time (only), whilst securely ensuring that the data is read only when it is ready.

8.6.2 The interrupt driven interface

The interrupt interface control scheme is useful where the speed of operation of the interface is considerably slower than the execution time of an instruction on the processor. Its advantages and disadvantages are almost entirely the reverse of the polled interface. It makes efficient use of the processor for these slow interfaces, but at the cost of an element of unpredictable behaviour which must be controlled carefully. The scheme works as follows.

Most conventional processors have one (or more) inputs on the control bus known as interrupt request (IRQ) lines. If an interface places a pulse on this input, indicating that it wishes to transfer data, the following sequence of operations takes place.

1 The processor finishes the instruction which it is presently carrying out, and then pushes the *address* of the next instruction in its current (background) program onto the stack, in a way comparable with a subroutine call (see Section 7.7).
2 The processor then asserts one of the lines on its control bus usually known as 'interrupt acknowledge' (IACK) to indicate that it has recognised the interrupt request.
3 The interface which generated the interrupt pulse watches the IACK line to see when its request is recognised. When it

detects that the processor has asserted IACK, the interface switches on a parallel input port dedicated to this task, which transmits a binary number known as an 'Interrupt Vector' to the processor via the data bus. The vector may have been previously set up on the interface using a bank of switches, or it may be loaded into a latch on the interface when the system is first switched on ('system initialisation'). The vector *uniquely identifies* the interrupting interface.

4 The processor picks the vector up from the data bus, and uses it to *construct* a unique address in memory to which it now branches, rather like a subroutine call. The processor is now executing a piece of code which deals with the interface's specific requirements, often known as the 'interrupt service routine' (ISR). IACK is switched off (the interrupt has been acknowledged and is now being dealt with).

5 The ISR will normally terminate with a 'return' instruction, similar to a subroutine. When the processor finds this instruction at the end of the ISR, it takes whatever it finds on the top of the stack (which should be the address of the next instruction in the background program, left there in step 1) and forces this into the program counter. Program flow therefore returns to the point in the background program where the initial interrupt request occurred, and continues on from there.

The whole process is rather similar to a subroutine call, except that instead of being initiated by a subroutine CALL instruction embedded in a piece of program code, it is initiated by a pulse generated by an external device such as an interface. This is where many of the difficulties associated with interrupt driven interfaces arise. This pulse can, in principle, arrive at any point in the execution of the background program. The ISR can therefore be launched at any such point. The interrupt arrives like a 'bolt out of the blue' as far as the background program is concerned.

Since it is not possible in general for the programmer to predict whereabouts in the program this will occur, it is doubly necessary to make sure that all of the processor's context (such as registers, etc.) are saved in a safe place such as the stack. This will avoid the possibility of the ISR interfering with the background program by overwriting useful data. Failure to do this will result in the appearance of apparently random program errors, which can be difficult to detect and deal with. The use of incorrectly set up interrupt driven interfaces can lead to very puzzling system behaviour because of this randomness.

Nevertheless, if correctly implemented, the interrupt driven scheme is very efficient for slow interfaces. To take the example of the SAD again, the background program would send out the 'start conversion' pulse, and then continue with *useful processing*. A polling loop would not be used. A little while later the 'busy bit' would indicate that the conversion value was ready, and this signal event would be used to drive the IRQ line. A vector would be fetched from the SAD and this would be used to activate the code (the ISR) which transfers the conversion value into memory. This conversion value would be placed by the ISR at an address in memory where the background program can gain access to it when the interrupt process is complete. The processor continues with useful computation at all times. This is important for slow interfaces where a large number of passes through a polling loop could not be tolerated.

It is likely in realistic systems that a number of interfaces will be interrupt driven, many of which could be active at the same time. Whilst it is possible for 'an interrupt to interrupt an interrupt' on a typical processor, the ISR would normally start with a 'disable interrupts' (DI) instruction. This de-sensitises the processor to other interrupts during the execution of the ISR, to avoid confusion through unwanted interaction between several interrupt sources. When complete, the ISR executes an 'enable interrupts' (EI) instruction to re-sensitise the processor.

The vector uniquely identifies each interrupting interface, so no confusion should arise in identifying the interface which requires service. However there remains the question of *which* of a number of simultaneously active interfaces should respond to the processor's interrupt acknowledge (IACK) cycle. This is dealt with in the system hardware by arranging the interfaces into a priority order. The active interface with the highest priority responds to the IACK cycle by providing its vector. All other active interfaces suspend, holding their requests as pending until the prioritising hardware gives them permission to supply their vectors in response to subsequent IACK cycles.

8.6.3 The direct memory access (DMA) interface

DMA interfaces are typically used where the speed of response of the interface is much greater than the execution speed of an individual instruction. In this case, it would be a considerable disadvantage to have *any* instructions (i.e. a program) involved in the data transfer, since this would slow the process down significantly. This is especially true where large blocks of data are to be transferred from relatively fast devices, for instance from a CD ROM or disk drive.

DMA transfers are typically carried out under the control of a direct memory access controller (DMAC). This is really a specialised processor designed for the purposes of shifting blocks of data around the computer system. It works in conjunction with the normal processor, and when it operates, it takes over the control of the system buses from the main processor. Significantly, because of its dedicated task, the DMAC does not need to carry out any instruction fetch-execute cycles in order to work out what it needs to do. Its function is built directly into its hardware design. These aspects of the DMAC design mean that data can be transferred around the system as fast as the buses will allow.

The DMAC's operation typically centres around a set of registers: a source address register (SAR), a destination address register (DAR), a transfer counter and a transfer register. There is also usually a control register whose contents determine the precise mode of operation of the DMAC. All of these registers normally can be written to or read from the main processor as simple parallel interfaces (in order to set the DMAC up), with the possible exception of the transfer register.

The DMA operation works as follows: When the DMAC is enabled by the main processor (e.g. by setting a specific bit in the control register), it takes over the control of the system buses and carries out a sequence of data transfer cycles. Each transfer cycle starts when the DMAC places the contents of the SAR onto the address bus. The memory read enable control bus signal (see Chapter 7) is then asserted by the DMAC, and the memory responds by placing the contents of the addressed memory location onto the data bus. The DMAC then places this data into its transfer register. The DMAC then switches the contents of the DAR onto the address bus and also the contents of the transfer register (i.e. the data item being transferred) onto the data bus. It then asserts the memory write enable (MWE) control bus signal (see Chapter 7). This operation writes the transfer item of data into a memory address specified by the DAR.

The operation of each cycle amounts to a direct memory to memory transfer via the DMAC.

Note that 'memory' in this context could be a memory-mapped interface, as well as a conventional memory location. It is therefore possible to use DMA to transfer information from memory to memory, from interface to memory, from memory to interface or from interface to interface.

Depending on the way in which the processor initialises the

control register, the other registers are updated in various ways by the DMAC at the end of each transfer cycle. For instance, if it is intended to shift the contents of a block of data from one place to another in memory, then both the SAR and the DAR might be incremented at the end of each cycle, so that they point to the source and destination addresses to be used for the transfer of the *next* data item on the next cycle. In another application, the DAR may be incremented, whilst the SAR remains fixed. This would correspond to shifting a block of data from a fixed address, typically an interface, into a block of memory locations. If the SAR is incremented, whilst the DAR remains fixed, then a block of data would be moved from successive memory locations to an interface. Finally, fixed SAR and DAR would correspond to data transfers directly between two interfaces.

At the end of each transfer cycle, the contents of the transfer counter are decremented. This counter normally will have been pre-loaded with the *number* of the items which are to be transferred. When the transfer counter reaches zero, the DMA operation is therefore complete, and the DMAC hands control of the system buses back to the main processor. It often also generates an interrupt to alert the processor that it can resume its normal operation.

The whole operation takes place without the direct involvement of the main processor, operating under the control of its relatively slow fetch-execute cycles.

At this point in our study of computer fundamentals, we have an adequate understanding of all of the main aspects of computer system operation. We will now look at some examples of specific computers and interfaces which are in widespread use in digital audio and music technology. The examples are a digital signal processor (DSP), the MIDI interface and a PC sound processing card.

8.7 The Texas Instruments TMS320C30 Digital Signal Processor (DSP)

The intention throughout this book is to describe techniques which apply generically to devices which are found in common use. It has not been our policy to present specific processors for instance, since they rapidly become out of date. However, this DSP illustrates many of the principles described earlier in this section of the book, and is a good example of devices in common use in digital audio.

The TMS320C30 is in reality a complete computer system on one integrated circuit package, one inch square in size. As can be seen in Figure 8.10, it has on-board memory (RAM and ROM), a controller (control unit), DMAC, serial ports and timers.

A timer is a hardware counter which can be incremented from some initial value (often zero) by a regularly occurring pulse (generated internally or externally), until it reaches a maximum value which can be pre-set under program control. It then generates a pulse on an external pin, or raises an interrupt to the processor to indicate that the timer period has expired. Some timers work by counting *down* from the preset value, and raise the interrupt when the count reaches zero. Clearly, the larger the pre-set value, the longer the counter will take to reach it, and the longer the timer period will be. Using a device such as this, the processor can set a timing interval, start the timer and get on with useful work, leaving the timer to monitor the passage of time autonomously. When the time interval expires, the processor is alerted via an interrupt, and it can take the action intended for this time interval. A timer could be used to set the sample rate in a digital audio system for instance.

Referring to Figure 8.10, the arithmetic unit in a typical DSP is optimised for the intensive numeric manipulation found in signal processing. The TMS320C30 has a number of ALUs which can therefore operate simultaneously and independently. The main ALU is attached to a 'barrel shifter'. It performs common arithmetic operations (add, subtract, etc.) on signed binary (integer) and floating point data in one processor cycle. The barrel shifter is an extension of this ALU, and can shift the individual bits of an ALU input left or right by up to 32 bits in a single cycle. This could be used for pre-scaling data for instance. The ALU has two sets of inputs, and one set of outputs, as we would expect. The multiplier block could also be considered as an extension to the ALU, although it is shown as a separate block in Figure 8.10. It can carry out an integer or floating point multiplication in a single processor cycle. There is no divider susbsystem. Division of two numbers is carried out in many of these processors by multiplying one number by the inverse of the other.

As we expect, the output of the ALU can be loaded into processor registers such as the data registers (R0–R7) and auxiliary registers (AR0–AR7). The auxiliary registers are mainly used as address registers. The contents of these registers can be routed around to the ALU for processing via the 'REG1' and 'REG2' buses.

There is a second set of arithmetic units (ARAU0-1). These

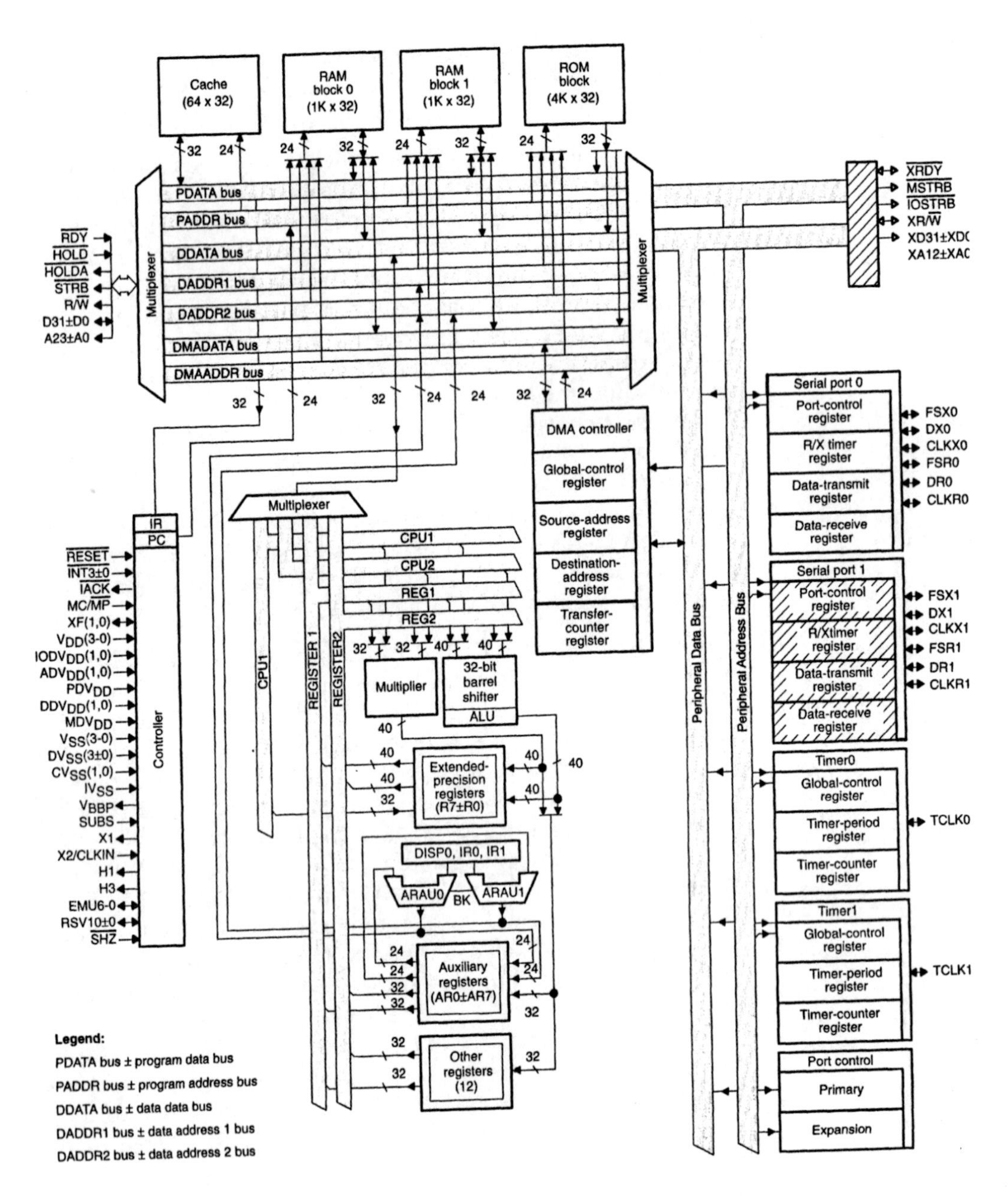

Figure 8.10 The TMS320C30 Digital Signal Processor.

are intended primarily to operate on the contents of auxiliary registers, to carry out any calculations on addresses contained in these registers required to locate operands in memory. The fact that there are two sets of ALUs means that the DSP can be calculating the next address to be used *simultaneously* with its processing of the present data item in the data ALU. This use of concurrent ('parallel') processing to enhance system throughput and performance is typical of a DSP.

Designers of processors can do much to enhance the speed of operation of a CPU, and the TMS320C30 is a good example of this. However the bus connecting the processor to memory soon becomes a bottleneck, and there is little the designer can do about this, since a typical memory device has only one address and one data bus port. The designers of the TMS320C30 have dealt with this *on the device* by providing multiple buses for accessing memory (PDATA, DDATA, DMADATA, etc.). The fact that there are multiple buses means that the processor can simultaneously deal with data from several sources – internal and external memory for instance, or from the four blocks of internal memory (cache, RAM and ROM). It may also carry out simultaneous DMA operations. This is another example of 'parallel' processing. (Unfortunately parallel processing has a different meaning in other aspects of computer literature, so the use of the term in relation to DSPs is a little misleading, although commonly used.)

The use of multiple buses external to the DSP is more difficult. The designer is ultimately constrained by the number of pins which can feasibly be attached to an integrated circuit. (The TMS320C30 has over 200!) This means that the PDATA, DDATA, etc. buses could not be extended off the chip, even if the external memory could be configured around them. To deal with this, the internal buses are switched one at a time to the external bus pins, as required during instruction execution, using a switching device known as a 'multiplexer'. There are, nevertheless, two sets of external bus connections in the TMS320C30: 'normal', at the top left of Figure 8.10, and the 'extended' (prefixed by an 'X': XD31-XD0) on the top right.

The final aspect of Figure 8.10 worthy of comment here is the memory unit known as the 'cache'. In a typical computer program, instructions are fetched from relatively localised areas in memory – programs do not on the whole fetch instructions from very disparate areas. The idea of a cache is to maintain a snapshot of the most recently used instructions (and data in some processors) in a place very local to the processor. This is done in the hope that the number of times the processor will need to go off-chip, to main memory to fetch an instruction, will be minimised. The cache contains a mechanism which determines whether the information required is contained in the cache, in which case it is provided from there, or whether the processor must go off-chip, to the main memory to fetch the information.

The use of an on-chip cache provides additional opportunities for optimised use of the parallel bus structure on the DSP to carry out multiple memory accesses. It also provides the opportunity for very fast memory accesses, since the location of the cache on the DSP chip itself means that accesses can be very quick. Most DSPs allow the speed of operation of external buses to be pre-programmed, so that the use of slower memory devices can be accommodated. If a cache is in use on the DSP, the use of these slower external memory devices may not be too much of a speed disadvantage.

8.8 The MIDI interface

MIDI (Musical Instrument Digital Interface) is an important interface for music technology and digital audio applications, and deserves a special mention in this chapter concerned with interfaces.

MIDI is an example of a *serial interface*. That is to say that the individual 1s and 0s constituting a MIDI message (see Chapter 4) are transmitted one bit at a time down a single wire or circuit. Information is typically transmitted to and from a processor as a parallel word (as described in Section 8.3 above), and it is the job of the transmitter interface to convert this parallel word into the serial bit-stream. In the receiver interface, this stream is converted back into a parallel format, ready for collection by the processor. In both the transmitter and receiver interfaces, this process is typically carried out by a 'UART' – a universal asynchronous receiver/transmitter.

The heart of both the receiver and transmitter sections of a UART is a synchronous digital device known as a *shift register*. In fact, a UART has two of these devices, one in the receiver, the other in the transmitter. The way in which these are used to transmit serial data from a source (the transmitter) to a destination (the receiver) is illustrated in Figure 8.11.

In Figure 8.11 a sample 4-bit pattern, 1011, is shown as being transmitted from a parallel input port in the transmitter to a parallel output in the receiver. In a MIDI system the data are often transmitted as 3 × 8-bit bytes. The transmitter processor loads the bit pattern to be transmitted as a parallel word into the parallel input of its shift register. It does this when the 'transmit buffer empty' status signal indicates that this shift register is clear of any previous data, and is therefore ready to transmit.

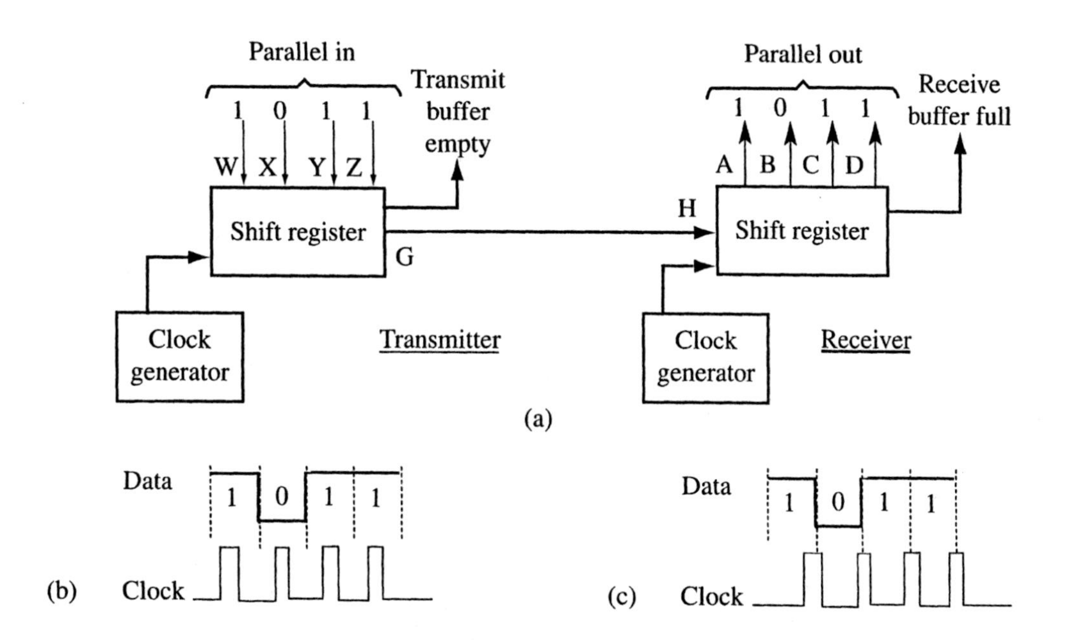

Figure 8.11 A serial interface system; (a) serial system overview, (b) data arriving at H with correct receiver clock phasing for the data input at WXYZ, (c) as in (b) but incorrect clock phasing.

Subsequent clock pulses in the transmitter move the bit pattern 1011 (in this example) to output G, one bit at a time. This bit pattern therefore arrives at point H in the receiver, as illustrated in Figure 8.11(b). Clock pulses have to be generated *in the receiver* in the phase relationship with the data shown in Figure 8.11(b). That is to say the clock pulses must be timed to be placed *within* the data bit intervals.

A phase relationship as shown in Figure 8.11(c) would not be acceptable, since each clock pulse straddles the boundary between each bit boundary, and the interpretation of the data pattern would probably be ambiguous. Assuming that the phase relationship is correct, the bit pattern (1011 here) is moved one bit at a time (per clock pulse), along the parallel outputs A–D in the receiver. When the bit pattern is in place in the receiver outputs, with the first bit input sitting at D, the UART asserts the 'receive buffer full', to indicate to the receiving processor that a data word has been received, and is ready to be transferred from the interface to the processor.

The problem of generating a clock source with the correct phase relationship is the classic problem in serial interfaces. It is not adequate just to have two clock generators (one in the receiver, one in the transmitter) which are very carefully tuned to have the same frequency. Eventually, over some length of time, the two clock generators would drift apart in phase, because of minute differences in frequency, just as the pendula of two accurately timed wall-clocks will gradually drift apart in phase, even if they start together.

There are various ways of dealing with this problem. In a UART (and therefore in MIDI interfaces), the actual data bits are normally preceded by a 'start' bit, and are followed by one or two 'stop bits'. In MIDI, the data is transmitted least significant bit first. When the start bit is detected at H, the clock generator in the receiver is started from rest. The frequency of the clock generator is typically 8 or 16 times the bit rate of the incoming data. Let us assume that the clock frequency is 8 times the incoming bit rate. The UART counts 4 pulses, and then checks that it is in the middle of the start bit. (This avoids spurious triggering due to noise). If all is well, it then counts in batches of 8 clock pulses. At the end of each batch, it is assumed that the next clock pulse will coincide with the *centre* of the next data bit, so this clock pulse is used to load the data bit then at H into the receiver shift register. After the last batch of 8 has been counted in this way (to load a final 'stop bit' into the receiver shift register), the receiver clock is 'put to sleep' by the UART, ready to be started up when the next start bit comes along. 'Receive buffer full' is then asserted.

Synchronisation of the clock generator is thus achieved by starting the clock generator up, from cold, on each data frame. This frame is typically 10 bits in extent: 1 start bit, 8 data bits and 1 stop bit. A typical MIDI 'note event' transmission occupies three such frames. Since the clock starts from cold on each frame, it only needs to remain in a suitable phase relationship for 10 bits, because it will be stopped and reset after this time. With a relatively accurate clock frequency this is quite easy to do, especially since the clock pulses are typically about one sixteenth the width of the data bits, and therefore there is a reasonable tolerance to clock pulse drift within the data bit. In MIDI transmissions, the data bits are transmitted at a rate of 31.25 Kbaud: that is, the maximum rate of transmission is 31 250 data bits per second.

Figure 8.12 shows the electrical layout of a typical MIDI interface. An electronic component known as a *transistor* is used in several places in Figure 8.12(a), and so one of these devices is shown in more detail in Figure 8.12(b). A transistor is a three-terminal device, where the terminals are labelled b (for base), c (for collector) and e (for emitter). A transistor is a current amplifier and a current switch. If a current, i_b, (see Figure 8.12(b)) is forced to flow into the base of the transistor shown in the figure, then current I_c flows collector to emitter (in the direction of the arrow) which is proportional to, but usually much larger than the base

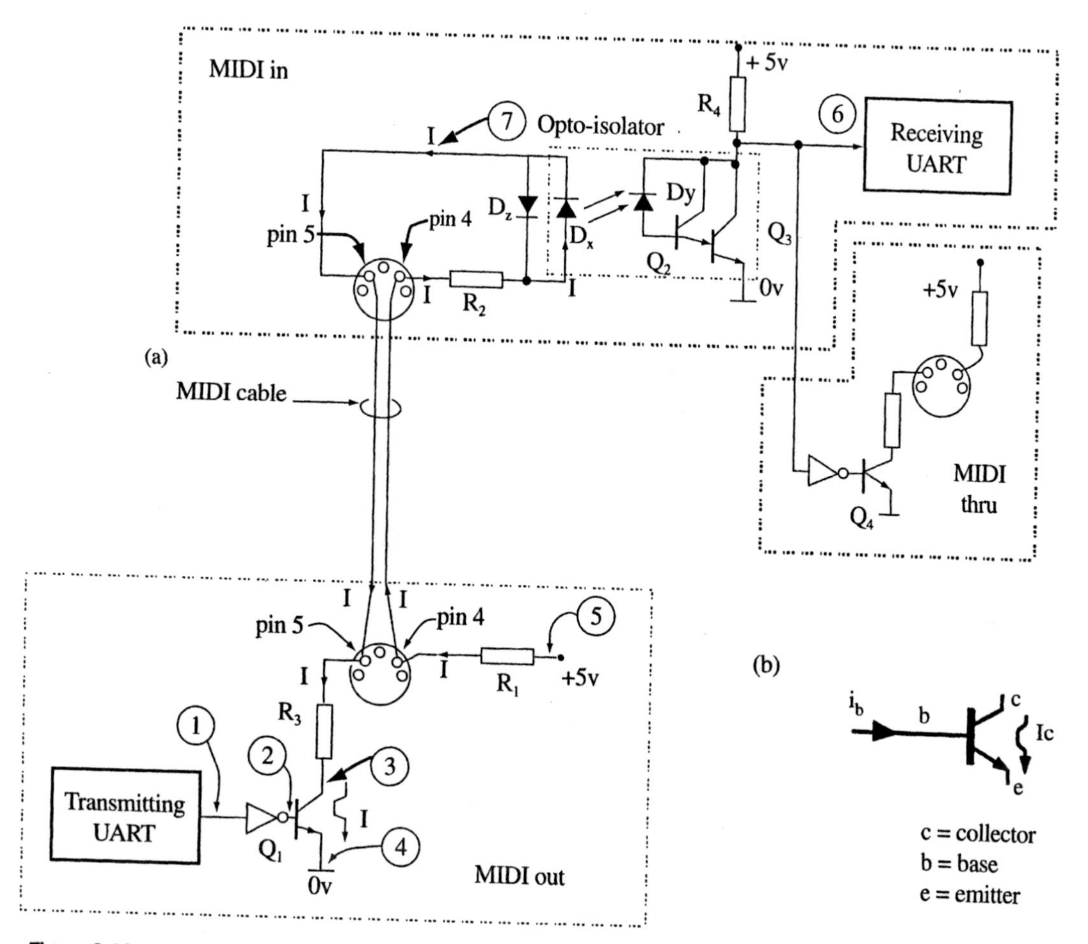

Figure 8.12 (a) Electrical layout of a typical MIDI interface; (b) detail of a transistor.

current. If the base current is sufficiently large then *any* current available at the collector is shunted down to the emitter, and the transistor functions as a simple switch.

We can now apply this knowledge to understanding the current flows in the MIDI circuit, Figure 8.12 (a). The inverter on the output of the transmitter UART converts a logic 0 at point 1 to a current at point 2, flowing into the base of the transistor Q_1. This makes Q_1 act like a switch, establishing a short-circuit between 3 and 4, allowing current to flow between these points. (A logic 1 at point 1 would open circuit the switch, blocking any current between 3 and 4.)

Assuming that Q_1 is switched on, current I flows from the power supply at 5, through resistor R_1, through the MIDI cable via pin 4 of the MIDI OUT plug, to the MIDI IN plug of the receiver. There it flows through R_2 and through the light emitting diode (LED),

D_x, of the opto-isolator. The LED lights up in response. Current I then flows back through the other conductor of the MIDI cable (pin 5), through resistor R_3, through transistor Q_1 to the 0V power supply return at point 4. This process forms the 'current loop' described in Chapter 4.

The opto-isolator turns the light from the LED into a tiny current flowing in diode D_y, which is greatly amplified by the dual transistor configuration Q_2/Q_3. This switches Q_3 on so that point 6 is connected to 0V. In summary: a logic 0 at point 1 causes a current to flow in the MIDI circuit and this results in 0V (a logic 0) at the input to the receiver UART. The converse is true: a logic 1 at point 1, stops current flowing in the MIDI circuit (because Q_1 is switched off), and the input to the receiver UART receives 5V (logic 1) via resistor R_4. When no MIDI signal is present, the MIDI line idles at 'current off'.

A logic 0 at point 6 causes the MIDI thru invertor to switch Q_4 on, so that the signal at Q_4 exactly follows the signal at points 6 (and hence 1).

The MIDI signal is communicated to the receiver via a light beam in the opto-isolator. This is used so that there is complete electrical isolation between the transmitter and the receiver. This is valuable because of the lengths of MIDI cable between the two. Such cables can act as antennae picking up large electrical noise signals from the mains circuits, which could cause damage and data errors in the MIDI circuits. The risk is exacerbated if the receiver device (e.g. a sound expander) and the transmitter device (e.g. a master keyboard) are run from different phases of the mains, as could easily be the case in a complex performance setup. Diode D_z prevents excessively large (circuit damaging) noise voltages building up at point 7 when no MIDI current is flowing.

8.9 PC sound cards

We complete this chapter by outlining the architecture of a typical PC sound card. An example of such a card would be a member of Creative Labs' Sound Blaster™ series, although the outline presented here is a fictitious (though hopefully representative) design, and is not modelled on any particular commercial product.

A description of such a design is appropriate here because these sound cards are very widely used in multimedia PC based applications, and some notion of their operation is therefore directly relevant to a book of this kind. Also, many of the techniques described earlier in this chapter find direct application in a

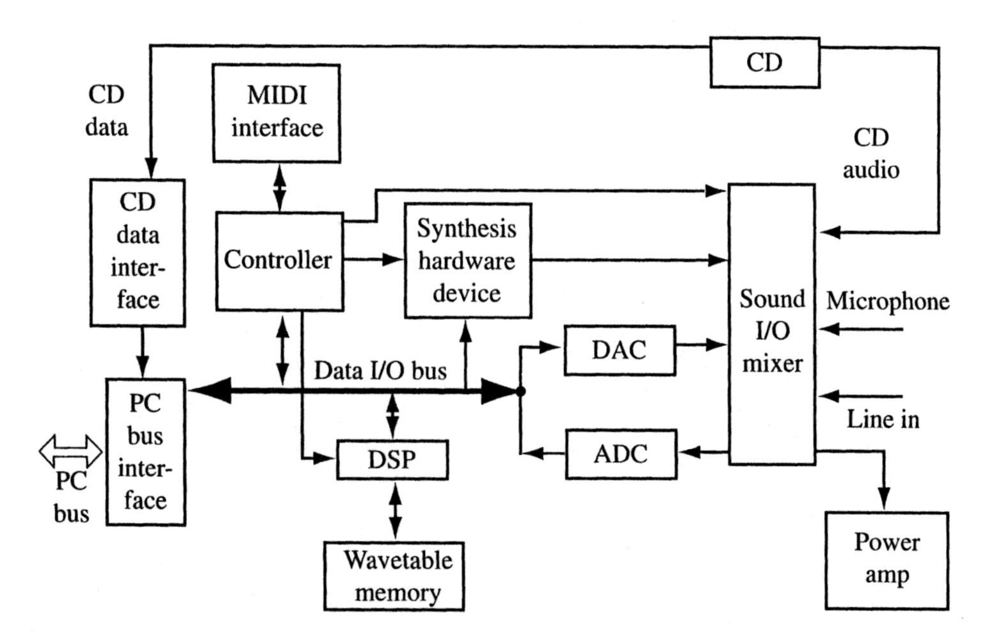

Figure 8.13 Outline of the architecture of a typical PC audio card.

typical sound card, and so a description of one of these devices forms a useful summary at this stage of the book.

Figure 8.13 shows the outline schematic of our sound card design. Such a card is designed to be plugged into the bus of a personal computer (PC), so that it can use the PC's facilities for storage of data, printing and file management and so on. However, much of the serious processing of audio data is carried out by dedicated microprocessors located on the card itself.

The first of these is the on-board controller. Its main job is to set up and control the other on-board components, and to handle the dialogue between these components and the user's application program running on the PC. This dialogue is carried out by messages exchanged between the controller and the PC over the card's PC bus interface. For example, the card can input audio signals and convert them into PCM data. These digitised PCM samples can be sent to the PC's disk for mass audio data storage. This will typically be done by DMA transfer to the PC, set up and managed by the controller. Similarly, audio data from the PC can be sent out from the PC under DMA control to the sound card, where it will be converted into analogue sound output via the DAC.

The controller also manages the on-board MIDI interface. This can be used as a general purpose MIDI device by any application running on the PC such as a sequencer or a multimedia interactive CD ROM encyclopaedia. It can also be used to provide MIDI control for other on-board devices such as the synthesis hardware or DSP. Transfer of MIDI data to and from the PC is handled by the controller, typically using interrupt driven mechanisms over the PC bus interface.

The on-board DSP can be used to process real-time 'effects' to be added to the sound samples which are to be output via the DAC. Examples of this might be filtering and reverberation, described in Chapter 5. The DSP also has an associated bank of fast RAM which can be used to store audio data to be processed by the DSP. For example, audio data fetched from the PC disk (perhaps originally input from a microphone, or 'line in' and sampled in by the on-board ADC) can be placed into this memory, to form the basis of a sampler managed by the DSP. Loop points could be set up, envelope parameters added to the waveform, and the resulting sound could be placed under the control of the MIDI interface, so that pitch and velocity information may be imposed on the sound output.

Many sound cards also have hardware synthesis units built into them, often based on FM (see Section 5.4). Taken together with the on-board MIDI interface, this unit allows many sound cards to emulate the behaviour and sound palette of common standalone MIDI sound expanders.

The sound mixer section is like a small, on-board 'mixing desk', which can be manipulated by the controller in response to signals received from the user's application running on the PC. A mix of line input, microphone, and CD can be used as sound input sources, and a mix of on-board synthesis hardware and DAC outputs sent to the on-board power amplifier.

The range of powerful audio processing facilities provided on a typical sound card illustrates the extent to which the world of computers and the world of multimedia are becoming ever more integrated. When linked to the content distribution facilities of the Internet, we can expect this integration to develop further, to yield exciting possibilities for the future evolution of audiovisual art.

8.10 Summary

In this chapter, we have gained an understanding of the way in which a computer is used as a component in a larger system. The operation of the MIDI interface has been presented in technical detail. We have considered the ways in which analogue signals such as sound can be converted into a binary number representation, suitable for manipulation within the computer. We have also considered the converse process, by which binary numbers can be converted into corresponding analogue signals. Finally, we have examined the ways in which these techniques are used in important processing devices used in music technology: the digital signal processor (DSP) and PC sound cards.

Suggested further reading

Chamberlin, Hal (1980) *Musical Applications of Microprocessors*, Hayden, ISBN 0-8104-5753-9.

Texas Instruments (1989) *Third Generation TMS320 User's Guide*, ISBN 2-86886-031-1.

Watkinson, John (1998) *The Art of Digital Audio*, Butterworth–Heinemann, Oxford, ISBN 0-240-51270-7.

Part 5

Programming for sound generation and processing

The processes of writing computer programs for audio, musical and multiple-media purposes are introduced in this section. It is quite feasible for the enthusiast to do this, so that he or she can contribute to the development of the subject in terms of technique and music, provided that a suitable introduction to programming is given. This is the purpose of Chapter 9 and examples are based on algorithmic composition for MIDI systems. Chapter 10 extends the programming techniques covered in Chapter 9 to show how sound can be synthesised and processed using the unit generator concept. The real-time control of signal processing and multiple-media applications is introduced. Examples of direct sound synthesis and manipulation are given using the MIDAS system as a framework.

These chapters are written for the newcomer. However, programming takes a good deal of study to carry out competently, so we recommend studying the art of programming the 'C' language using one of the many well-known books or on-line web tutorials.

9 Computer programming for musical applications

Overview

The main purpose of this chapter is to explain why there is a need to study computer programming in the subject areas of music and multimedia. It introduces the concepts which programming languages have in common, and sets them in a musical context. It is not a course on a specific programming language since many excellent texts already exist purely for this purpose. Instead we present each code example in terms of its 'algorithm' – the idealised list of instructions in the program which describes how a task should be done. The intention is to give the reader some understanding of how common musical programs such as sequencers operate, so that the reader may experiment with these ideas and contribute to the future development of music technology.

By the end of this chapter you should appreciate why programming is necessary, that different sorts of computing languages are available, and be familiar with some of the concepts that are required for programming with musical systems.

Topics covered

- The need for computer programming.
- Tasks involved in programming.
- Types of programming language.
- Structured programming for musical purposes.
- Programming for music with MIDI.

9.1 The need for computer programming

9.1.1 The prevalence of computers

Computers are all around us in many different guises. There are the familiar typewriter keyboards, screens and 'mice' which are used, amongst other things, for word-processing and music editing. There are also the hidden (or 'embedded') computer systems inside everyday objects such as car engines, washing machines and synthesisers. All of these systems are driven by *software* – or computer programs. As computers become increasingly prevalent in society, so engineers and musicians should become more aware of the software that makes these machines work.

9.1.2 The art of programming

Programming is a creative *art*, not just a technique. The purpose of writing a program is to get a computer system to do what you want it to do. This may sound straightforward, but it involves the following issues:

- *specifying* the task completely,
- *designing* a series of logical operations which will carry out that task,
- *coding* these operations into a form that a computer can understand – the computer program,
- *testing* the computer program to see if it meets the specification.

Notice that coding is only part of the process. The specification and design stages are very important and are often the most difficult parts of the procedure. The basic problem that computer designers face is bridging the gap between the communication methods of human beings and computers (see Figure 9.1).

People have goals and feelings and they communicate with verbal language and physical gestures. Computers work at the level of numerical code, blindly following a list of primitive instructions. The designers of computer systems have the formidable task of building a bridge between these vastly different worlds.

9.2 Types of programming language

There are many different tools available that help designers to bridge the gap between a human's intentions and the computer's internal instructions. This section outlines the various types of

Figure 9.1 The gap between humans and computers.

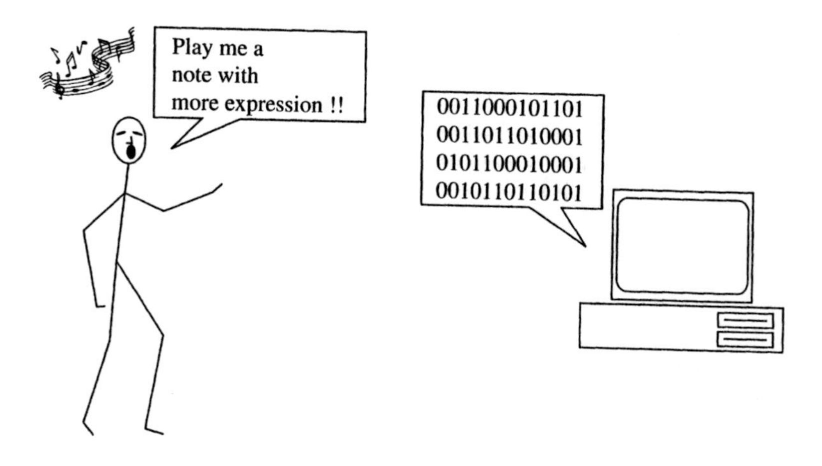

programming language. Readers wishing to explore this subject from a more technical background should refer to Chapter 7.

9.2.1 Programs

All computer programs consist of two basic elements – *instructions* and *data*. The instructions tell the program what to do, and the data are the pieces of information that are processed. Computer programs are classified according to how they treat and represent these two elements.

9.2.2 Low-level code

This section recaps the information presented in Chapter 7 in a form that is appropriate for comparing the different types of programming language. A computer processor follows a list of numerical instructions stored as binary code (lists of 1s and 0s). Early computer systems were given their instructions at this level; programmers had to flick dozens of switches either on or off. This was a tedious process and prone to errors, so various forms of 'short-hand' were developed to give the computer its commands.

One form of short-hand is called *assembler* code. This consists of a series of simple instructions that can be converted into the computer's corresponding binary code. Consequently the program-

Figure 9.2 A small extract of low-level code. Each line can be directly converted into a computer instruction.

```
load    a1,  d0
load    a2,  d1
add     d1,  d0
store   d0,  a3
sub     d0,  #243
store   d0,  a4
```

mer deals with words and numbers (rather than binary), but is still working at the *low-level* of simple computer instructions. Figure 9.2 shows a typical fragment of assembler code.

Assembler languages produce fast, compact programs, and they are still widely used today when there is a rigorous requirement for high speed of operation or small programs.

The main problem with assembler code is that programmers have to spend much of their time dealing with the detailed mechanics of each instruction as it interacts with the computer. As humans we can only hold a limited number of concepts in our minds at any one time. Therefore with low-level programming it is very difficult 'to see the wood for the trees' and to get a clear picture of the *structure* of the program.

9.2.3 High-level languages

High-level programming languages allow designers to work at a higher conceptual level, i.e. closer to the way in which a human would think, and further removed from low-level primitive computer instructions.

High-level programs often resemble a simplified form of written (English) language. They enable programmers to shape the overall structure of the program. Figure 9.3 shows what a piece of high-level code might look like.

Figure 9.3 An imaginary piece of high-level code. Instructions are carried out from top to bottom.

```
repeat 3 times :
        input_data()
        process_data()
        output_data()

if (user_requires_help)
        printout (help_page)

printout (menu_options)
```

There are many varieties of high-level language, each with a different emphasis according to how the instructions and the data are represented. Three common forms of programming language are used: procedure-oriented, object-oriented, or logic-based. In the computing world there is much discussion over which type of language is best for a particular job, and the arguments are quite complex.

There now follows a brief overview of each of these categories.

Procedure-oriented languages

Procedure-oriented programming languages place the emphasis on organising the program's instructions. They focus on the procedure (or 'algorithm') for carrying out the task. Figure 9.4 shows that a procedure is used to transform input data into output data. Examples of this sort of language are 'C', 'BASIC', 'FORTRAN' and 'COBOL'.

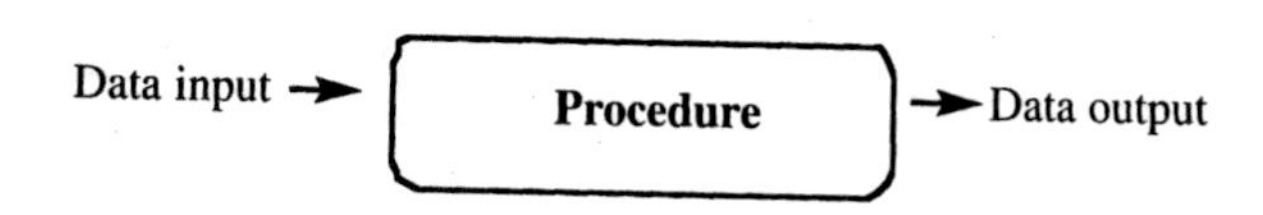

Figure 9.4 Procedure-oriented languages.

Object-oriented languages

Object-oriented languages provide an alternative way for computer programmers to look at the coding problem. The programmer must start by identifying all the *data* that is needed for the task, then encapsulating the data into packages called 'objects'. Every program needs instructions, but in an object-oriented language the instructions are 'packaged up' inside the data objects. For some tasks this makes the program easier to construct, particularly where there are many different types of information to deal with. Examples of this sort of language are 'Java', 'Smalltalk', and 'C++'.

Logic-based languages

Logic-based programming languages provide the programmer with yet another way of defining the problem, this time by emphasising the logical *relationships* between data elements. Chess-playing computers are usually programmed using logic-based languages. The programmer specifies the *rules* by which the data objects interact so that the program can look for the best solution to an logical problem. Examples of this language type are 'Prolog' and 'Lisp'.

In this text, however, we concentrate on the more widespread procedure-oriented languages where the 'algorithm' (an ordered set of instructions) is the basic building block.

9.3 Structured programming for music

The material in the remainder of this chapter is presented in the form of *algorithms* that demonstrate the ideas behind musical programming. This section explains why algorithms provide a useful way of learning about programming, and introduces some of the basic concepts necessary for writing musical computer programs.

9.3.1 Algorithms

An algorithm is a precise list of instructions of *what* has to be done in order to achieve a particular task. We have seen (Section 9.1.2) that the art of programming includes:

1 Determining exactly what the task is (*specification*), and
2 Producing a series of logical instructions to carry out the task (*design*).

The usual output of the design phase is a set of algorithms. They cannot actually 'run' on a computer in this form, as they need to be put into specific computer *code.* In other words algorithms need to be coded and embedded within a particular computing language. The language is then converted (*compiled*) into the low-level numerical code required by the computer hardware.

This chapter concentrates on algorithms because their structure is relatively easy to follow. If you want to hear these examples running on a computer, you will first need to code them into a programming language. Many people using computers for audio and musical applications use the 'C' programming language because it runs at a fast speed on most machines. Therefore in this text we show the algorithms in a simplified version of 'C', so that you would not have much translation to do should you wish to put them into 'C' code.

9.3.2 Functional elements of musical programming

There is a set of fundamental tasks which most musical computer programs have in common. Figure 9.2 showed how *data input* is given to a *procedure* which then produces *data output*. The data has been *transformed* by the computer. This concept forms the basis of practically all computer programs.

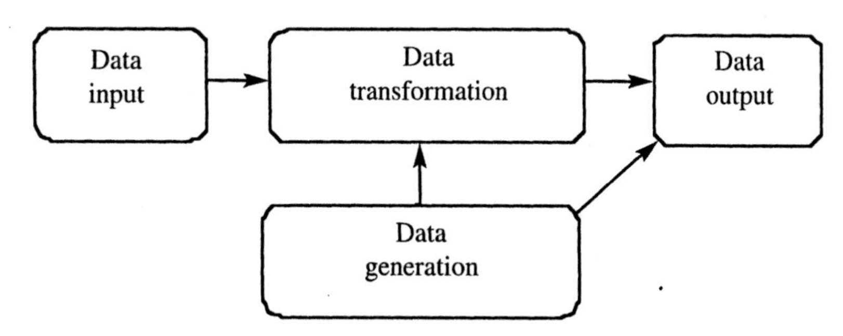

Figure 9.5 The basic functional elements for musical computer programs.

The basic functional elements required by musical programs are:

- data input (e.g. recording)
- data output (e.g. playback)
- data transformation (e.g. processing)
- data generation (e.g. synthesis, composition).

Figure 9.5 shows that data is transformed by the computer, but may also be *generated* from within the program. In fact our first algorithms (Section 9.5) generate musical output *without* requiring any input from the user.

The diagram shows that the generated data may be also transformed by the user's input. An example of this is a person playing a keyboard synthesiser. The synthesiser *generates* musical sounds, but the pitch is controlled (*transformed*) by a user (*data input*) on a piano-type keyboard.

9.3.3 Programming for data at various speeds

It is important to consider the *rate* at which data is to be generated and transformed. Not only does this have implications for the computing power required to run the program, but also the style in which the program is written.

In the following sections, we will consider the input, transformation, generation and output of data at different rates.

9.4 Programming at control rate

In this section we will examine the meaning of control rate, and describe how the MIDI standard can be used to make interactive musical programs. We then introduce a series of code examples for generating and processing musical notes using MIDI.

9.4.1 Definition of 'control rate'

Control rate is a term that refers to the speed at which humans operate, i.e. the speed at which people can *control* equipment. To get an idea of how fast this is, try to imagine all the information you generate when you play an electronic keyboard. There is a limit to the number of notes you can play in any one second. If you take your hand off the keyboard to move the pitch-bend lever, then you cannot play as many notes, but you are producing an additional stream of pitch-change information.

A typical human player tends to produce control rates consisting of between 1 and 100 pieces of information per second.

9.4.2 MIDI as a control rate medium

The most widely accepted standard for musical control data is MIDI (see Chapter 4 for full details). It allows computers to take musical data in from keyboards, transform that data, and output it to produce sound on another keyboard (or sound module). In this way MIDI enables people to write interactive musical computer programs. MIDI deals with musical note events, and provides *control* of those events. It does *not* cover the methods by which the sounds themselves are *generated* – that is left to *audio-rate* processing code within the sound modules.

9.4.3 MIDI programming libraries

A programming 'library' is a set of functions, made available within a computer system, which can be used by programmers to ease the task of writing code. Library *functions* can be incorporated into your code for a variety of purposes. On most multimedia computer systems nowadays there will be a MIDI library which offers a range of functions for interacting with MIDI-based keyboards and sound modules.

In this text we base our examples on a simplified MIDI library that has the following functions:

- **midi_note**( ) This function sends a MIDI note message from the computer to a sound module. On receipt of such a message the module will play a note, and will keep playing it until it is told to turn it off. The information about *which* note is to be played is placed within the function's brackets.
- **pause( )** This function causes the computer to wait for a certain time. It is used for controlling the timing of the MIDI notes being sent out, so they do not all play at once. Notice that this is not specifically a 'MIDI' function (as it does not

send a MIDI message) but is included here as it is needed by many of the following examples.

Other functions in our library will be introduced as we use them. All of the tasks that are referred to in this chapter are possible on a modern multimedia computer system, but you need either access to a suitable library or to write your own!

9.5 Data generation and output

We now introduce a series of code examples for playing musical notes using MIDI. Many interesting musical features can be created by this method.

When learning a foreign language, it is often easier to speak it than it is to listen. In the same way, the generation of data from within a program is easier than processing data input from the outside world. So we begin with a series of algorithms that simply generate MIDI notes.

9.5.1 Notes, chords and tunes

Figure 9.6 shows an algorithm for playing a single note for one second. Even though this is a simple algorithm it will need some explanation, as the format of each command will probably be unfamiliar to you. The formatting is very similar to that used in the C programming language.

- Anything within these marks /* . . . . */ is a *comment*. It is not actually part of the algorithm – it is just there to explain something to the reader. All good code contains plenty of comments.
- Each separate command finishes with a semicolon ;
- The `midi_note()` function has three numbers in its brackets. These are the pieces of information (or 'parameters') which it needs to carry out the task of sending the note message to the sound module. The first number is the *pitch*, the second is the *channel*, and the third is the *velocity*. (For a full explanation of these concepts, please refer to Sections 4.6 and 4.6.1.)
  - A pitch of **60** corresponds to 'Middle C' on a keyboard. (MIDI pitch numbers go from 0 to 127.)

Figure 9.6 A simple algorithm for playing one MIDI note.

```
midi_note (60, 1, 127);        /* Play a note */
pause (1000);                  /* Wait */
midi_note (60, 1,0);           /* Turn note off */
```

- MIDI channels are used to distinguish between data intended for separate instrumental parts. (MIDI channel numbers go from 1 to 16.)
- A velocity of **127** means 'play the note very loud'. (MIDI velocity values go from 0 to 127.)

- The `pause()` function has just one number in its brackets. This is the length of time in milliseconds (1000 ms = 1 second) that the computer is to wait for, before going onto the next command.
- The second `midi_note()` function is similar to the first, but the *velocity* number is **0**. A zero velocity has the effect of turning a note *off*.

In summary, this first algorithm sends MIDI messages to a sound module to play a note of 'Middle C' very loudly for one second (on whatever instrumental sound is set on channel 1).

Figure 9.7 shows a second algorithm. It is similar to the first, but sends more messages.

Notice that three `midi_note()` messages are sent before the `pause()`. These will be heard *simultaneously* as a musical *chord*. The notes of the chord are determined by the 'pitch' number inside the brackets. Each key on a keyboard has its own MIDI pitch number. The pitch number increases by one for each semitone up the keyboard. Therefore E is represented by a pitch number of 64 since it is 4 semitones higher than C (pitch = 60).

Notice also that all three notes are turned off (after the `pause()` command). If you forget to turn a note off in MIDI, it will carry on playing indefinitely (a so-called 'stuck note').

The first algorithm played a single note. The second one played a chord. Try Example 9.1 to extend these ideas into playing a melody (a series of notes in sequence).

Figure 9.7 An algorithm for playing a chord (many notes at the same time).

```
midi_note (60, 1, 127);     /* Play Middle C */
midi_note (64, 1, 127);     /* Play E */
midi_note (67, 1, 127);     /* Play G */
pause (2000);               /* Wait */
midi_note (60, 1, 0);       /* Turn off C */
midi_note (64, 1, 0);       /* Turn off E */
midi note (67, 1, 0);       /* Turn off G */
```

Example 9.1 Write an algorithm that plays a simple well-known tune. Base it on the algorithms in Figures 9.6 and 9.7. Consider carefully which notes to use, where the pauses are placed and where each note is turned off.

9.5.2 Musical repetition

Music is a fundamental human form of expression that involves arranging sound in time. It consists of a fine balance between change and repetition. Too much repetition becomes tedious, while too much change leaves the listener unable to follow the piece.

So far, we have seen how to make a computer play a series of notes in order to play a tune or a chord. Now we consider how to make patterns that repeat. Figure 9.8 shows an algorithm containing a *loop*. It plays a two-note theme in the bass, and repeats this four times altogether.

Notice the following features of this algorithm:

- At the top of the code is a *variable* called 'count'. A variable is a named entity which can temporarily store data. In this case the data is an **int** (an integer, or whole number). The line `count=0` sets the initial value of 'count' to zero.
- The line `while (count < 4)` denotes the start of a 'loop'. All the code between the curly brackets { and } will be continually repeated *while* the variable 'count' contains a value less than 4.

Figure 9.8 A loop is used to repeat a section of code. In this case it produces a repeating bass-line.

```
int  count;                   /* Variable called count */
count = 0;                    /* Set count to '0' */
while (count < 4) {           /* Start of loop */
  midi_note (48, 1, 64);      /* Play a bass C */
  pause (250);                /* Wait quarter of second */
  midi_note (48, 1, 0);       /* Turn C off */
  midi-note (43, 1, 64);      /* Play a G */
  pause (250);                /* for quarter of second */
  midi_note (43, 1, 0);       /* Turn G off */
  count = count + 1;          /* ADD one to count */
}                             /* } denotes End of loop */
```

- The line `count = count + 1;` adds 1 to the value of count each time around the loop. In this way 'count' starts off as 0, increases to 1, then 2, then 3. At this point the loop has gone around four times. As count increases to 4, the line `while(count < 4)` is no longer true and so the loop cannot go round again. Thus 'count' can be considered to be a *loop control variable* as its value determines how long the loop should run for.

Example 9.2 How would you change this code so that it played a four-note bass-line repeated eight times?

Figure 9.9 extends the concept of the loop by getting the loop variable to *change* a musical value as it counts.

Notice how the variable 'count' now appears in the line `midi_note(count, 1, 64)`. This causes a new note to be played each time the value of 'count' increases. Thus the MIDI pitches 0 to 127 will be played, increasing one semitone at a time, and each one lasting a tenth of a second. The program will play a chromatic scale.

```
int  count;                      /* Variable */
count = 0;
while (count < 128) {            /* Start of loop */
  midi_note (count, 1, 64);      /* Play note - use count */
                                 /* to specify its pitch */
  pause (100);                   /* Tenth of a second */
  midi_note (count, 1, 0);       /* Turn note off */
  count = count + 1;             /* Add one to count */
}                                /* End of loop */
```

Figure 9.9 A loop that plays an ascending semitone scale.

Example 9.3 The standard five-octave MIDI keyboard covers the range of notes from 36 to 96. How could you change the code so that the loop only played the pitches 36 to 96 inclusive?

Example 9.4 How could you make this into a *descending* scale?

Example 9.5 How could you change the code so that it played a *whole-tone* scale (where there are two semitones between each step of the scale)?

9.5.3 Scales and arrays

A scale of 'C major' consists of the following notes: C, D, E, F, G, A, B, C. These do *not* have the same number of semitones between them (for example 'D to E' is two semitones, but 'E to F' is one semitone). There is a regular pattern of tone (T) and semitone (S) gaps which is the same for *every* major scale (T,T,S,T,T,T,S). Therefore we cannot use a loop with a simple incrementing 'count' to play a major scale. Instead we must somehow *store* the pattern of tones and semitones so that we can calculate which note is to be played next. Let us store it as the series of numbers 2,2,1,2,2,2,1. This defines how many semitones there are between each step of the scale.

Figure 9.10 shows how it is possible to allocate a section of computer memory just big enough to store this pattern. As we give the computer an instruction to allocate memory space for these seven integers, it usually can be asked to give back to us a *pointer* that points to the start of the series of numbers in memory.

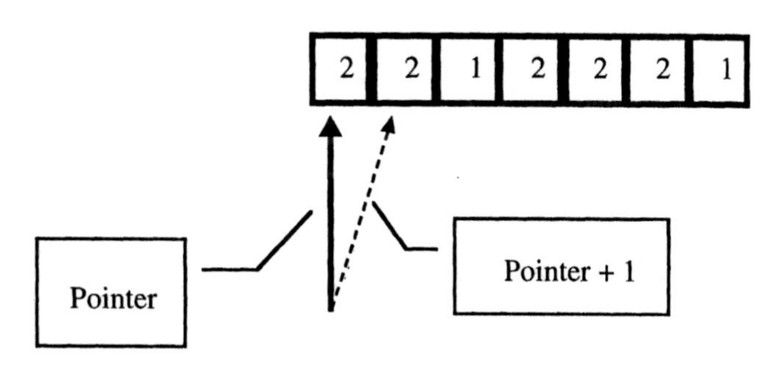

Figure 9.10 The tone-tone-semitone series stored as numbers in computer memory. A 'pointer' allows us to look at each element when we need to use it.

We use the pointer to access each element of the number series. It always points to the first element of the series. If we look at (pointer + 1) we are looking at the next element along, (pointer + 2) is the next, and so on.

Most high-level languages give programmers a concise shorthand for dealing with series of numbers in memory like this. An *array* is a set of memory elements that can be used for just this purpose. Figure 9.11 shows an algorithm that uses an array to play a C major scale.

This algorithm is more difficult to understand as it involves some cross-referencing between different variables. It is worth reading through the following paragraphs several times (refer-

```
int scale[7] = {2,2,1,2,2,2,1};/* Scale stored in array */
int index;                      /* Variable called 'index' */
int pitch;        /* Variable to hold current note value */
index = 0;
pitch = 60;                     /* First pitch is Middle C */

while (index < 7) {
  midi_note(pitch, 1, 64);                  /* Play note */
  pause(250);
  midi_note(pitch, 1, 0);               /* Turn note off */
  pitch = pitch + scale[index];  /* Calculate new pitch */
  index = index + 1;
```

Figure 9.11 Use of an array to play the relative semitone steps of a major scale. The algorithm performs one octave of C major, ascending.

ring to the code) because this sort of technique is used a lot in programming. Notice the following features of the algorithm:

- The array is called 'scale' and consists of seven integer elements. Our series of semitone spacings is stored in the array. This is all done in the first line of the code. In this example we have called the loop control variable 'index'. This is because it is used to *index* the array (i.e. it is the number added to the pointer to access a particular array element). It starts off at 0 and increases until it is 7, at which point the loop will stop running (because the statement `while (index<7)` will no longer be true).
- The array is accessed each time around the loop to get the value of the next scale step. An individual array element can be accessed as follows: scale[0] is the first element of the 'scale' array, scale[1] is the second, etc. (This is actually a short-hand for 'look at the memory element one step on from the *pointer* called "scale"'.) In our example we use the variable 'index' within the square brackets [ ] to look at a specified element of the array. The line `pitch = pitch + scale[index]` is used to increase the 'pitch' variable by the value read from the array.

Example 9.6
What changes would you need to make to make this algorithm:

1 Play a scale of 'D' instead of 'C'?
2 Play 2 octaves (i.e. 14 notes in total)?
3 Play a *minor* scale?

9.5.4 Random generation of musical notes

Not all music is prescribed and written down in a 'score'. Much music from around the world involves *improvisation*. 'Jazz' and 'blues', for example, rely on the players inventing much of the music as they play. There is usually an agreed *structure* (e.g. a repeated 12-bar set of chords), but the note-by-note music is not written down.

The algorithms so far have generated notes from sequences of commands (to play chords and tunes) and loops (to play scales). In this section we will introduce a new function for making *random* choices about which note to play. In this way the computer can sometimes generate unexpected musical output.

We will use the following function to generate random numbers:

```
int random_number (min, max)
```

This function requires two numbers as its inputs – the minimum and maximum numbers it is allowed to generate. The word 'int' at the start indicates that this function 'returns an integer'. This means that the function generates a whole number and places it into any variable that is written to the left of the function. As an example, the line:

```
pitch = random_number(60, 72);
```

generates a random integer (whole number) between 60 and 72 inclusive, and places it into the variable called 'pitch'.

Figure 9.12 shows an algorithm that combines the concepts of arrays, scales, loops and random number generation. Notice the following features:

- The `blues[7]` array stores seven pitch values which correspond to the notes of a blues scale. Can you work out what they are?
- Each time around the loop, one of the blues notes is picked at random. The variable 'choice' is assigned a random number between 0 and 6. This is used to fetch one of the elements of the blues array for use in the `midi_note( )` function.

```
int blues[7] = {60,63,64,65,67,70,72);                    /* Blues Scale */
int choice;                              /* Variable to hold random value */
int count = 0;                     /* Counter variable, set to 0 at first */

while (count < 64) {                              /* Loop this 64 times */
  choice = random_number(0, 6);                 /* Random number 0 to 6 */
  midi_note(blues[choice], 1, 64);                         /* Play note */
  pause(250);
  midi_note(blues[choice], 1, 10);                     /* Turn note off */
  count = count + 1;
```

Figure 9.12 An algorithm to generate random streams of notes selected from a blues scale.

- The loop goes round 64 times. Hence this algorithm generates a 64-note random blues tune.

Example 9.7 Change the algorithm so that it produces a bass note *and* a tune note each time around the loop. (Hint: the bass line might need its own special array of notes).

Example 9.8 The current algorithm produces a stream of notes at a constant velocity with a fixed repetitive rhythm. How could you expand it so that:
a) the velocity varied with every note?
b) the timing of each note was different?
c) some notes were silent (i.e. a musical 'rest')?

9.6 Data input and transformation

The examples so far have involved the generation of data from within the computer. This section deals with the input of MIDI data (from a keyboard, for example) and its subsequent transformation within the computer program. We will introduce a new function to 'catch' the MIDI messages from the keyboard so that they can be manipulated by the program. The newly transformed notes are sent out via MIDI to be played on the sound module.

```
int get_midi_note( &pitch, &channel, &velocity )
```

This function *receives* MIDI note messages from an external keyboard. In this way a keyboard player can enter information into the computer by simply playing it in. Each incoming MIDI note carries values of pitch, channel and velocity. The & sign is a coding short-hand which indicates that the variables pitch, channel and velocity will be updated (changed) by this function. The 'int' again shows that the function *returns* an integer. If it returns a 1, then a new MIDI note has arrived (and the variables pitch, channel and velocity will have been updated with the new note data). If it returns a '0', then no action is necessary, since no note has been detected.

9.6.1 Real-time transformation

Many different types of transformations are possible in real-time using MIDI. Examples are transposition, inversion, and harmonisation.

Figure 9.13 shows an algorithm which plays another note for every note played in by the user at the MIDI keyboard.

The first line within the loop calls the function `get_midi_note()` which places a value of 1 or 0 into 'new_note'. An **if** statement is used to determine whether a new MIDI note has been received. It does this by checking to see if the variable 'new_note' is equal to 1 `if (new_note == 1)`. If this is true, then the two `midi_note()` lines within the **if** statement (i.e. within the { } braces) will be called. They immediately send the same MIDI data off to the sound module, **and** send another note whose pitch is seven semitones higher (a musical '*fifth*'). When the keyboard player *releases* the note, the same thing happens but the velocities are all zero, so both notes are turned off.

Figure 9.13 A real-time algorithm for reacting to incoming MIDI notes. A 'harmony' note is produced for each note that is received.

```
int pitch, chan, veloc;          /* 3 MIDI variables */
int new_note;        /* Variable to test for new data */

while (TRUE) {                        /* Keep on looping */
  new_note = get_midi_note (&pitch, &chan, &veloc);
  if (new_note == 1) {
    midi_note (pitch, chan, veloc);
    midi_note (pitch+7, chan, veloc);
  }
}
```

9.6.2 Sequencing

Some of the most popular types of musical program act as MIDI sequencers. A sequencer is like a 'word-processor' for music. Notes can be played in, edited, put together with other musical parts, and played back as a finished composition.

From a programmer's point of view, a sequencer needs to be able to:

- receive MIDI notes
- store the information in an organised way
- allow the user to interact with the stored data
- play back the completed music.

Figure 9.14 shows a block-diagram representation of how a sequencer works. It portrays the main data structures that need to be handled by the program code in order to implement the basic functionality of a sequencer.

As the keyboard is played it produces MIDI note data. This is received by a 'record' function (using something like `get_midi_note()` to capture each note event). The information from the note (pitch, channel and velocity) is stored in a data structure, along with a record of *when* the note was played. Each new event is stored in its own structure, but the individual structures are linked together into what is known as a *linked list*. A linked list is like a chain of structures. Each structure contains a pointer to the next structure in the list.

Figure 9.14 The data structures for a typical sequencer program.

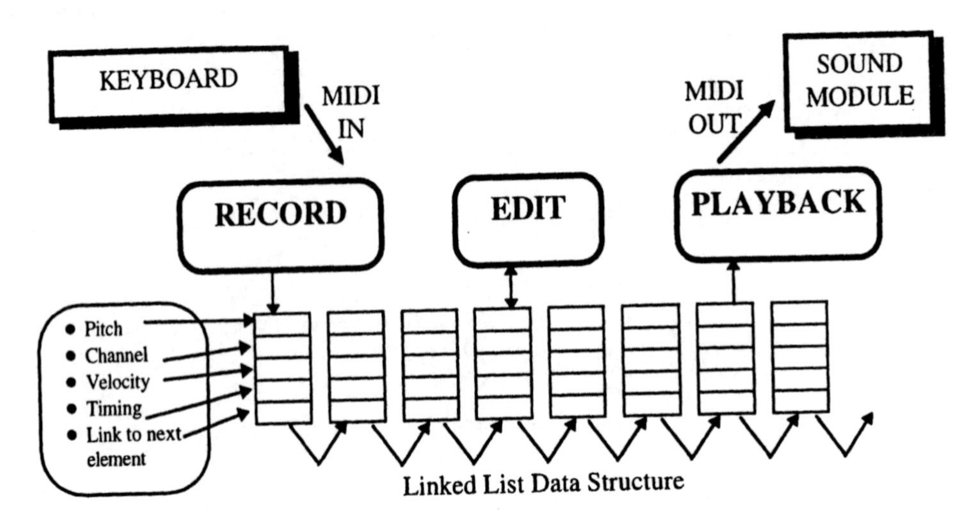

The data can be changed using various 'edit' functions. Typical editing operations include:

- *transposing* a series of notes, or 'correcting' a wrong note,
- *quantising* the rhythms (changing the timings of the events so that they occur at regular intervals),
- *deleting* a note event (in which case one item of the linked list is removed and the list re-connected).

The series of notes (the 'sequence') can be heard at any time by using the 'playback' function. This converts the stored data back into MIDI messages which it sends out to a sound module.

Example 9.9 Produce the outline of a program (based on the previous examples) which would function as such a sequencer. You are provided with a function:

```
int time()
```

which returns an integer number representing the instant in time when the function is called.

9.7 Summary

The art of programming is necessary in the world of modern media technology. Data is processed on digital computers that are controlled by software. These computers exist in various forms, often 'embedded' into everyday devices.

There are a variety of programming languages available, and you may be required to learn a specific language for a specific task. Different media require different rates of data handling, and this affects the way in which programs are put together. However, there is a set of basic requirements (input, output, generation and processing) that is common to almost all programming situations.

Suggested further reading

The web-site for this book (http://www.york.ac.uk/inst/mustech/dspmm.htm) contains sound examples for each of the pieces of code outlined here. It also contains links to further information and tutorials about programming.

Roads, Curtis (1995) *The Computer Music Tutorial*, MIT Press,

Cambridge, Massachusetts, ISBN 0-262-68082-3 (especially Chapters 18 and 19 on algorithmic composition).

There are thousands of books about programming (in various languages, including C). They are written for readers from a wide variety of backgrounds, and we advise that, where possible, you look at a book before purchase to see if it is in a suitable style and at an appropriate level for your particular background. We list below three examples which take very different approaches.

Kelley, Al and Pohl, Ira (1998) *A Book on C*, Addison-Wesley, ISBN 0201183994.

Kernighan, B. and Ritchie, D. (1988) *The C programming language*, Prentice Hall, Englewood Cliffs, NJ, ISBN 0131103628.

Perry, Greg (1994) *Absolute Beginner's Guide to C*, Howard W Sams & Co., ISBN 0672305100.

10 Programming for audio and visual synthesis

Overview

This chapter is intended as a tutorial guide for those wishing to write computer programs for audio and visual synthesis applications. It explains how computer programs can generate and manipulate sound and images, and expands on the material covered in Chapter 9 by introducing a series of synthesis examples.

The concept of 'unit generators' is introduced as a way of defining audio processes with the help of computer programming libraries. This concept is then developed to include the generation of graphical objects. This leads to a description of how sound and images may be synchronised as part of a multiple media work.

The University of York's MIDAS system is used as an example of an audio-visual programming environment, and more detailed information about this may be found at the web-site given in the References.

Those readers not wishing to embark upon the tutorial at this point, but desiring an overview, are advised to read Sections 10.1, 10.2, 10.6 and 10.7.

Topics covered

- Programming libraries for audio applications.
- Unit generators for sound synthesis.
- The MIDAS programming library.
- Audio synthesis and signal processing examples.
- Additive and subtractive synthesis.
- Frequency and amplitude modulation.
- Programming for audio-visual synthesis.
- Audio-visual examples.

10.1 Programming at audio rate

10.1.1 Definition of 'audio rate'

The phrase 'audio rate' refers to the speed at which a computer has to work in order to generate soundwaves. Sound can be represented by numbers (details in Chapter 3) and thus can be stored and manipulated on a computer. Audio signals can only be faithfully reproduced by using many tens of thousands of sound 'samples' per second. Since it requires several computer instructions to create or manipulate each sound sample, a typical audio program needs to make hundreds of thousands (or perhaps even millions) of calculations per second. This is a considerable load for any computer system, and so audio-rate programs need to be written with this in mind.

10.1.2 Structure of audio rate programs

Most of the programming techniques covered in Chapter 9 (control rate processing) are also applicable to audio rate situations. The difference in audio-rate situations tends to be the way that the overall program is constructed. Control rate programs generally follow a structure that can be easily seen in the layout of the code (for example a loop that generates ten MIDI notes). Audio rate programs are often split up into 'chunks' of code that are run continuously one after another. These chunks are known as *unit generators*.

10.2 The unit generator concept

10.2.1 History and definition of unit generators

A unit generator is piece of computer code that a has a simple, well-defined role, usually concerned with the generation or processing of one sound sample. They can be considered as the 'building blocks' from which audio signal processing applications are built.

The first unit generators were produced by Max Mathews (see Section 1.9) in his pioneering work generating digital sound. He wrote code for generating soundwaves, for filtering the audio signal and for sending the signal to the loudspeaker. The user of the program could then specify how these units were to be connected together to carry out a particular audio task. This was achieved by the user typing in a text-based 'orchestra' file.

10.2.2 Types of unit generator

A computer music synthesis system allows the composer to construct *networks* of unit generators. A typical set of unit generators provided by such a system would include:

- sound input (e.g. from microphones),
- sound output (e.g. to amplifiers and loudspeakers),
- sound-*file* reading and writing to and from a computer disk (sound *stored* for later use),
- oscillators and waveforms (sound *generation*),
- filters, delays, echo and reverberation (sound *processing*),
- table-reading (e.g. for providing amplitude envelopes),
- mathematical operations (+, –, /, ×, exponential, sine, etc.).

A series of unit generators can be connected together to form a complex signal processing operation.

10.3 Programming libraries

A programming *library* is a collection of functions that can be used by a programmer. They are useful for the following reasons:

- They hide unnecessary details (such as machine specifics) from the programmer, thus making the task of coding easier. This also tends to make the code more readable.
- They aid portability between machines. The programmer's code does not have to be altered in order to get it to run on a different machine since it is written using standard functions in the library.

In fact the C programming language relies on the use of libraries to achieve most of its functionality. One of the reasons that 'C' is so widespread is that the vocabulary of the language (its built-in commands) is rather small. This enables it to be easily rewritten for a variety of computers. Its command set includes symbols (such as { } + – and ;) and words (such as int, while, and if). Most commonly used 'commands' in C, such as printf (which prints messages on the computer screen), are functions within a standard C library.

Other libraries exist to give the programmer easy access to a set of ready-made functions. In Section 9.4.3 we considered the use of a MIDI library to send and receive MIDI messages. In this chapter we introduce a specific audio-visual library which can give you access to functions that produce sound and images.

10.4 MIDAS

MIDAS is an audio-visual toolkit that has been designed at the University of York (UK) to run on a variety of computing systems. It is included in this text as an example of a signal processing environment. More information is available from the web-site (at end of chapter) which should allow you to try out some of the sound and image examples for yourself.

10.4.1 Overview of MIDAS

MIDAS is an acronym for the 'Musical Instrument Digital Array Signal-processor'. It allows an array of musical instruments and computers to be connected together to form a real-time signal processing engine (see Figure 10.1).

Devices that are connected to the MIDAS network communicate with each other by means of specially defined messages known as *protocols*. The main purpose of these protocols is to allow the user to set up networks of unit generators. These networks are

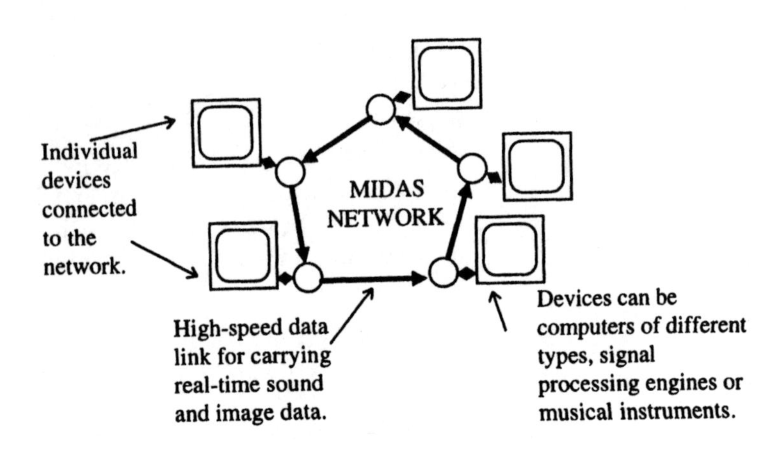

Figure 10.1 The MIDAS concept. A series of devices are connected together by a high-speed data link, enabling real-time signal-processing operations to be carried out.

generally built to run real-time tasks involving sound, image and human interaction.

10.4.2 Unit generators on MIDAS

In MIDAS the unit generators are called Unit Generator Processes (UGPs) and can consist of any operation that a computer can do (e.g. play sound, draw an image, or work out what the mouse is pointing at).

The MIDAS web-page (which can be found as a link from this book's web page) describes all the UGPs that are available. They fall into one of the following categories:

- audio UGPs (for making and processing sound),
- data UGPs (for storing, printing and displaying information),
- graphical and user-interface UGPs (for producing images and interacting with the user),
- maths and logical UGPs (for making calculations),
- MIDI handling UGPs (for sending and receiving MIDI messages),
- musical structure UGPs (for handling 'score' information).
- signal manipulation and flow control UGPs (for making decisions, scaling and manipulating data).

As an example, let us look at the documentation for the sinewave oscillator UGP. This is shown in Figure 10.2.

This shows that the name of this type of UGP is **OSCIL**. It has two inputs and one output. The inputs are the amplitude and frequency of the oscillation, and the output is the oscillating sinewave signal. The inputs and output are labelled with names (OSC_AMP, OSC_FREQ, and OSC_OUT), so that they can be referred to easily in your program code.

If the UGP that you need does not exist, you can write your own.

Figure 10.2 The documentation for the **OSCIL** UGP, taken from the web-site.

```
UGP name: OSCIL

Description: Simple integer sinewave oscillator.

Inputs:
0: (OSC_AMP): amplitude of output signal.
1: (OSC_FREQ): Frequency in Hz.
Outputs:
3: (OSC_OUT): Oscillating output.
```

UGPs are simply small sections of C code written in a special way that defines their inputs and outputs. Details of how to write UGPs are on the web.

10.4.3 The MIDAS C programming library

There are several ways of controlling MIDAS UGPs, but one of the most flexible is to write computer programs using the MIDAS C programming library. As with all programming libraries, the MIDAS library provides a set of functions that you can call from your program. With the MIDAS library you can construct and play a network of UGPs. Your program is compiled, and then *linked* with the MIDAS library to form the executable code. This is now a program that you can run on your computer.

10.4.4 Functions in the MIDAS programming library

The MIDAS library provides a set of functions that you can call from your C program in order to set up and run UGP networks. The main functions that we shall use throughout this chapter are listed below. You may wish to refer to these descriptions as you work through the audio examples in Section 10.5.

- initialise_midas(); This must always be the first function call to MIDAS as it prepares the computer to receive MIDAS messages.
- set_sampling_rate(); This sets the audio system to work at the sampling rate specified in the brackets. The higher the rate, the better the quality, but more processing power is needed. For example set_sampling_rate(8000); means that all audio UGPs will be run 8000 times every second. This is lower than 'compact disc' quality of 44 100 times, so the sound quality is poorer, but the computer does not have to work as hard and so it can support more complex networks in real-time.
- create( ); This creates a UGP in computer memory. The data that you put in the brackets explains what type of UGP it should be. For example create(OSCIL, 1); will create an OSCIL UGP. Every single UGP in a network must be assigned its own unique identification number, and in this case our OSCIL will be known as UGP number 1.
- connect(); This is used to connect one UGP's data output into another UGP's input. Thus data can flow from one to the other. This is the basis of forming a network of UGPs. As an example, the command connect(1, OSC_OUT, 2, OSC_FREQ); means connect the output of UGP number 1 to the frequency input of UGP number 2.

- declare_input(); This sets up an initial value on a UGP's input. For example declare_input(1, OSC_FREQ, 440); initialises the frequency input of UGP number 1 to be 440 Hz.
- input_data(); This function sends data to the input of a UGP while it is running. It can thus be used to update the data values in the network (for example to make the oscillators play notes of different frequencies). Its format is similar to declare_input().
- network_defined(); This informs MIDAS that you have now fully set up your network of UGPs.
- midas_play(); This causes the current MIDAS network to be run for the number of seconds specified in the brackets. For example midas_play(0.5); will run MIDAS for half a second. At a sampling rate of 8000 samples per second, this implies that the MIDAS network will run 4000 times.

10.5 Audio synthesis and signal processing examples

The purpose of this section is to lead you through a set of audio synthesis and signal processing tutorial examples. Each example demonstrates a different concept that can be built from unit generators. In theory these examples could be implemented to some degree on a variety of systems (such as Csound). We will be using the MIDAS system, and the C programming code required to operate MIDAS is provided where appropriate. All code and sound examples are available from the web site.

The tutorial begins with the creation of a single tone, and gradually builds up to an interactive audio system for hearing the effects of amplitude and frequency modulation.

10.5.1 Making sound with an oscillator

In this first example (Figure 10.3) a complete piece of C code is shown. The MIDAS library header is included with the line #include <midas.h>. The main body of code is contained between the curly brackets { and }. It consists entirely of function calls to the MIDAS library.

Example 10.1 What parts of the code would you need to change in order to make it play:

- a note one octave higher (double the frequency)?
- a quieter note (e.g. one with half the amplitude)?
- a note that plays for twice as long?

```
#include <midas.h>                     /* Headers & definitions for MIDAS */
void
main()
{
 initialise_midas();                   /* Must be first call to MIDAS library */
 set_sampling_rate(8000);              /* 8kHz samples */

 create( OSCIL, 1);                    /* Create OSCIL UGP no. 1 */
 declare_input( 1, OSC_AMP, 16000);    /* Set UGP no. 1's amplitude to 16000 */
 declare_input( 1, OSC_FREQ, 440);     /* Set its frequency to 440 Hz */

 create( AUDIO_OUT, 2);                /* Create AUDIO_OUTPUT UGP */
 connect( 1,OSC_OUT, 2, SEND_AUDIO);/* Connect OSCIL to AUDIO_OUT */

 network_defined();                    /* Network is now complete */

 midas_play(2.0);                      /* Run whole network for 2 seconds */
}
```

Figure 10.3 A MIDAS network program consisting of a single sinewave oscillator connected to the audio output. It produces a constant tone for 2 seconds.

Try to work through the code in order to understand what is happening. Refer to the MIDAS function list (Section 10.4.4), and study the comments on the right hand side of the code.

Figure 10.4 is a graphical representation of the network of UGPs that is created by the first example. Sometimes a network is most clearly portrayed by graphical symbols.

10.5.2 Playing a series of notes

The previous example played a single tone. The program is now extended so that it plays four notes altogether (Figure 10.5).

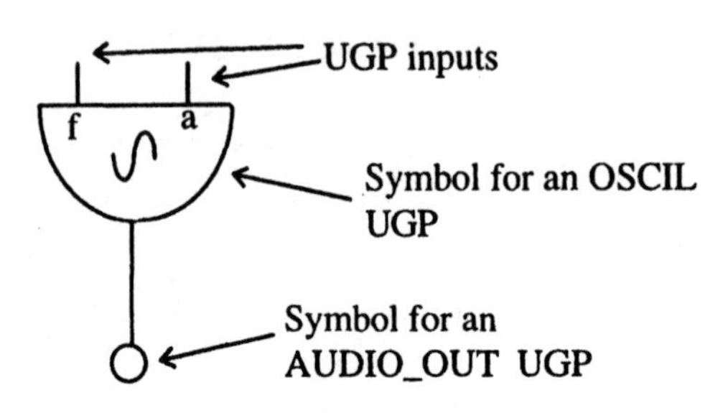

Figure 10.4 A graphical representation of the UGP network created from the code in Figure 10.3.

Figure 10.5 The MIDAS network defined in Figure 10.3 is extended to play three more notes after the first. This is done by sending input_data() messages, then running MIDAS again.

```
network_defined();                       /* Network complete */

midas_play(0.5);                         /* Play first note */

input_data( 1, OSC_AMP, 13000);          /* 2nd Amplitude */
input_data( 1, OSC_FREQ, 660);           /* 2nd Frequency */
midas_play(0.5);                         /* Play note */

input_data( 1, OSC_AMP, 17000);          /* 3rd Amplitude */
input_data( 1, OSC_FREQ, 550);           /* 3rd Frequency */
midas_play(0.5);                         /* Play note */

input_data( 1, OSC_AMP, 14000);          /* 4th Amplitude */
input_data( 1, OSC_FREQ, 660);           /* 4th Frequency */
midas_play(2.0);                         /* Play note */
```

Notice how the new amplitude and frequency data is sent (using the input_data() function call), and then midas_play() is called again to run the network in order to play the note.

The one piece of C code is acting as both the *orchestra* (definition of instruments) and the *score* (notes to be played). In many systems, such as Csound, the two concepts are kept separate.

10.5.3 A simple sequencer

We now extend the code yet again to implement an 8-note sequencer. The network of UGPs remains the same (one OSCIL, one AUDIO_OUT), but we modify the way in which data is sent to the oscillator.

Figure 10.6 shows the code required for the sequencer. The pitches and amplitudes are set up at the top of the code in two *arrays* (see Section 9.5.3 for a description of arrays). The function set_up_oscil_network( ) is a function that you would have to write. It would contain most of the code shown within the main() function of Figure 10.3 (except midas_play()). In other words it initialises the MIDAS system and creates and connects the required UGP network.

As the code goes around the **for** loop, the values from these arrays are read out one by one and are sent to the amplitude and frequency inputs of the oscillator (using the input_data() function).

This program causes MIDAS to keep playing the series of eight

```
int pitches[8]={200, 300, 400, 200, 800, 100, 200, 100};          /* Arrays */
int amps[8]={20000, 19000, 18000, 17000, 20000, 30000, 20000, 25000};
int i; /* Counter variable used in the 'for' loop at the end of this code */

 set_up_oscil_network();         /* Your own function to set up the network */

 while(TRUE) {          /* Keep on looping until the user stops the program */

   for(i=0; i<8; i++) {         /* Play through the 8 notes in the array */

     input_data( 1, OSC_AMP, amps[i]);                  /* Send Amplitude */
     input_data( 1, OSC_FREQ, pitches[i]);              /* Send Frequency */
     midas_play(0.2);                     /* 0.2 seconds = 5 notes per second */
  }
}
```

Figure 10.6 This code performs the function of a simple sequencer, by sending pitch and amplitude values to the network defined in Figure 10.3.

notes. The sound of each note is a simple sinewave tone. In the next section we will expand this code to make the sound more complex.

Example 10.2 To ensure that you understand the essential elements of the code in Figure 10.6, could you change it so that:

- the sequence plays at twice the speed?
- there are 12 notes in the sequence?
- the sequence plays for 10 times only, then stops?

10.5.4 Additive synthesis

If we add a series of sinewaves together we can make a more complex tone. This is called *additive synthesis*. The following example explains how we can do this within MIDAS.

Figure 10.7 shows how the outputs of two oscillators can be added together by connecting them to a 4-input MIXER UGP. The unused mixer inputs are set to zero.

Figure 10.8 demonstrates how the additive synthesis network could be used to play the same sequence of notes as in the previous example, but with tones that contain two partials. Data is sent to the inputs of the oscillators so that one plays the funda-

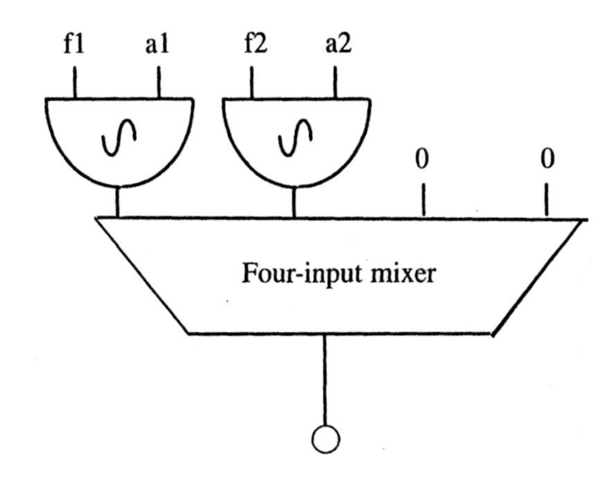

Figure 10.7 A MIDAS network with two OSCILs (of different frequencies and amplitudes) added together using a MIXER. The unused inputs are set to 0.

Figure 10.8 MIDAS code for playing a sequence of notes on two OSCILs. The second one is sent frequency data three times that of the first.

```
#define OSCIL1 1                    /* Define names for UGPs */
#define OSCIL2 2
#define MIXER 3
#define OUT 4

create( OSCIL, OSCIL1);             /* Create individual UGPs */
create( OSCIL, OSCIL2);
create( MIXER_4, MIXER);
create( AUDIO_OUT, OUT);
connect( OSCIL1, OSC_OUT, MIXER, MIX_IN_1);              /* Connect UGPs */
connect( OSCIL2, OSC_OUT, MIXER, MIX_IN_2);
connect( MIXER, MIX_OUT, OUT, SEND_AUDIO);

while(TRUE) {                                            /* Keep looping */

  for(i=0; i<8; i++) {                               /* Play the 8 notes */

    input_data( OSCIL1, OSC_FREQ, pitches[i]);           /* Frequency 1 */
    input_data( OSCIL2, OSC_FREQ, pitches[i]*3);         /* Frequency 2 */

    input_data( OSCIL1, OSC__AMP, amps[i]);              /* Amplitude 1 */
    input_data( OSCIL2, OSC__AMP, amps[i] - 2000);       /* Amplitude 2 */

    midas_play(0.2);                               /* 5 notes per second */
  }
}
```

mental tone and the other plays the third harmonic (three times the frequency) at a lower amplitude.

Notice how the names OSCIL1 and OSCIL2 have been defined by the user (at the top of the code) to make it easier to identify each UGP in the code.

The additive synthesis network could be extended to include many more OSCILs (using a MIXER with more inputs if necessary). Complex instruments often consist of many tens of partials. A new OSCIL is needed for each extra partial. Additive synthesis therefore requires a lot of computing power because it employs a large number of oscillators. Its advantage is that each harmonic is independently controllable, which allows great precision in creating a specific tonal quality.

10.5.5 Additive synthesis with inharmonic shifting

Let us assume that we have already created a network of ten OSCILs whose outputs are mixed (added) together. The section of code in Figure 10.9 can be used to send frequency data to each OSCIL.

The frequencies sent are in the form 220*i. This means that all the partials are an integer multiple of the first frequency (220 Hz, 440 Hz, 660 Hz, 880 Hz, etc.) and hence the partials are harmonically related. However, an offset is added to all frequencies (220*i) + offset and this offset increases as time passes. This has the effect of gradually raising the pitch, but it also destroys the relationship between the partials. They are no longer harmonically related.

As an example, take the frequencies 220 and 440. One is exactly double the other, so they are harmonically related. If we add an offset of 20 to both frequencies we get 240 and 460. Notice that 460 is no longer an exact multiple of 240. We are said to have *inharmonically* shifted the frequencies.

Figure 10.9 A fragment of MIDAS code to send increasingly inharmonically related frequencies to a set of OSCILs.

```
int offset=0;                  /* Variable to be added to frequency */
while(offset<1000) {           /* Loop around with increasing offset */
  for (i=1; i<=10; i++) {      /* For each of the 10 OSCILs */
    input_data(i, OSC_FREQ, ((220*i)+offset) );     /* Send frequency data */
  }
  midas_play(1.0);             /* Run MIDAS for 1 second */
  offset=offset+20;            /* Increase the offset - i.e. shift frequency */
```

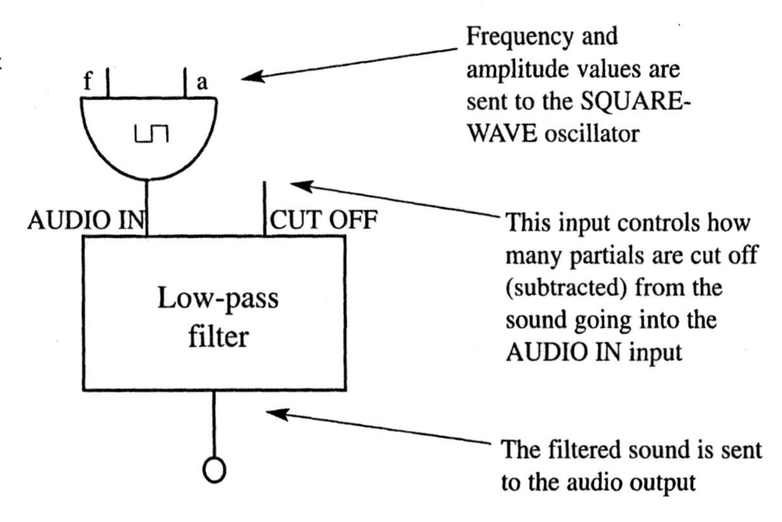

Figure 10.10 A network for generating a complex (square-wave) signal, then filtering that signal. This is an example of subtractive synthesis.

When the code in Figure 10.9 is run the listener will notice the increasing disturbance of the harmonic relationship. What starts out sounding like a single rich tone gradually begins to sound increasingly harsh and consisting of many tones.

10.5.6 Subtractive synthesis and filtering

Figure 10.10 shows a UGP network that generates a signal that is rich in harmonics (in this case a squarewave) and then passes this signal through a lowpass filter. This has the effect of removing the upper harmonics and thus simplifying the sound.

Figure 10.11 This code emulates the 8-note sequencer seen earlier, but also adjusts the cut-off frequency of the lowpass filter. Each note is richer in harmonics than the last as the cut-off frequency is raised.

```
while(TRUE) {

  for(i=0; i<8; i++) {

    input_data( 1, OSC_AMP, amps[i]);                          /* Amplitude */
    input_data( 1, OSC_FREQ, pitches[i]);                      /* Frequency */
    filter_val = (filter_val+50);        /* Increase the filter cutoff value */
    if (filter_val>3999) filter_val=0;     /* but set to zero again when it */
                                            /* gets close to sample_rate/2 */
    input_data( 2, LOWPASS_CUTOFF, filter_val);               /* Cutoff Freq */
    midas_play(0.2);
  }
}
```

The code in Figure 10.11 uses the sequencer code from Figure 10.6, but this time playing the network described in Figure 10.10. The cut-off frequency of the filter (so called since harmonics above this frequency are removed) is changed for each note played.

Example 10.3 In electronic dance music there is often a bass sound whose timbre changes while the note plays. Typically this is a use of subtractive synthesis where the filter is opened and closed (i.e. its cut-off frequency is raised and lowered).

How would you re-write the above code so that the timbre changes *during* each note?

The filter cut-off is increased by 50 Hz each time which has the effect of gradually letting through more harmonics thus making the sound brighter and louder. Notice that when the cut-off frequency exceeds 3999 Hz it is reset to zero. If it were allowed to go higher than this it would let through frequencies greater than half the sampling rate of 8000 Hz. This would cause the *aliasing* distortion described in Chapter 3.

10.5.7 Frequency modulation

Frequency modulation (FM) is the act of changing (or modulating) the frequency of one signal by using another signal. A simple, but very effective, implementation of this can be built from two OSCILs (see Figure 10.12). The oscillator that produces

Figure 10.12 A network of two OSCILs performing frequency modulation. The modulating OSCIL varies the frequency of the carrier OSCIL. The creation and connection code is shown to the right.

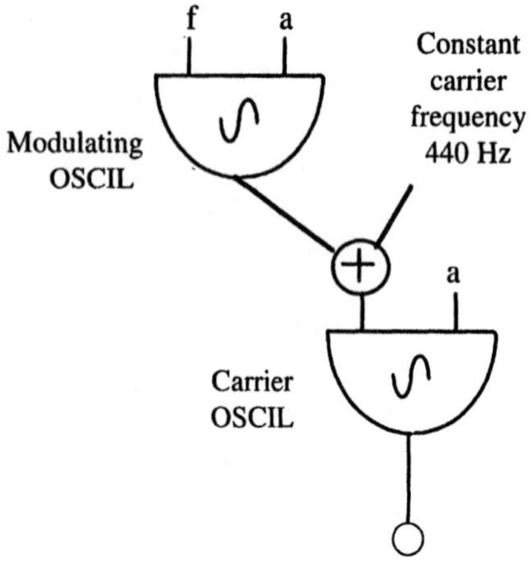

f
a
Modulating OSCIL
Constant carrier frequency 440 Hz
a
Carrier OSCIL

```
create( OSCIL, MODULATOR);
create( OSCIL, CARRIER);
create( ADDER_2, ADD);
create( AUDIO_OUT, OUT);

connect( MODULATOR, OSC_OUT, ADD, ADD_IN_1);
declare_input( ADD, ADD_IN_2, 440);
connect( ADD, ADD_OUT, CARRIER, OSC_FREQ);
declare_input( CARRIER, OSC_AMP, 10000);
connect( CARRIER, OSC_OUT, OUT, SEND_AUDIO);
```

```
input_data(MODULATOR, OSC_AMP, 10);        /* Modulator Amp */
input_data(MODULATOR, OSC_FREQ, 7);        /* Modulator Freq */
midas_play(5.0);                           /* Run MIDAS for 5s */

input_data(MODULATOR, OSC_AMP, 10);        /* Modulator Amp */
input_data(MODULATOR, OSC_FREQ, 20);       /* Modulator Freq */
midas_play(5.0);                           /* Run MIDAS for 5s */

input_data(MODULATOR, OSC_AMP, 20);        /* Modulator Amp */
input_data(MODULATOR, OSC_FREQ, 100);      /* Modulator Freq */
midas_play(5.0);                           /* Run MIDAS for 5s */

input_data(MODULATOR, OSC_AMP, 100);       /* Modulator Amp */
input_data(MODULATOR, OSC_FREQ, 500);      /* Modulator Freq */
midas_play(5.0);                           /* Run MIDAS for 5s */
```

Figure 10.13 A code fragment to send data to a two-OSCIL FM network. In this case four sets of data are sent to the modulating OSCIL and each tone played for 5 seconds.

the pitched tone that we hear is called the carrier OSCIL. Its frequency is varied by connecting its frequency input to the output of another oscillator, called the modulating OSCIL.

FM is used in musical synthesis because it generates sounds with complex timbres using relatively few oscillators. The code example in Figure 10.13 is used to send data to the two-OSCIL FM network.

The first line of code sets a modulating amplitude of 10. This means that the carrier frequency (the pitch that we hear) varies from 440 to 450 Hz. This is because the modulation output (which is varying from 0 to 10) is added to the carrier frequency of 440 Hz. The second line sets the modulating frequency to 7 Hz. This means that the carrier's frequency will be varied seven times per second. This will be heard as a 'vibrato' effect – a subtle variation in pitch of the kind that acoustic instrumentalists often use.

Subsequent lines of code vary the modulation by different amounts. By the time the modulation amplitude is 100 and its frequency is 500 Hz, we no longer hear a single vibrato tone, but rather a complex, harsh-sounding note. Each change in modulation amplitude or frequency produces a significant change to the sound.

Example 10.4 The above code produces step changes in modulation amplitude. How could you change the code so that it moves slowly from no modulation to a high modulation amplitude?

10.5.8 Real-time control of signal processing networks

Up until this point the MIDAS code has acted as both an orchestra and a score. Pre-coded data has been sent out to the UGPs in order to make them play a tune, or change timbre. We now introduce the concept of real-time (or 'performance') control. The user can interact with the parameters (and effectively 'play' the instrument) by entering data *as the network is run.*

There are several ways that the user can send information in real-time. These include the user playing a MIDI device (such as a keyboard, or an ultrasonic MIDI beam) which can be connected to the computer running MIDAS. The incoming MIDI messages can be decoded by MIDI UGPs. In this way human gestures can be converted into control values for synthesis or signal processing.

An alternative method of entering data in real-time is to use the computer's mouse to send variable data to the UGPs. To achieve this in MIDAS we need to use a WINDOW UGP. Its job is to open a graphical 'window' on the computer screen and send messages back to the program denoting where the mouse is, and whether its buttons are pressed. The window is also a useful screen area for producing text and images, and we will be using this facility in Section 10.6.

Figure 10.14 shows the code needed to set MIDAS up for graphical interaction. The code in Figure 10.15 uses the current values of x and y to send data to the modulating OSCIL. As the mouse is moved the effect of the change in modulation can be heard instantly.

Whenever you create a system that receives data from a user's

Figure 10.14 Two sections of code used for creating an interactive graphical environment for MIDAS. The WINDOW UGP is the entity that creates the interactive area on the computer screen.

At the top of the code ...

```
#include <midas.h>              /* Standard headers & definitions for MIDAS */
#include <graphlib.h>     /* Graphics library to allow a window to be used */
#define SCREEN_WIDTH 600       /* Defining how big we want the window to be */
#define SCREEN_HEIGHT 600      /* measured in 'pixels' - dots on the screen */
```

Within the main() function ...

```
create(WINDOW, WIN);            /* Create a WINDOW UGP for mouse interaction */
 declare_input(WIN, WIN_X, SCREEN_WIDTH);             /* Sets width of window */
 declare_input(WIN, WIN_Y, SCREEN_HEIGHT);           /* Sets height of window */
```

```
extern int x, y;        /* External variables for Mouse Position - they are
                           defined within the graphics library which is 'exter-
                           nal' to this code */

while (TRUE) {                                   /* Keep running the network */

  window_manager();     /* Reads the current mouse position into variables x
                           and y */

  input_data(MODULATOR, OSC_AMP, x*20); /* Send scaled x position to Mod Amp */
  input_data(MODULATOR, OSC_FREQ, y);      /* Send y position to Mod Freq */
  midas_play(0.05);                       /* Run MIDAS for 1/20s each time */
```

Figure 10.15 This code reads the position of the mouse every 1/20th of a second, and sends the data to the modulating OSCIL.

input device you must consider whether the values that come out of that device are appropriately scaled to be sent directly to your UGP network. Notice how the mouse's 'x' position (i.e. how far it is across the window) has been multiplied by 20. This is to ensure that the modulation amplitude receives values in the range 0 to 12000 rather than the range of 0 to 600 which the mouse produces (because the screen width has been defined as 600 in the program example in Figure 10.14).

10.5.9 Amplitude modulation

Another form of modulation that is often used in synthesis is known as amplitude modulation (AM). In its simplest form it also consists of two OSCILs (see Figure 10.16) but the modulat-

Figure 10.16 A MIDAS network demonstrating amplitude modulation. Notice how the carrier's basic amplitude is 16000, but this is varied by adding on the output of the modulating OSCIL.

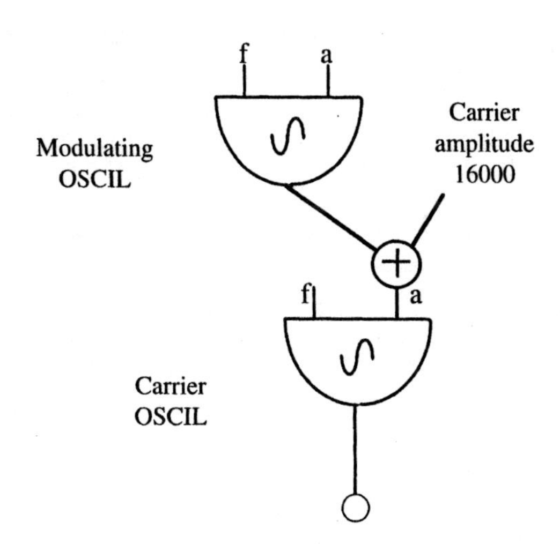

ing OSCIL controls the *amplitude* of the carrier OSCIL. When the modulation has a low amplitude and frequency it is heard as a 'tremolo' effect – with the volume of the note pulsing repeatedly louder and softer. At higher levels of modulation the sound's timbre becomes more complex. Again, it is possible to use the mouse position to control the modulation parameters in real time. The code to do this would be very similar to that in Figure 10.15.

10.6 Programming for audio-visual synthesis

10.6.1 Overview

The term 'multimedia' is so widely used that it can mean different things to different people. A system is usually considered to be 'multimedia' if it uses a variety of system output methods. Typical systems combine one or more of the following:

- text
- images
- animation
- sound
- control messages (e.g. MIDI).

Multimedia systems also tend to stress the importance of *user interaction*, possibly using a variety of input devices (e.g. computer keyboard, 'mouse', 'joystick' or touch-screen). The emphasis is on *engaging* users by allowing them to interact with a variety of media using a range of techniques.

10.6.2 Different rates of operation

The programmer's task is to ensure that all of these different forms of input and output work together seamlessly. This is not a straightforward task because each of the different media has its own *rate* of operation.

- TV images flash large amounts of data onto the screen at 25 or 30 frames per second (depending on which country's television system you are using).
- Audio data is typically generated at 44 100 stereo sound samples per second.
- MIDI messages are produced for each note – typically around 10 small messages per second.
- Text appears on the screen as the user types it, a few characters each second.

All of these media, running at their different rates, need to be *synchronised* so that events which are meant to happen together *appear* to the user as if they are together.

10.6.3 Graphics unit generators

Unit generators were originally invented to generate and process sound samples. However the concept of a 'unit task' can be extended to other media. Many MIDI systems allow the user to connect networks of processing objects together (see 'Max', Section 1.16.1). The University of York's MIDAS system uses unit generators to control and synchronise graphics, audio, MIDI and any other processing task that the computer can be used for.

A graphics unit generator is a chunk of code that generates an image object. Image objects range in complexity; from primitive graphical forms (dots, lines, rectangles, circles, etc.) to more elaborate structures such as 'kaleidoscopes' and 'clouds'.

10.6.4 Audio-visual networks

It is possible to combine the various media into a single unified network that synchronises all the different processing rates.

Graphics unit generators need only run when they are required to generate an image. There is no point in calculating the position of a rectangle several thousand times a second when the screen image is only updated 25 times each second. However, the graphics unit generators can be *incorporated into networks running at a higher rate.*

This is done by using *data buffers* on the input of every unit generator. When a 'slow' process (e.g. MIDI input) is connected to a fast process (e.g. an audio oscillator), the data buffer holds the latest value of the slow input, and the fast process repeatedly reads it. On the other hand, when a fast process is connected to a slow process the buffer is overwritten several times before the slow process reads the data. When processes are at the same rate, then the data is passed without any loss or repetition.

Even very simple interconnections of audio and graphics unit generators can yield interesting results. Figure 10.17 shows a network of three unit generators which is used to generate complex graphical patterns.

As this network is run, the oscillators produce data which controls the position of the circle on the screen. The frequency of oscillation in the 'x' (horizontal) direction can be different to that

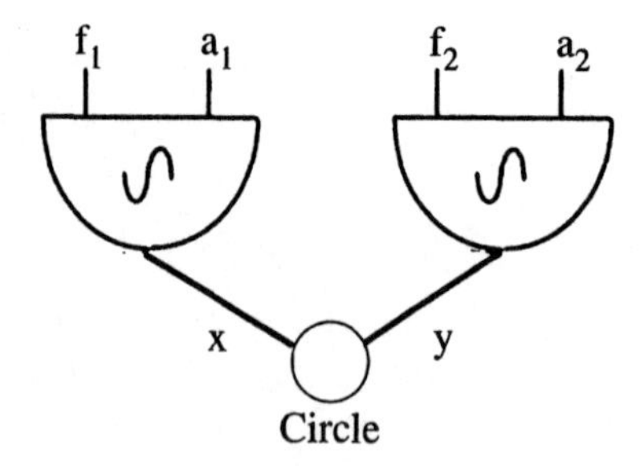

Figure 10.17 An audio visual network. Two 'audio' oscillators are used to move a graphical circle around the screen.

in the 'y' (vertical) direction. Therefore the circle moves around the screen in a complex trajectory. Figure 10.18 shows the result of running this network for several seconds. In this example the circle unit generator has been instructed to leave a 'trace' of all its previous positions on the screen.

10.6.5 Real-time interaction with audio-visual networks

The situation becomes more interesting when the user is watching the pattern emerge dynamically, and can influence the development of the pattern by interacting with it (using the computer mouse for example, or a MIDI keyboard).

In the following example (see Figure 10.19) a mouse-reading UGP outputs x and y position values to a 'mirror' UGP. The mirror UGP draws a triangle, but also 'reflects' the x and y values through an imaginary mirror thus producing a total of four symmetrically arranged triangles.

The user controls the position of the first triangle by using the computer mouse. As the mouse is moved, the four triangles are

Figure 10.18 A screen-shot produced by the simple audio-visual network shown in Figure 10.17.

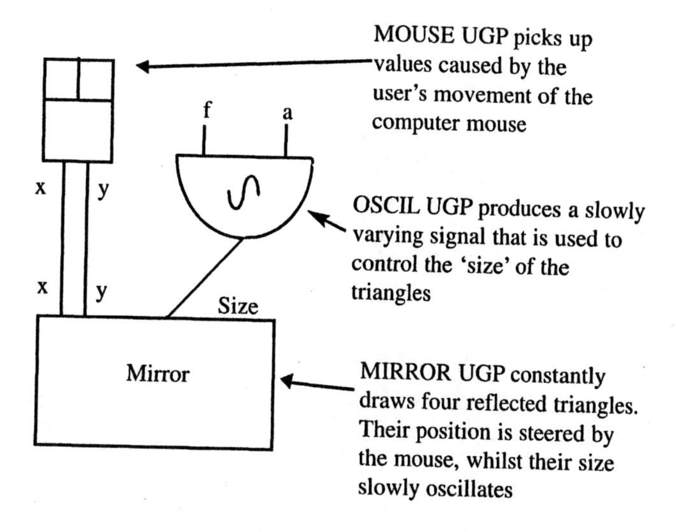

Figure 10.19 A three unit generator network for producing complex images. The complexity comes from the mirror's ability to generate symmetrical shapes, coupled with the OSCIL's slow change of image size, and the user's overall control using the computer mouse.

seen to move in opposite directions around the screen (as if you were looking down a child's kaleidoscope). An extra element of complexity is added by an OSCIL which is used to slowly vary the size of the triangles.

The effect is likened to having an 'active' drawing tool. The user is in overall control of the shape, but the picture that emerges is a *combination* of the user's input and the computer algorithm. A snapshot of this image is shown in Figure 10.20 after running the network for just a few seconds.

Figure 10.20 The graphical output of the network shown in Figure 10.19. The final picture is the result of a few seconds of interaction between a human user and a simple computer algorithm.

On careful inspection you might see that the pattern is not completely symmetrical. This is because the MIRROR UGP has another input which allows a degree of asymmetry to be specified.

10.6.6 Simultaneous control of sound and image

The previous two examples used graphical and OSCIL UGPs (traditionally used to produce audio) to produce a solely graphical output. By reconfiguring the UGPs we can produce a network that produces images and sounds that are linked together by a common algorithm.

Figure 10.21 is an expanded form of the network shown in Figure 10.17. Two oscillators still drive a CIRCLE unit generator to produce a complex image on the screen (similar to that shown in Figure 10.18). Notice how a MOUSE UGP is used to control the relative frequencies of those oscillators. The user controls this in performance and sees instant changes in the trajectory of the circle.

The outputs of these two OSCILs have a secondary connection. The same outputs are used to drive a third OSCIL – this time for making sound. You can tell it is used for sound generation because it is connected to an AUDIO_OUT (shown as a small

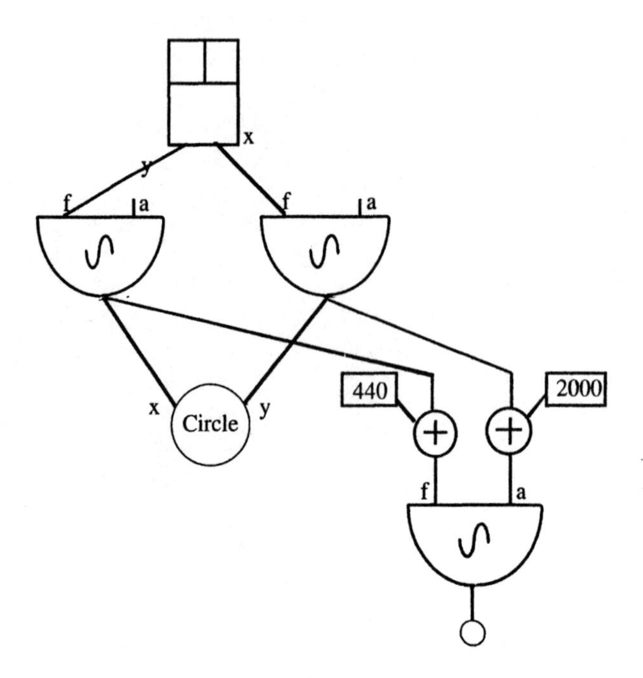

Figure 10.21 An audio-visual MIDAS network. The user moves a mouse to control the frequencies of two OSCILs. These simultaneously control the circle's graphical trajectory *and* the audio modulation which determines the sound's timbre.

circle in the diagram). The first two oscillators *modulate* the audio oscillator. The outputs are *added* to a value of 440 (for the frequency input) and 2000 (for the amplitude input). FM and AM take place simultaneously thus producing a complex sound.

The most important thing to notice is that the first two OSCILs direct *both* the graphical trajectory and the timbre of the sound. These OSCILs are in turn controlled by the human user's movement of the mouse. This creates an audio-visual 'mouse instrument' which the user can learn to play. Players of this instrument often comment on how the sound becomes 'harmonious' at the same time that the graphics appear 'pleasing'. This is because the two media are linked to a common algorithmic source.

10.7 Summary

Computer programs can be used to generate sound and image, and to allow the user to control the development of the audio-visual artefact in real time. This requires the use of programming libraries which provide a set of pre-written functions that can be used in your code. The libraries themselves are pieces of computer code that can be edited or expanded as required.

The unit generator concept allows the user to concentrate on connecting together a series of well-defined unit operations for sound generation or processing. In recent years this concept has been expanded to also include the generation and manipulation of image, MIDI messages and any other operation that a computer is capable of performing. Computer programming of unit generators allows the designer to construct interactive audio-visual compositions and environments.

Suggested further reading

The Internet site associated with this book (http://www.york.ac.uk/inst/mustech/dspmm.htm) gives more detail on the MIDAS system, and multiple-media applications. Some books which you may find useful are listed here:

Begault, D. (1994) *3-D sound for virtual reality and multimedia*, Academic Press, Boston, ISBN 0120847353.

Gurewich, Ori and Gurewich, Nathan (1994) *Easy Multimedia: Sound and Video for the PC Crowd*, Windcrest, ISBN 0070252572.

Heller, Steven and Drennan, Daniel (1997) *The Digital Designer: The Graphic Artist's Guide to the New Media*, Watson-Guptill Pubns, ISBN 0823013464.

Part 6

Interface design for the future

One of the objectives of this book is to assist the committed enthusiast (lay and professional) to contribute to the development of music technology and audio processing systems of the future. This section deals with the way in which people interact with such systems and considers a 'human factors' approach to their design. It concludes with a case study (Section 11.6) which describes an outline design of a new audio-visual instrument that the reader can take on and develop further.

Chapter 11 can be read without prior knowledge. The design exercise in Section 11.6 can be treated as something to read and think about, or it can be used as the basis of a practical exercise for the reader. To successfully carry out the exercise you would probably want to have studied most of this book in some detail.

11 Designing the musician–machine interface

Overview

It is said that a computer system is only as good as the people who use it. However, it is often the *interface* – the means by which people communicate with the computer – that is the limiting factor. This is equally true of computer music systems. This chapter considers the different ways in which people interact with machines.

It introduces the subject of human–computer interaction (HCI) and explains the variety of ways in which computer systems are operated. It then describes the common models by which interfaces to electronic musical instruments are designed, and considers their relative advantages and disadvantages. This knowledge leads to a series of guidelines for the development of electronic musical instruments of the future.

The aim is for you to think about how computer-based musical tools are currently designed for use by humans, and how this might change in the years to come, maybe with your assistance!

Topics covered

- Introduction to human–computer interaction.
- Different styles of interface.
- How computer designers think about the user interface.
- User interfaces in computer music.
- Guidelines for future musical instrument development.
- A design exercise for a musical system.

11.1 Human–computer interaction

Human–computer interaction (HCI) is the study of how human beings use computer systems. It has emerged as a subject area of increasing importance as computing technology has developed.

11.1.1 The early days of computing

The engineering tasks associated with the first two decades of computing history (the 1950s and '60s) were primarily directed towards getting computers to work and to keeping them operational. Computers tended to be vast devices which consumed large amounts of energy and constantly needed maintenance. Users were assumed to be engineering operators who knew the machine operation intimately. Data was entered in a variety of formats – from binary codes stored on punched card or paper tape to the setting of control registers by spin-wheels. All computer operation was by 'batch' process; users submitted their data and were informed of the result at a later date. There was no need to study the *interface* between machine and operator because it was assumed that users would be happy to code their programs and data into whatever form the computer could process.

11.1.2 Personal computers

The development of cheap personal computers for home and office use in the early 1980s forced manufacturers to consider the 'man–machine interface' (MMI) as a major part of the design phase of each new product. The MMI was what customers came into contact with and so concepts such as 'ease of use' and 'speed of response' became a focus for commercial rivalry. Whilst manufacturers designed hardware and software primarily to sell products, certain academic groups became more interested in the long-term implications of different user interfaces. By 1990, the name of the subject had changed to human–computer interaction (HCI) and the number of people studying it had grown enormously.

11.2 Styles of interface

Computer designers make certain assumptions about the users of their systems. These have given rise to a number of different styles of computer interface. Some of the most influential styles are explored here.

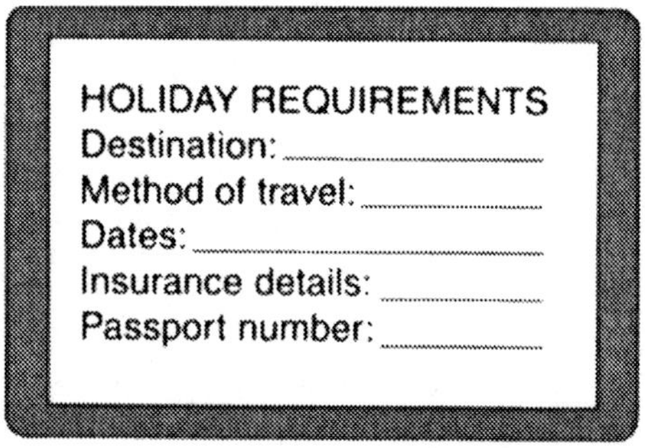

Figure 11.1 A form-filling interface in a travel-agent's booking office. The clerk types in all the information required while talking to the customer.

HOLIDAY REQUIREMENTS
Destination: ____________
Method of travel: ____________
Dates: ____________
Insurance details: ____________
Passport number: ____________

11.2.1 Form-filling

Perhaps the simplest type of interaction consists of the user being required to answer questions or fill in numbers in a fixed format, rather like filling out a form. There is no free dialogue, so this is not a flexible interface style. However, it is one which favours those applications where the user is simply the provider of sets of information (e.g. travel-agent booking systems – see Figure 11.1).

11.2.2 Command line

One of the most prevalent interface metaphors throughout the 1970s and early 1980s was the 'command line'. Here (Figure 11.2) the computer gives a 'prompt' to the user (for example an on-screen **A:>**) and awaits the user's instructions. These instructions are usually in the form of a set of tightly defined letter commands (for example `mv file1 file2` which under the UNIX operating system will rename 'file1' to 'file2').

The advantages of a command-line environment are ease of programming, no restriction on the choice of the next command, and a fast way of issuing instructions for those who have become accustomed to the system by using it for a long time. The disadvantage is the apparently cryptic and unforgiving set of keyword

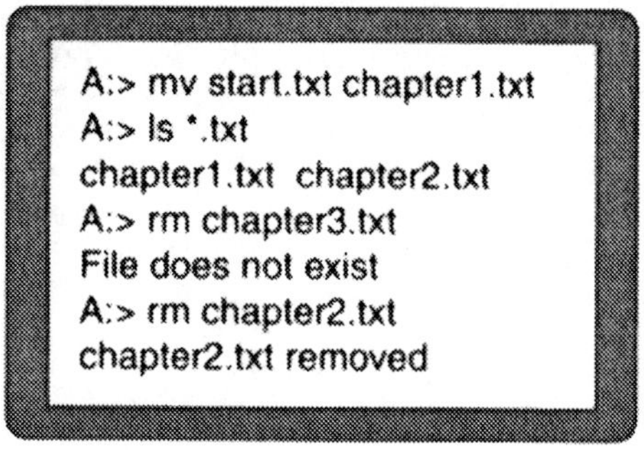

Figure 11.2 A command-line interface showing a series of commands typed by the user (next to the A:> prompt) and the responses by the computer.

```
A:> mv start.txt chapter1.txt
A:> ls *.txt
chapter1.txt  chapter2.txt
A:> rm chapter3.txt
File does not exist
A:> rm chapter2.txt
chapter2.txt removed
```

commands which a user has to know before being able to effectively use the system.

11.2.3 Windows, icons, menus, pointers (WIMP)

The most influential of all current interaction environments is the 'WIMP' interface. The origins of graphical windowing systems go back to the 1960s, although the ideas were developed in greater detail by the Xerox corporation during their pioneering 'Dynabook' project in the 1970s. Their goal of producing a book-sized personal communication system was not achieved at the time, but their idea of a user interface that could be used by anyone was to have a dramatic effect on all subsequent interface developments.

The fundamental goal behind such designs is to give the user a meaningful way of working with the computer. For example, the computer screen could show a picture of an office or a 'desktop' representation with filing cabinets, waste-paper baskets and in-trays. This is often easier for users to relate to than the arbitrary code words of command-line interfaces.

The user points at 'objects' on the screen to access their functionality. This is traditionally done using a 'mouse' pointer, but there are a variety of other devices available such as tracker-balls, joysticks, graphics tablets and 'concept keyboards', each of which has its own ergonomic advantages and disadvantages.

Data of different types are arranged in 'windows'. These are intended to mimic a set of paper pages on a desk, and can be moved around with the pointing device to get the best arrangement for the user. Not all computer operations will fit into this idea of a desktop, so the concept of the *menu* emerged.

A menu is a list of choices from which a user must choose. Their advantage is that new users can search for their required option without having to remember the command (unlike command-

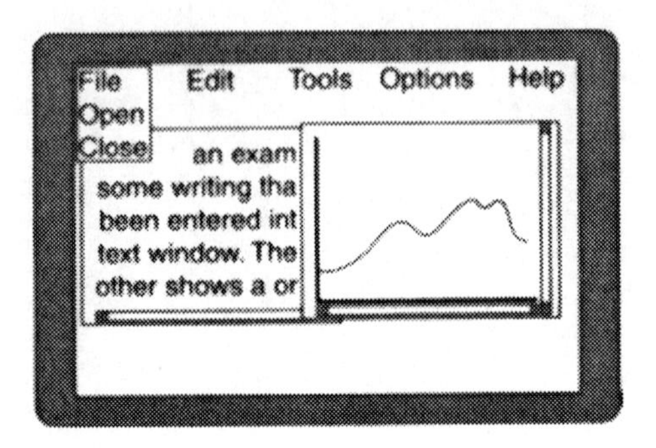

Figure 11.3 A 'WIMP'-style interface. The user is currently selecting a menu option with the mouse. Data is displayed in different 'windows'.

line systems). The corresponding disadvantage is that advanced users are forced to navigate the set of menus to find the item they are looking for.

11.2.4 Direct manipulation

Ben Shneiderman coined the term 'direct manipulation' in the 1980s when he became aware that certain computer systems seemed to engage the user's attention more actively than others. He noticed that such interfaces captivated human users by presenting them with a series of objects (usually on-screen) which could be moved, edited, explored and activated by the actions of pointing, clicking, pressing and dragging. The objects responded with immediate and continuous feedback in terms of visual display or sound. Video games are cited as commercially successful examples of this genre.

The advantage of direct manipulation systems is that beginners can explore the functionality by simply 'trying out' the interface, as most actions are reversible. Indeed the early 'WIMP' interface designs were based on these same fundamental ideas in the quest to find an interface which would suit both new and experienced users.

However, direct manipulation interfaces are not easy to program. They require a lot of computing power and not every action can be represented by the direct manipulation of an object. The result is that WIMP interfaces have had to rely increasingly on menus (which are not 'direct' interfaces).

11.2.5 Computer supported co-operative work (CSCW)

Much of the world-wide focus of HCI researchers has shifted towards managing people-to-people communication using computer systems as intermediaries. This is hardly surprising due to the academic and media focus on the world-wide communications network (for example the World Wide Web – WWW). Much work is being done to manage the complexities of video conferencing systems. This focus is likely to get stronger as the amount of information being exchanged around the world increases, and more people become connected to the network.

11.2.6 Virtual reality (VR)

Another major growth area is the field of virtual environments which aim to allow users to experience *virtual reality*. This means

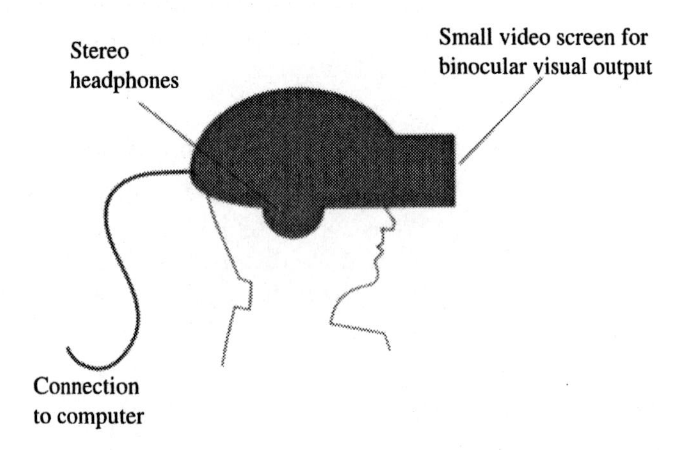

Figure 11.4 A virtual reality headset. Sound and three-dimensional vision are experienced by the user.

that users are given the means of interacting with the computer in the ways that they would interact with objects in the real world. It is often associated with 'immersive' interfaces, where the user puts on a helmet (for stereo audio and visual output) and gloves (as input devices).

There is a disagreement in the academic world as to whether VR offers a way forward or in fact is a retrograde step. Many industrialists believe that the combination of the real world with the virtual world of a computer offers the best solution to numerous technical problems. Others believe that by putting goggles on the user's head we are blocking out the real world and moving further inside the world of the computer, rather than making computers integrate better with our everyday lives. Whatever the future for VR systems, the fundamental questions of how the user should interact with the system still remain.

Section 11.3 examines the different ways in which people have thought about the human–computer interface.

11.3 Models of user interaction

Throughout HCI's brief history academic researchers have attempted to gain a better understanding of the complex issues involved where computers meet people. To aid this understanding, they have developed a range of 'models' that help computer designers to think about the interaction between machine and human. In the following sections the main ways of thinking about HCI are outlined in historical order.

11.3.1 The human as information processor

Between the mid-1970s and mid-1980s there was an increasing need to speed up systems involving computers and people. During this time a notion was developed which was to prove extremely influential, and still is even today. Designers of computer systems needed to predict the behaviour of their intended human users. To do this they imagined the user to be a *processor of information*. The computer displays a choice, the human thinks about it, chooses a response, and the computer then stores that information. Card, Moran and Newell in 1983 called this interaction between humans and computers the 'recognise–act' cycle (Figure 11.5). Human beings are seen as the information bottleneck, taking measurable times to read, interpret and respond to symbols displayed on the computer screen.

In this model the computer is totally in charge of the interaction. Users can make decisions but based only on the options presented on the screen. The time taken to make these decisions and to take the appropriate action is the metric by which human performance is measured and subsequent improvements made in the interface design.

11.3.2 Scripts for human tasks

Card *et al.* (1983) also outlined a method of analysing the tasks which a human wishes to perform with a computer system. It is based on the fundamental assumption that the user has a series of tasks to do in order to achieve his or her goal, and that these tasks can be broken down into a series of simpler tasks resulting eventually in a list of actions to be taken. An 'action' is defined as a task which needs no further problem solving (e.g. 'press the return key').

Figure 11.5 The 'recognise–act' cycle. The computer displays a question. The user then has to think about it, and type a reply. This all takes time which can be measured.

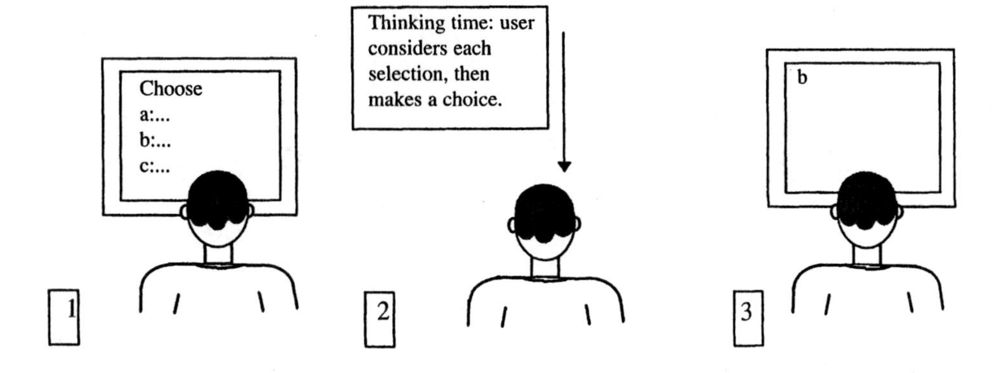

Using this method, designers try to predict the precise series of actions required for everything a user is likely to want to do with a computer system. These lists are called 'scripts'.

11.3.3 Mental models

The development of larger and more complex systems (such as hospital administration programs, where comprehensive task analysis would be impractical) prompted researchers to try to understand more about what was going on inside a user's mind when using a computer. The idea is that users develop their own internal representations of the world (including the computing system they are using). The closer the user's internal representation is to the designer's model of the program, the better the user can operate the computer.

The following two sections outline important differences in the way that users can think about the systems they are operating.

Structural versus functional

Users are thought to develop two distinct types of mental model about any device they are in contact with: *structural* and *functional*.

The key issue is whether users are required to know how the system *actually operates* internally (a structural model), or simply to know *how* they use it (a functional model). It is assumed that the better a device's design, the less the user is concerned with the mechanics of its operation. As an example, a person can drive a car using only a functional model of how the car works ('this pedal stops the car', 'this wheel turns it left and right'). However, a structural model of the car is required in order to repair it (you need to know *how* the engine works in order to diagnose a fault and mend it).

Automatic versus controlled

A further distinction is made as to whether the computer operator uses *automatic* or *controlled* processes. An automatic process is defined as an action that has been learnt by the user, and thus can be used automatically in response to a certain stimulus. The most obvious example of this is a car driver who slams a foot on the brake pedal in response to an object unexpectedly veering into the road. In contrast a controlled process is one which demands a certain amount of conscious control and cognitive effort (e.g. reading through a list of menu options and selecting the most appropriate one).

11.3.4 Humans as 'actors'

A more recent train of thought rebels against the idea of modelling humans as information providers. In 1991 the term 'human actors' was coined, meaning 'those who initiate actions'. The idea is that human beings should be regarded as independent 'agents' who can control their own behaviour, rather than being a purely reactive part of a human–computer system.

This model is not yet widely accepted, yet potentially offers a more realistic interpretation of the ways that people operate in an increasingly computerised world. It is very easy for engineers to design systems that make all the decisions, pausing occasionally to ask the user for a carefully structured morsel of input data. This does not necessarily make the best interface to a computing system. The 'actor' model of HCI design implies that systems should not direct the order of the interaction – but rather that the user does this.

11.4 User interfaces in computer music

If we are to come up with new methods for interacting musically with computers it is important to understand what current methods exist, along with their relative advantages and disadvantages. There have been many different ways in which musicians have been expected to interact with computer systems over the years. This section identifies a series of categories that describe these different methods.

11.4.1 The 'computer programmer'

Users of Max Mathews' digital computer music programs and the many systems derived from them (Section 1.9) take on two distinct roles:

1 'designer of algorithmic instruments'; the user selects and connects pre-compiled signal processing blocks into networks known as 'orchestras',
2 'specifier of number streams'; the user types lists of numerical data into a file known as a 'score'.

The joint effect of these roles is to turn the composer into a computer programmer. People working in this manner are almost indistinguishable, to casual onlookers, from others in the same room using computers for programming in conventional languages such as C, Pascal, BASIC and FORTRAN. The computing tools are based around a computer screen and involve text editors and command-line operations.

Some composers still prefer to work in this manner. People who think more easily in terms of lists and numbers might find this a creative medium in which to compose, but for many it is a barrier to exploring electronic music.

11.4.2 The 'conductor'

The direction of Max Mathews' research efforts, having spawned a whole new approach to making music, changed dramatically in the 1970s. His new emphasis was on enabling users to conduct music, the score of which was stored on the computer. Mathews felt that the human expression found in live performances of acoustic instruments and orchestras had to be harnessed for the new sound world of the digital computer.

Users conduct a pre-stored score and generally steer the musical expression – inflecting the tempo, volume and perhaps the pitch. This is a different role altogether from playing an acoustic instrument, where the player is controlling all the sound parameters in real time. However, conducting programs are a useful way to synchronise a complex computer accompaniment with a live human player.

11.4.3 'Graphic sound and score designer'

The spread of microprocessor-based computing systems in the late 1970s contributed to the increasing popularity of the visual display unit (VDU) as the primary source of feedback to the typical computer user. While early VDUs almost exclusively displayed text information, they were gradually developed so that data could be displayed in a variety of formats (graphs, shaded areas, sliders, etc.) This makes good use of the highly developed human visuo-spatial abilities. Even today, when many other types of input and output devices are being developed, it is assumed that the primary source of feedback to the user should be visual. Consequently a VDU is still a standard part of most computing systems.

The Fairlight and Synclavier computer music systems popularised the idea of editing sound parameters via visual feedback to the user (Sections 1.11.1 and 1.11.2). A mouse, joystick or arrow keys are used to move a cursor around a screen. Graphical displays of synthesis parameters (or conventional score notation, or even sample data) can be manipulated by the action of the cursor.

Nearly all modern musical computing systems follow this paradigm. It is generally thought to be more flexible and satisfying to use than the numerical manipulation of a set of score parameters. The very popular concept of the graphical MIDI-based sequencer relies upon the combination of visual displays (usually with the choice of several types of editor) and direct manipulation of notes on a MIDI keyboard (or other instrument).

11.4.4 'Push-button controller'

Another style of user operation appeared with the growth of the digital keyboard industry in the early 1980s: that of adjusting parameters by displaying them on a small embedded liquid crystal display (LCD).

Prior to this, keyboard synthesisers had presented the user with an array of knobs, buttons and sliders for sound control. Each of these controlled a single parameter of the synthesis process (for example filter 'cut-off frequency', output volume or pitch offset) and thus gave the user direct continuous control of any of the available parameters. The user still had to think in terms of 'parameters' but these parameters were controllable by physical actions which could be learnt. Thus editing was made intuitive and immediately correctable, with the added bonus that the same interface could be used to alter the sound in the course of a live performance.

Since the advent of the digital age the knobs and sliders have been mostly replaced by a small parameter screen. The manufacturers put most of their microprocessing effort into the sound generation process and its real-time triggering from the keyboard. A small part of the software is devoted to the user interface. Keyboard players who wish to edit sounds have to navigate the complexities of a hierarchical menu structure of cryptic alphanumeric parameter values. This is usually done via a liquid crystal character display and a pair of 'up/down' buttons. This causes the keyboard player to regard the control of sound as intellectually divorced from (and on a different plane to) the directness of triggering the sound by playing the keyboard.

This has resulted in the affirmation by the commercial music industry that the purchase of pre-made sounds is the way forward for musical creativity, and that sound production should be left to the experts. It is no wonder that a large gulf emerged between the academic community (whose focus was on the detailed creation of new sounds) and the more public music technology industry (with its focus on 'quick and easy to use' sounds and speed of production).

Various manufacturers attempted to re-introduce devices covered in sliders and knobs, but ironically many users found them difficult to use and reminiscent of the 'old way of doing things'! Such is the entrenchment of user-interface ideas that users now *expect* a small parameter editing screen for a product to look up to date.

Manufacturers seem to have no choice but to produce those things which will sell. However, this does not mean we have to accept the current way of doing things as being the *best* way of doing things. More recently many keyboard manufacturers have developed higher-resolution displays that allow a limited form of graphical interaction, coupled with a small set of sliders which allow real-time control of a few parameters. This, at least, is an improvement on the 'two-line LCD with up/down buttons' interface.

11.4.5 'Instrumentalist'

As can be seen from the historical overview in Chapter 1, much of the design effort for the musician–machine interface has been centred on producing better editing and recording options. In contrast, live performance demands strict real-time operation and flexibility of response. Designers of new musical instruments need to work very hard in order to produce a device that gives the user a good sense of control.

Once again the piano-type keyboard dominates the real-time control market. The majority of MIDI instruments are keyboards, as the MIDI v1.0 specification was effectively designed around the requirements of keyboard players. As mentioned earlier (see Section 1.4.4), the keyboard is not a good way of continuously controlling musical events after they have been triggered (a requirement of much music).

The keyboard is not the only form of interface for live electronic music as demonstrated by the range of non-keyboard based MIDI controller devices which now exist. In the future (looking beyond MIDI) the question of the *form* of the performance interface will become even more important. New instruments need to be developed that have comparable sound quality and flexibility of control as conventional acoustic instruments. This is the challenge for the artistic engineer.

11.5 Suggested areas of change

Huge strides have been made in the development of computer interfaces in the last three decades. It is important to remind ourselves of just how recent a subject area this is. Even in a static subject, we would not expect to have discovered everthing in such a time span. However, the technological advances towards the end of the twentieth century have kept the topic area moving and changing at an incredible pace. Some of the basic tenets of HCI will probably hold true throughout this change, whereas others will become obsolete, or at least less relevant. It is the purpose of this section to highlight those areas of computer-based musical systems and instruments which may need to change or develop.

11.5.1 Dynamic systems

The phrase 'dynamic systems' is used in this context to describe those systems where a user needs to be in constant control of a complex device. A car is a good example, as it cannot drive itself for very long due to its requirement for constant multi-parametric input from its driver in order to make it move and to keep it going along the correct path. Musical instruments are also good examples of dynamic systems.

An acoustic instrument, such as a clarinet, does not make sound unless a human physically activates it. Even then, it needs constant control and fine balancing to keep the note sounding. The player is continuously, and subconsciously, listening to the sound produced and taking appropriate subtle or gross action to maintain the required note.

If people are to be in charge of computer-based musical systems (and not the other way round), then the interface must be designed to enable this to happen. It is too easy for designers to opt for the simplest human–machine interface. Not only are complex, responsive interaction tools difficult to design and program, but there seems to be pressure from industry to produce 'easy-to-use' interfaces. However a *simple* interface may not be the best way of maximising the potential of the human-computer interface.

11.5.2 Ease of use

We all know what it is like to be annoyed by a poorly designed computer interface. Most people would expect an ideal computer system to be easy to use and easy to learn. However, nobody expects learning to drive a car to be easy. No musician would ever say that a violin or a clarinet was easy to use. Yet the subtlety of control and real-time interaction which can be shown by car drivers and instrumental performers is often astounding.

By making the assumption that interfaces should be *easy* we are in danger of undervaluing the adaptation capabilities of human operators and thus limiting the potential human–computer interaction to a lowest common denominator of 'easy to use' commands.

People are actually very good at adapting to new situations, by learning and by practice and repetition. Users of traditional dynamic systems, such as musical instruments, put themselves through hours (and years!) of rehearsal in order to achieve control intimacy with their instruments. It was mentioned above that a clarinet player must continuously control many aspects of the instrument simultaneously and creatively in order to play a note. This is a difficult task which takes a great many hours of practice. Strangely enough, when an electronic device is made where the playing interface is made easy, objections are immediately raised by listeners who complain that the sound is 'flat', 'boring', 'lifeless' and 'nothing like the real thing'. A musician who has to work hard to produce a good sound is likely to be rewarded with appreciation from an audience. In contrast 'easy-play' instruments tend to produce 'cheap-and-easy' musical results.

11.5.3 Information and menus

A car driver is not usually thought of as 'giving information to the car', or a violinist of 'transferring data to the violin', but rather, in both cases, of directly interacting with the dynamic system in question. We need to be careful, when designing musical computer systems, to use the model of direct control rather than that of information gathering. It is a subtle shift of thinking, but without it the computer interface is likely to default to a 'WIMP' system with point-and-click menu options, without even questioning if this is the correct style of interface for this situation.

It has been shown that users of menu-based systems do not usually learn the menus, but instead rely on continuous traversal and interpretation. This means that users are constantly interpreting menu systems – keeping them at the level of 'controlled' processes (see section 11.3.3), and are unable to learn them as 'automatic' processes. If musicians are busy navigating menu systems they are not concentrating on making music, or looking elsewhere (e.g. at the score, the conductor or fellow performers).

11.5.4 Learning an instrument

Players of musical instruments have always required considerable dedication and commitment to hard work and rehearsal in order to learn how to play well. Young musicians typically amass thousands of hours of practice on their instrument before they do their first major concert.

Therefore, we should perhaps assume that if a computer interface demands more than a surface level of operation, users should be expected to spend long periods of time learning the dynamics of how to 'drive' it. Many computing interfaces assume, however, that you need no instruction, since you navigate the menu system and interpret it accordingly each time you want to access a certain function.

A conventional acoustic musical instrument is an inanimate object which relies on the innate 'processing power' and creativity of the human player in order to bring it to life. However, it is the nature of this intimate interaction between performer and instrument which is the very essence of the performance. There must be a radical shift of thinking in the design of computer systems if computer instruments are to be accepted as long-term, worthwhile tools to which people are willing to dedicate large amounts of time and effort.

11.5.5 Configurable instruments – moving the goal-posts

Many computer systems advertise their 'flexibility' in that they can be configured to produce different sounds, notes, responses and functions at the touch of a button. This flexibility is often very useful. For example the authors' *MidiGrid* live performance software (Section 1.15) can be customised by the user to produce individual 'grid patterns' of different sizes and complexity, containing whatever layout of performance material is required. This flexibility of configuration has been responsible for MidiGrid's successful use in schools and for various music therapy situations. Teachers and therapists can restrict the musical material that is available to the end-user. In doing so they form customised musical environments where the tonality, instrumentation and physical layout of the notes (and thus the type of hand gestures used to play them) are defined for a particular music/client combination.

However, there is a danger that players will never learn to control the instrument beyond a surface level of exploration because the 'goal-posts' are constantly being moved. Players of traditional acoustic instruments undergo a good deal of configuration themselves in the process of learning to control their instruments!

Generally, if we allow system interfaces to be continually reconfigured, we are perhaps in danger of removing any reason for human operators to work hard at learning to control the system interactively.

11.5.6 Use of graphical interfaces in live performance

One aspect of a developing level of control shown by a traditional instrumentalist is a decreasing reliance on visual cues. Users of electronic and computer equipment often label sliders and buttons (for example writing on sticky tape on a mixing desk – to label which instrument is on each track). This is analogous to the labelling of a piano keyboard with the letter names of the notes on the stave. The problem is that while you are reading the labels you are not looking at the music, or your fellow performers or the conductor, and will probably not even be *thinking* about the music. Observation of competent pianists will quickly reveal that they do not need to look at their fingers, let alone any annotation which may be associated with the keys.

As users develop their musical performance ability on a particular instrument, they rely increasingly on tactile feedback, and decreasingly on graphical information. There is a general lesson here for the designers of human–computer interfaces in high-performance systems. Graphics are a useful way of presenting information (especially to beginners) but are not necessarily the primary channel which humans use when fully accustomed to a system.

11.5.7 Physical control versus interface dialogues

With an acoustic instrument there is no 'interface dialogue' between musician and instrument; instead the instrument responds instantly to the player's hand movements. It is then up to the player to interpret the sound produced and to use this as feedback to the musical process.

This should also be true for a computer-based system in performance. Traditional computer interfaces are based on the exchange of information within a user-interface dialogue. The users spend much of their time and cognitive energy in navigating the menu systems and deciding between sets of choices.

The computer should not dictate the conversation or insist that the user selects from a set of pre-defined options, but should instead provide an environment for creative exploration. This means that the user should be able to directly trigger and control the artistic medium being used. For example in a drawing program the user should be able to move a pen around the screen or a 'graphics tablet' and instantly see the result on the screen. Perhaps mistakes could be erased by using the opposite end of the pen. Compare this to the command-based equivalent – using a set of menu options or remembering the commands to 'draw a line from point (x^1,y^1) to point (x^2,y^2)'.

The command-based dialogues keep the user one step removed from the actual process of creation. The user does not 'get the feel' of the system and therefore it does not become an automatic, fully-learned process.

Dialogues are good in those situations where the user *wants* to be guided through a set of questions or options, but the necessity for these should diminish as the player becomes more competent at physically operating the system.

11.6 Design exercise

In this section we give an outline of a design for a new computer-based audio-visual instrument. This will bring together and summarise many of the ideas presented earlier in the book. We also hope that it will encourage you to experiment with the design of new instruments – one of the stated intentions of our book.

11.6.1 Overview of the instrument

All electronic instruments require methods:

- for the human player to control the formation of sound
- to create (or synthesise) the sound signals,
- to amplify those sound signals so that they can be heard.

This situation is portrayed in Figure 11.6. In this example we consider the design of an instrument which can produce *visual* output as well as sound. Both sound and image will be controlled by the human player.

The following issues need to be addressed when designing such an instrument:

- What physical form will the playing interface take?
- How are the audio signals amplified and heard?
- How are the images produced and displayed?

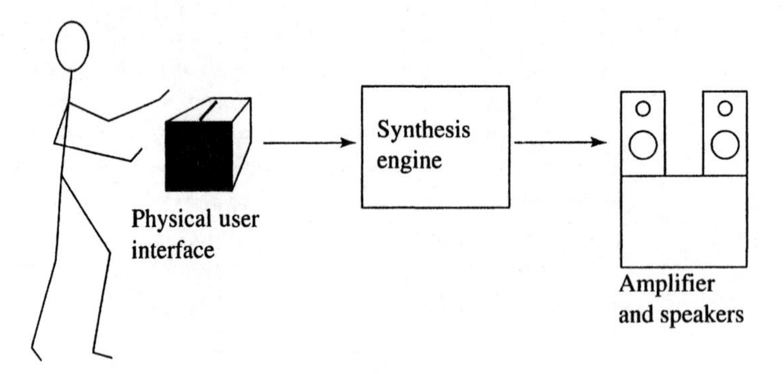

Figure 11.6 Overall schematic diagram of the instrument.

- By what electronic method does the playing interface communicate with the synthesis engine?
- What synthesis methods does the engine use to create audio signals and images?
- How are the signals from the playing interface interpreted by the synthesis engine? In other words, how are the signals used by or connected to the synthesis algorithms? (This is sometimes referred to as the *mapping* between the interface and engine).
- How can we guarantee that the entire process of taking the user's actions and producing the audio-visual output happens seamlessly from the point of view of the human player?

In the following sections we examine each of these issues in turn.

11.6.2 Physical form of the playing interface

We need to decide on what our human instrumentalist will actually do in order to operate the instrument. Even though it is possible to construct electronic instruments that are played by several people at once, let us presume that we require a single human player.

Our instrument will be controlled by physical gesture (i.e. it is not programmed with a score in advance, see Section 11.5.7). Many acoustic instruments require co-ordinated input from two hands (consider the piano, guitar, clarinet and violin), so it would seem sensible to build in this requirement to our interface design. Different hands often control separate parts of the sound (for example a violinist's left hand controls the pitch, whereas the right hand holds the bow which controls volume and instru-

Figure 11.7 The placement of the instrument's sensors. The left hand moves in front of a capacitance sensor. The right hand moves in front of an ultrasound beam, and squeezes some conductive foam rubber.

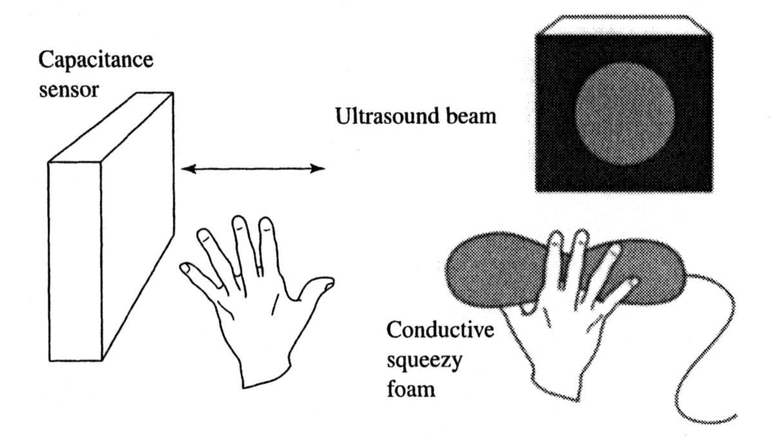

mental timbre). Let us adopt this model, but experiment with the way that each hand controls the sound.

Figure 11.7 shows an example physical interface for our instrument. It uses three different types of sensor.

1 An ultrasonic position detector. This device emits pulses of ultrasound which are used to determine the distance to the nearest object in the path of the sound waves (see Section 1.15.3). A typical range of such a device is over a metre or so.
2 A capacitance proximity detector. This device senses how close a body is to it – typically over the range of a few centimetres.
3 A clump of variable resistance foam rubber which can be squeezed in the hand. When a voltage is applied across it this produces a variation in a small electrical current passing through the foam proportional to the pressure of the squeeze.

In Figure 11.7 the player's left hand is used to move to and from the capacitance sensor over a range of a few inches. Let us presume for now that this somehow controls the *volume* of the sound **and** the *size* of the images on the screen. The right hand moves back and forward in front of the ultrasound beam. This could be used to control the *pitch* of the notes. However, this hand is also holding a piece of foam rubber which can be squeezed. Let us choose to alter the note's *timbre* as it is squeezed. At the same time the *colour* of the images on the screen will get brighter and lighter and their dynamic pattern or trajectory will change.

Production of the audio signals

The placement of the speakers is an important consideration in any musical set-up. Sometimes it is appropriate for the output sound to be routed through a mixing desk and played through a loudspeaker system. In our system we might suggest that we have the speakers built in to the body of the instrument. This helps the sound to be heard (and felt) by the player. It also makes it more natural to incorporate the instrument into an ensemble of acoustic instruments, as the sound has its own definite location in the room.

Production of the images
At the back of the instrument we will place a computer screen on which the synthesised images will appear. An alternative video output would usually be provided for the benefit of an audience or other players. For example, the images could be projected onto a wall or be played over a 'video-wall' system (involving an enlarged image spread out over a grid of television monitors).

General appearance of the instrument
Other considerations about the appearance and feel of the instrument are left to the reader to decide. Often what an instrument looks like and how it responds to touch are highly important factors for the people who will be playing it. The shape of the instrument and its colour make it attractive to different sections of the population. Its size, shape, weight and durability affect who can physically play it – and who will be attracted to play it. The above discussion is just one way of specifying how the instrument should work. You are encouraged to think of your own ways. Should the sound and image have separate controls? Should there be more controls, fewer controls? Could we use a different arrangement of sensors? Do we require the use of two hands (or should it be just one finger? two hands and two feet? head and eye movement?). What about different types of sensor? The discussions in Chapter 1 should help you to explore a range of possible ideas.

11.6.3 Communication with the synthesis engine

We have now outlined what the instrument might look like and what the components of the playing interface could be. These components must be able to communicate with the 'engine' which synthesises the sound and image. In other words we must now define how the sensors pass on their information to the machine that will make the sounds and create the images.

We will use MIDI as the communication method for linking the sensors with the synthesiser engine. Chapter 4 describes MIDI in some detail and discusses its limitations. MIDI will be useful to us as long as we can work within the constraints of its specification. If we can limit its use to the passing of *control-rate* information (see Section 9.4), then it will be a usable interface for our purposes. The advantage of using MIDI is that it is well-defined and supported on many computers and music technology devices.

The sensors themselves do not produce MIDI – they simply output a voltage proportional to 'proximity' or 'squeeze pressure' etc. (depending on the type of sensor). What is needed is a method for converting these voltages into MIDI messages. We will use the device mentioned in Section 1.15.3 – the University of York's MidiCreator box (see Figure 11.8). MidiCreator can be programmed so that the MIDI messages that you specify can be sent out when there is a change of voltage on each sensor. More information about this device can be found on the web site for this book.

For our purposes the MIDI 'control change' message (Section 4.6.1) is the most versatile message to use. There are many control change messages to pick from, and for want of a better choice we will select number 7 – the 'volume control' message. MidiCreator will be configured so that each sensor sends out a MIDI volume control message on its own individual MIDI channel. For our purposes here we are *not* using the volume control message to carry volume information. Instead we are using an already well-defined MIDI message and re-employing it as a way of carrying data from each sensor to the synthesis engine. As the synthesis engine receives each MIDI message it will extract the MIDI *channel* (so it knows which sensor originated the information) and the *data* value (so it knows the posi-

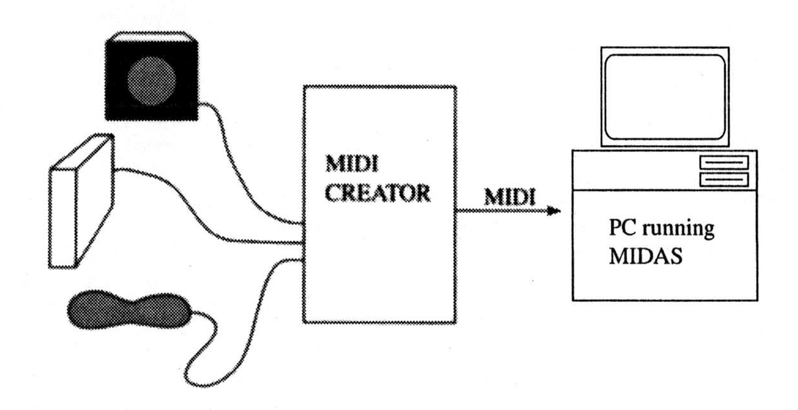

Figure 11.8 The MidiCreator box is used to convert the sensors' voltages into MIDI messages. These messages are then sent to a PC running the MIDAS system (the synthesis engine).

tion of the sensor on a scale from 0 to 127). The synthesis engine will then treat this data according to how we choose to set up the synthesis algorithm.

There are many synthesisers that will receive MIDI information and turn it into sound. For this instrument we will be using the MIDAS system (outlined in Section 1.16.2, and described in more detail in Chapter 10). It gives us the flexibility of defining exactly how the MIDI messages get converted into sound, and it also allows interactive generation of images.

Sections 10.4.1 and 10.4.2 describe how MIDAS enables the instrument designer to construct networks of sound and image synthesis algorithms known as UGPs (unit generator processes). One type of UGP receives MIDI 'control change' messages (on a specified channel) from the computer's MIDI IN port and extracts the information from the messages. This data can then be passed on to the synthesis UGPs in order to create sound and images.

11.6.4 Synthesis methods for image and sound

There are many different sound synthesis methods and these have been outlined in Section 5.2. We will use a simple audio-visual design that has already been described in Section 10.6.6 (you may find it helpful to revise this section now). Here, two oscillators are used to control the position of a circular shape on the screen, and simultaneously to modulate the sound of an audio oscillator.

Figure 10.21 shows a computer mouse being used to control the frequency inputs of the oscillators. Instead of the mouse we will use the MIDI UGPs mentioned above to extract the data from the incoming MIDI messages. The outputs of the MIDI UGPs will be connected to the synthesis UGPs, so the incoming data will affect the sound and images being synthesised. The manner in which this connection is done is very important to the instrument's response and the degree of control given to the human player. For this reason it is discussed in more detail in the next section.

11.6.5 Mapping

We now consider how the signals from the playing interface are interpreted by the synthesis engine. This process is sometimes referred to as the *mapping* between the interface and engine. Referring back to Section 11.6.2 we see that there are three sensors in our instrument. In Section 11.6.3 we decided to use

MIDI 'volume control change' messages to transmit the data, and that each sensor would be represented by a specific MIDI channel. Therefore we must have three UGPs for extracting the MIDI data (one tuned to MIDI channel 1, and the others to channels 2 and 3), and each looking out for volume control change messages. The data on the outputs of these three UGPs are numbers in the range 0 to 127 which represent the current value of each sensor.

These numbers could be fed *directly* into the inputs of the synthesis UGPs. However, it is better that we ask some questions first:

- Are the sensor data outputs in the correct range? (i.e. is 0 to 127 an acceptable range for a particular synthesis parameter, or does it need to be scaled?)
- Is it the *position* of the sensor that's important or its *movement*?
- Does each sensor directly control a single synthesis parameter? (or can the sensor contribute to the control of many parameters?)
- Can more than one sensor control or influence a single synthesis parameter? (This is known as cross-coupling).

Let us consider the output of each of the sensors in turn.

Ultrasound sensor

The ultrasonic beam is controlled by the player's right hand (see Figure 11.7). In Section 11.6.2 we stated that a movement of the player's hand back and forth in the beam would control the *pitch* of the notes.

Therefore we must connect the output of the MIDI UGP (for Channel 1 – the channel we have allocated for the beam sensor) to the frequency input of the audio oscillator. However, the data output of the MIDI UGP will be in the range 0 to 127, whereas we may wish the frequency to have a range of 50 to 2000 Hz, for example. Thus the connection must go via a 'RANGER' UGP. This allows you to specify the range of the input data and the desired range of the output data. When MIDAS is running the data is then automatically scaled to fit the desired range.

We have just mapped the sensor's position directly onto pitch. Subsequent experiments with the instrument might give rise to discussions about whether this is the best way for the pitch mapping to work. It might be that you want the instrument to play discrete pitches, tuned to a particular scale. This could be done by reconfiguring MidiCreator to send out a scale of MIDI 'note' messages or perhaps by changing the mapping within MIDAS by writing your own UGPs.

Capacitance sensor

The capacitance sensor is operated by the player's left hand (again see Figure 11.7). In Section 11.6.2 we decided that a movement of the player's hand to and from the sensor would control the *volume* of the sound and the *size* of the images.

We could simply make the *distance* from the sensor proportional to the volume. However, let us take a moment to consider how acoustic instruments are played louder. Pianos and drums need to be hit harder, violins and cellos have the bow moved faster and with more pressure, and an acoustic harmonium needs to be pedalled faster and more furiously to inject more air into the bellows. In nearly all cases an increase in volume corresponds to an increase in the energy that the player puts into the instrument. This activity seems to continuously involve the player in 'keeping the instrument going' and serves to make the experience an engaging one (see Section 11.5.1). It also has the effect that the instrument stops making any sound if you stop interacting with it – thus putting the human player very much in control (see section 11.3.4).

Therefore we will use the *rate of change* of hand position (i.e. the speed at which the hand is moved) in order to control the volume. This can be done in MIDAS by feeding the output of the MIDI UGP (for channel 2) into a 'SPEEDO' UGP. This produces an output value which is proportional to the speed at which its input data changes (the movement of the hand in front of the capacitance sensor).

We now have a measure of how fast the left hand is being moved. This needs to be connected to the amplitude input of the audio oscillator *and* the 'size' input of the circle drawing UGP. It is quite probable that two inputs will need different ranges (e.g. audio amplitudes are usually in the range 0 to 32 767, whereas the diameter of a circle in pixels is likely to be in the range 10 to 200). Therefore we need to connect the output of the SPEEDO UGP to *two* RANGER UGPs. Each RANGER scales the values accordingly so that the desired effect on the sound and image occurs. In practice we would need to experiment with the settings of these scaling UGPs, perhaps running the network and making adjustments several times until we are happy with the dynamic response of the instrument.

Conductive squeezy foam sensor

The foam is squeezed by the right hand. In Section 11.6.2 we suggested that the pressure of the squeeze should control the sound's timbre *and* the pattern and colour of the circles.

We can achieve this by connecting the output of the MIDI UGP for channel 3 (the foam sensor's allocated channel) to the frequency inputs of the modulating oscillators. In other words we will use this sensor to take the place of the mouse in Figure 10.21).

We will need to make an additional connection from the MIDI UGP to the 'colour' input of the circle. Once again, each connection needs its own RANGER UGP which would be fine-tuned during early trials of the instrument.

You may also wish to consider whether any of the parameters are cross-coupled. For example, as the foam is squeezed it may be that the rate of change of pressure also affects the pitch. In many acoustic instruments there are examples of such cross-coupling. For instance, a change in dynamic (volume) may have effects in timbre and pitch. There is a large number of combinations of connections and cross-couplings which could be made. There is scope here for much experimentation.

The network connections described above are shown in Figure 11.9. This is the basis for the algorithm that MIDAS will carry out in order to bring our instrument to life.

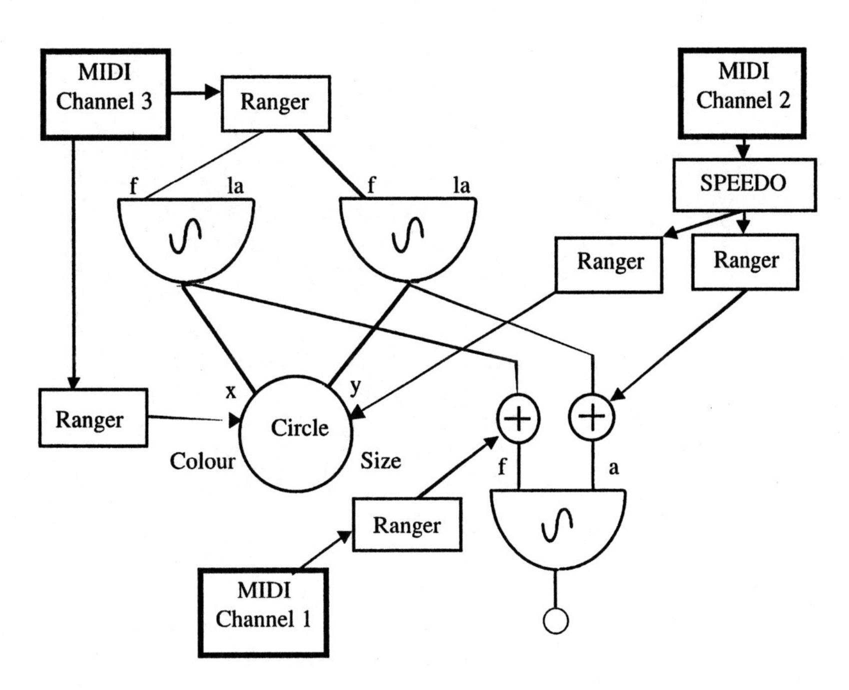

Figure 11.9 The final network diagram for the instrument. It is a modified form of Figure 10.21, and you may wish to compare the two diagrams to study the differences.

11.6.6 Performance considerations

When the instrument is physically constructed, and the sensors are connected to the MIDAS system which runs the finished network, there are still two major factors to be considered.

The response of the instrument

A musical instrument needs to react immediately to its human player. Any time lag between a performance gesture and the sound produced is very confusing and should be avoided where possible. Having said that, human beings are very good at adapting to difficult situations, and cathedral organists often have to compensate for a relatively large time lag between pressing a key and hearing the sound. Nevertheless, such a time-lag should be minimised in our instrumental design.

In a real-time computer music system the sound data is generated as samples and these are sent to an audio *buffer*. This allows the synthesis engine to produce sound samples in bursts, as the samples can queue up in the buffer, waiting to be output. The digital to analogue converter (DAC) reads the samples at a regular rate (the *sample rate*) from this buffer, and the analogue signal at output of the DAC is amplified and sent to the loudspeaker.

The aim is to keep the buffer well-stocked with sound samples. If too *many* samples are produced (for example 2 seconds worth), then there will be a noticeable time lag between the performer's gesture, and the audio response to that gesture. This is because the sound samples relating to that gesture will be waiting in the buffer for that time. As a general guide, the response of the system from human gesture to perceived sound needs to be 10 ms or less. At a sampling rate of 44 100 samples/sec this corresponds to 441 samples. This is therefore the maximum number of samples which should, on average, be in the audio buffer.

The human ear is very sensitive to change, and it is therefore very important that the instrument can work without 'glitching'. Audio glitching in this case occurs when the sound samples are temporarily not available at the audio output. This happens when the computer synthesis engine is not able to produce samples fast enough to keep up with the DAC and the buffer is emptied. There are several remedies.

- Simplify the synthesis network so that it has less work to do in order to generate each sound sample. The side-effect of this is that the synthesis technique is less complex, and this might be perceived as a thinner, weaker sound.

- Reduce the sample rate of the system so that the current synthesis algorithm can produce enough samples for the DAC. However, this has the result of degrading the overall sound quality, and carries the risk of aliasing distortion (see Chapter 3).
- Reduce the 'polyphony' of the music. If more than one note is playing at the same time, it might be possible to decrease the number of simultaneous instrumental parts. This reduces the load on the processor but means that the music has to be simpler.
- Re-code the algorithms making up the instruments (the unit generator in this case) so that they run more efficiently on the current processor. This is a common way of 'tweaking' a real-time system to get the best timing response. The downside of this is that code becomes more specialised for one processor, and is likely to be more difficult for other programmers to understand at a later stage.
- Use more processing power. MIDAS has been developed to allow the same network of UGPs to be run over a variety of physical processors. This might mean adding a DSP card onto the computer, or perhaps linking several computers together.

The response of the player

A new player of this instrument will be faced with a challenge in the form of two questions: 'What can it do?' and 'How do I control it?'. The instrument must be interesting and attractive enough to entice the player to try it in the first place. It then must allow the player to do *something* fairly quickly in order to secure their interest. Admittedly not all acoustic instruments allow novice players to make a sound immediately (have you ever tried a clarinet for the first time?). However in those cases there is an established body of musical works, recordings and accomplished performers to inspire the newcomer that it will be worth their while to invest a lot of effort in learning the instrument. Sections 11.5.2 and 11.5.4 discuss the trade-off which can occur between making an easy-to-use system and one which will bring long-term rewards for hard work and practice.

Your instrument will undoubtedly need to go through a series of 'design-test' cycles before you are happy to let people use it. There may be some changes required that do not show up until users have been trying it for some while. This is hardly surprising since well-known instruments such as the violin are the result of the cumulative labour of countless craftsmen, composers, performers and audiences over a large number of years. You will need to experiment with your new instrument!

11.7 Summary

Humans have been interacting with acoustic musical instruments for a lot longer than they have been using computer systems. There is a great deal of work to be done to improve the interfaces of computer music systems in the future – particularly if they are going to be truly useful to artists and musicians.

Particular emphasis needs to be given to developing computer systems into devices that can be freely controlled in real time, without making heavy mental burdens on users by forcing them to make constant choices. Computer instruments also need improved physical interfaces that offer the artist the same type of flexibility as an acoustic instrument.

These are tough challenges for engineers and musicians, but the final results should be worth the effort. It may even be that you, the computer music instrument designer, may show the way forward in the development of new computer instruments. If you, the reader, feel that you can make a contribution in this way then the book has achieved its objective!

Suggested further reading

Details about the equipment described in the design exercise, and more information on human–computer interaction can be found at the web-site associated with this book (http://www.york.ac.uk/inst/mustech/dspmm.htm. Some other books of interest are listed below:

Card, S.K., Moran, T.P. and Newell, A. (1983) *The Psychology of Human–Computer Interaction*, Lawrence Erlbaum Associates, Hillsdale, NJ, ISBN 0898592437.

Preece, J. *et al.* (1994) *Human–computer interaction*, Wokingham, Addison-Wesley, ISBN 0201627698.

Roads, Curtis (1995) *The Computer Music Tutorial* (especially Part V), MIT Press, Cambridge, Massachusetts, MIT Press, ISBN 0-262-68082-3.

Shneiderman, Ben (1997) *Designing the User Interface: Strategies for Effective Human-Computer Interaction*, Addison-Wesley, ISBN 0201694972.

Index

Printed in the United Kingdom
by Lightning Source UK Ltd.
104114UKS00003B/43-122